# Algorithms for Scheduling Problems

# Algorithms for Scheduling Problems

Special Issue Editors

**Frank Werner**
**Larysa Burtseva**
**Yuri Sotskov**

MDPI • Basel • Beijing • Wuhan • Barcelona • Belgrade

**MDPI**

*Special Issue Editors*

Frank Werner
Otto-von-Guericke University Magdeburg
Germany

Larysa Burtseva
Universidad Autonoma de Baja California
Mexico

Yuri Sotskov
United Institute of Informatics Problems
Belarus

*Editorial Office*
MDPI
St. Alban-Anlage 66
Basel, Switzerland

This is a reprint of articles from the Special Issue published online in the open access journal *Algorithms* (ISSN 1999-4893) in 2018 (available at: http://www.mdpi.com/journal/algorithms/special_issues/Scheduling_Problems)

For citation purposes, cite each article independently as indicated on the article page online and as indicated below:

LastName, A.A.; LastName, B.B.; LastName, C.C. Article Title. *Journal Name* **Year**, *Article Number, Page Range.*

ISBN 978-3-03897-119-1 (Pbk)
ISBN 978-3-03897-120-7 (PDF)

# Contents

# About the Special Issue Editors

**Frank Werner**, Apl. Prof. Dr. rer. nat. habil., studied Mathematics from 1975–1980 and graduated from the Technical University Magdeburg with honors. He defended his Ph.D. thesis on the solution of special scheduling problems in 1984 with "summa cum laude" and his habilitation thesis in 1989. In 1992, he received a grant from the Alexander-von-Humboldt Foundation. He currently works as an extraordinary professor at the Faculty of Mathematics of the Otto-von-Guericke University Magdeburg (Germany). He is an editor of one monograph and an author of three further books, as well as of more than 250 publications in international journals. He is on the Editorial Board of 14 journals, and in particular is an Associate Editor of the *International Journal of Production Research* and the *Journal of Scheduling*. He was a member of the Program Committee of more than 50 international conferences. His research interests are Operations Research, Combinatorial Optimization and Scheduling.

**Larysa Burtseva**, Dr., graduated from the Rostov State University (Russia) in Economic Cybernetics (1975) and defended her Ph.D. thesis on Control in Technical Systems (Technical Cybernetics, 1989) at the Radioelectronics University of Kharkov (Ukraine). Since 2000 she has worked as a full professor at the Engineering Institute of The Universidad Autónoma de Baja California, Mexicali, Mexico. She has led the Laboratory of Scientific Computation since 2005. She has more than 70 publications in international journals and conference proceedings, as well as six book chapters. Her research interests are Discrete Optimization, particularly Combinatorial Optimization, Scheduling, and also packing problems for different applications. She is a member of the National System of Researchers of the National Council for Science and Technology of Mexico (SNI CONACYT) and is a regular member of The Mexican Academy of Computation (AMEXCOMP).

**Yuri N. Sotskov**, Prof. D. Sc., finished secondary school with a gold medal in 1966 and graduated from the Faculty of Applied Mathematics of the Belarusian State University in Minsk in 1971. In 1980, he defended his Ph.D. thesis at the Institute of Mathematics of the National Academy of Sciences of Belarus. In 1991, he defended his D.Sc. thesis at the Institute of Cybernetics of the National Academy of Sciences of Ukraine. He has the title of Professor in Application of Mathematical Models and Methods in Scientific Research (Russian Academy of Sciences, 1994). He currently works as a principal researcher at the United Institute of Informatics Problems of the National Academy of Sciences of Belarus. He has more than 350 publications: five scientific monographs, two text books, more than 150 papers in international journals, books, and conference proceedings in English on applied mathematics, operations research, and scheduling. He is on the Editorial Board of four journals. He has been a supervisor of eight Ph.D. students (defended). In 1998, he received the National Prize of Belarus in Science and Engineering.

# Preface to "Algorithms for Scheduling Problems"

Optimal scheduling is an important area of operations research, as it has both theoretical and practical aspects. It is clear that efficient algorithms are most desirable for practical applications, since most real-world scheduling problems have large sizes but require fast decisions. Due to this, practitioners often prefer to use rather simple algorithms which provide schedules that may be far from optimal with respect to their quality.

This book is based on a Special Issue entitled "Algorithms for Scheduling Problems". In the Call for Papers, we invited prospective authors to submit their latest research in the area of the development of scheduling algorithms. We were looking for new and innovative approaches to solving scheduling problems exactly or approximately. High-quality papers were solicited to address both theoretical and practical issues of scheduling algorithms. Submissions were welcome both for traditional scheduling problems, as well as for new applications. We mentioned that potential topics included, but were not limited to, sequencing in single-stage systems and multi-stage systems with additional constraints such as setup times or costs, precedence constraints, batching, lot sizing, resource constraints, etc., also including single- or multi-criteria objectives and a broad range of scheduling problems in emerging applications such as sports, healthcare, and energy management.

In response to the Call for Papers, we selected eleven submissions, all of which are of high quality, reflecting the stable and growing interest in the area of effective and efficient algorithms to solve problems for real-world production planning and scheduling. All submissions have been reviewed by at least three experts in the operations research area. Each chapter of this book contains one of these selected papers. We hope that practical schedulers will find some interesting theoretical ideas in this Special Issue, while researchers will find new practical directions for implementing their scheduling algorithms. In the following, we give some short comments on the particular chapters.

The first three chapters deal with single-machine problems. Chapter 1 presents a new greedy insertion heuristic with a multi-stage filtering mechanism for energy-efficient single-machine scheduling. The authors wrote that in order to improve energy efficiency and maintain the stability of the power grid, time-of-use (TOU) electricity tariffs have been widely used around the world, which bring both opportunities and challenges to energy-efficient scheduling problems. Although methods based on discrete-time or continuous-time models have been suggested for addressing these problems, they are deficient in solution quality or time complexity, especially when dealing with large-size instances. For such problems, a new greedy insertion heuristic algorithm with a multi-stage filtering mechanism including coarse granularity and fine granularity filtering is developed in this paper. To show the effectiveness of the proposed algorithm, a real case study is provided, and detailed computational results are given.

Chapter 2 is devoted to a stability approach to scheduling problems with uncertain parameters. The stability approach combines a stability analysis of the optimal schedules with respect to possible variations of the job processing times, a multi-stage decision framework, and the solution concept of a minimal dominant set of schedules, which optimally covers all possible scenarios (i.e., for any fixed scenario, this minimal dominant set contains at least one optimal schedule). In contrast to fuzzy, stochastic, and robust approaches, the aim of the stability approach is to construct a schedule which remains optimal for the most possible scenarios. If there exists a schedule dominating the other ones for all possible scenarios, then this schedule remains optimal for each scenario, which may be realized. This may be possible if the level of uncertainty is not high. Otherwise, a decision-maker must look for a schedule which provides optimal—or close to optimal—objective function values for the most possible scenarios among other schedules. To this end, the desired schedule must dominate a larger number of the schedules. This may be possible if the schedule has the largest optimality (stability) box. The authors address a single-machine scheduling problem with uncertain durations of the given jobs. The objective function is the minimization of the sum of the job completion times. The stability approach is applied to the considered uncertain scheduling problem using the relative perimeter of the optimality box as a stability measure of the optimal job permutation. The properties of the optimality box are investigated and used to develop algorithms for constructing job permutations that have the largest relative perimeters of the optimality box.

Chapter 3 addresses a scheduling problem where jobs with given release times and due dates must be processed on a single machine. The primary criterion of minimizing the maximum lateness of the given jobs makes this problem strongly NP-hard. The author proposes a general algorithmic scheme to minimize the maximum lateness of the given jobs, with the secondary criterion of minimizing the maximum completion time of the given jobs. The problem of finding a Pareto optimal set of solutions with the above two criteria is also strongly NP-hard. The author states the properties of the dominance relation along with conditions when a Pareto optimal set of solutions can be found in polynomial time. The proven properties of the dominance relation and the proposed general algorithmic scheme provide a theoretical background for constructing an implicit enumeration algorithm that requires an exponential running time and a polynomial approximation algorithm. The latter allows for the generation of a Pareto sub-optimal frontier with a fair balance between the above two criteria. The next three chapters deal with flow shop and job shop scheduling problems as well as their hybrid (flexible) variants, often inspired by real-life applications.

In Chapter 4, the maximization of the number of just-in-time jobs in a permutation flow shop scheduling problem is considered. A mixed integer linear programming model to represent the problem as well as solution approaches based on enumerative and constructive heuristics are proposed and computationally implemented. The ten constructive heuristics proposed produce good-quality results, especially for large-scale instances in reasonable time. The two best heuristics obtain near-optimal solutions, and they are better than adaptations of the classic NEH heuristic.

Chapter 5 addresses a scheduling problem in an actual environment of the tortilla industry. A tortilla is a Mexican flat round bread made of maize or wheat often served with a filling or topping. It is the most consumed food product in Mexico, so efficient algorithms for their production are of great importance. Since the underlying hybrid flow-shop problem is NP-hard, the authors focus on suboptimal scheduling solutions. They concentrate on a complex multi-stage, multi-product, multi-machine, and batch production environment considering completion time and energy consumption optimization criteria. The proposed bi-objective algorithm is based on the non-dominated sorting genetic algorithm II (NSGA-II). To tune it, the authors apply a statistical analysis of multi-factorial variance. A branch-and-bound algorithm is used to evaluate the heuristic algorithm. To demonstrate the practical relevance of the results, the authors examined their solution on real data.

Chapter 6 is devoted to the effectiveness in managing disturbances and disruptions in railway traffic networks, when they inevitably do occur. The authors propose a heuristic approach for solving the real-time train traffic re-scheduling problem. This problem is interpreted as a blocking job-shop scheduling problem, and a hybridization of the mixed graph and alternative graph is used for modeling the infrastructure and traffic dynamics on a mesoscopic level. A heuristic algorithm is developed and applied to resolve the conflicts by re-timing, re-ordering, and locally re-routing the trains. A part of the Southern Swedish railway network from the center of Karlskrona to Malmö city is considered for an experimental performance assessment of the approach. A comparison with the corresponding mixed-integer program formulation, solved by the commercial state-of-the-art solver Gurobi, is also made to assess the optimality of the generated solutions.

Chapter 7 deals with a generalization of the job shop problem. It is devoted to a formalization of the resource-constrained project scheduling problem (RCPSP) in terms of combinatorial optimization theory. The transformation of the original RCPSP into a combinatorial setting is based on interpreting each operation as an atomic entity that has a defined duration and has to reside on the continuous time axis, meeting additional restrictions. The simplest case of continuous-time scheduling assumes a one-to-one correspondence between the resources and operations and corresponds to a linear programming problem setting. However, real scheduling problems include many-to-one relations that lead to an additional combinatorial component in the formulation of the RCPSP due to the competition of the operations. The authors investigate how to apply several typical algorithms to solve the resulting combinatorial optimization problem: an enumerative algorithm including a branch-and-bound method, a gradient algorithm, or a random search technique.

The next three chapters deal with complex manufacturing systems and supply chains, respectively. Chapter 8 considers a number of geographically separated markets, with different demand characteristics for different products that share a common component. This common component can either be manufactured locally in each of the markets or transported between the markets to fulfill the demand. However, final assemblies are localized to the respective markets. The decision-making challenge is whether to manufacture the common component centrally or locally. To formulate this problem, a newsvendor modeling-based approach is considered. The developed model is solved using a Frank–Wolfe linearization technique along with Benders' decomposition method.

The authors of Chapter 9 write that the current literature presents optimal control computational algorithms with regard to state, control, and conjunctive variable spaces. The authors of this chapter first analyze the advantages and limitations of different optimal control computational methods and algorithms which can be used for short-term scheduling. Second, they develop an optimal control computational algorithm that allows the optimal solution of short-term scheduling. Moreover, a qualitative and quantitative analysis of the scheduling problem arising in the manufacturing system is presented.

Chapter 10 is devoted to a graph model of hierarchical supply chains. The goal is to measure the complexity of the links between different components of the chain (e.g., between the principal equipment manufacturer and its suppliers). The information entropy is used as a measure of knowledge about the complexity of shortages and pitfalls in relationship to the supply chain components under uncertainty. The concept of conditional entropy is introduced as a generalization of the conventional entropy. An entropy-based algorithm is developed, providing an efficient assessment of the supply chain complexity as a function of the supply chain size.

Finally, Chapter 11 deals with an image processing work-flow scheduling problem on a multi-core digital signal processor cluster. It presents an experimental study of scheduling strategies including task labeling, prioritization, and resource selection. The authors apply the above strategies as executing the Ligo and Montage application. A joint analysis of three conflicting goals based on the performance degradation provides an effective guideline for choosing a better strategy. A case study is discussed. The experimental results demonstrate that a pessimistic scheduling approach works better than an optimistic one, and leads to the best optimization criteria trade-offs. The pessimistic heterogeneous earliest finish time (PHEFT) scheduling algorithm performs well in different scenarios with a variety of workloads and cluster configurations. The developed PHEFT strategy also has a lower time complexity in comparison with earlier versions, known as HEFT (heterogeneous earliest finish time first).

The editors of this book would like to thank the authors for submitting their high-quality works, the reviewers for their timely constructive comments, and the editorial staff of the MDPI AG for their assistance in promptly managing the review process and for their continuous support in the preparation of this book whenever needed. We hope that the reader will find stimulating ideas for developing new approaches and algorithms for different types of scheduling problems. Due to the large interest in this subject, we plan to prepare a Volume 2 of the Special Issue "Algorithms for Scheduling Problems". The submission deadline (in autumn 2019) will be announced on the website of the journal *Algorithms* later in this year.

**Frank Werner, Larysa Burtseva, Yuri Sotskov**
*Special Issue Editors*

*algorithms*

MDPI

*Article*

# A New Greedy Insertion Heuristic Algorithm with a Multi-Stage Filtering Mechanism for Energy-Efficient Single Machine Scheduling Problems

**Hongliang Zhang \*, Youcai Fang, Ruilin Pan and Chuanming Ge**

School of Management Science and Engineering, Anhui University of Technology, Ma'anshan 243032, China; ahutfangyoucai@163.com (Y.F.); rlpan9@ahut.edu.cn (R.P.); 13083209689@163.com (C.G.)
\* Correspondence: zhanghongliang_17@126.com; Tel.: +86-555-231-5379

Received: 25 December 2017; Accepted: 6 February 2018; Published: 9 February 2018

**Abstract:** To improve energy efficiency and maintain the stability of the power grid, time-of-use (TOU) electricity tariffs have been widely used around the world, which bring both opportunities and challenges to the energy-efficient scheduling problems. Single machine scheduling problems under TOU electricity tariffs are of great significance both in theory and practice. Although methods based on discrete-time or continuous-time models have been put forward for addressing this problem, they are deficient in solution quality or time complexity, especially when dealing with large-size instances. To address large-scale problems more efficiently, a new greedy insertion heuristic algorithm with a multi-stage filtering mechanism including coarse granularity and fine granularity filtering is developed in this paper. Based on the concentration and diffusion strategy, the algorithm can quickly filter out many impossible positions in the coarse granularity filtering stage, and then, each job can find its optimal position in a relatively large space in the fine granularity filtering stage. To show the effectiveness and computational process of the proposed algorithm, a real case study is provided. Furthermore, two sets of contrast experiments are conducted, aiming to demonstrate the good application of the algorithm. The experiments indicate that the small-size instances can be solved within 0.02 s using our algorithm, and the accuracy is further improved. For the large-size instances, the computation speed of our algorithm is improved greatly compared with the classic greedy insertion heuristic algorithm.

**Keywords:** energy-conscious single machine scheduling; time-of-use electricity tariffs; greedy insertion heuristic; coarse granularity and fine granularity filtering mechanism; concentration and diffusion strategy

---

## 1. Introduction

Driven by the rapid development of global economy and civilization, there will be a consistent growth in energy consumption in the years ahead. According to a survey of the International Energy Agency (IEA), the world-wide demand for energy will increase by 37% by 2040 [1]. Non-renewable energy resources such as coal, oil, and gas are diminishing day-by-day, which is threatening the sustainable development of many countries. Meanwhile, greenhouse gas emissions generated from inappropriate usage of fossil fuels have taken a heavy toll on the global climate as well as the atmospheric environment [2]. Therefore, how to save energy and then improve the environment quality has become a pressing matter of the moment.

As the backbone of many countries, the industrial sector consumes about half of the world's total energy and emits the most greenhouse gases [3,4]. Hence, energy saving in the industry sector has priority in promoting sustainable economic development. As we all know, most of the energy is converted into the form of electricity that numerous industrial sectors use as their main energy [5,6].

Nevertheless, electricity is hard to store effectively, and thus, must be produced and delivered to its customers at once [7]. In addition, the electricity demand is always uneven, which leads to an increase in generating cost owing to the utilization of backup power facilities during peak hours [8]. In order to maintain balance between electricity supply and demand, electricity providers usually implement demand-side management programs [9], which are an essential component of realizing the goals of a smart grid and rationalizing the allocation of power resources [10].

One of the demand-side management programs is time-of-use (TOU) electricity tariffs, which have been widely used around the world. Usually, a common TOU tariff scheme can be divided into three types of periods: off-peak, mid-peak, and on-peak periods. The basic nature of the TOU scheme is that the retail prices set by electricity providers vary hourly throughout the day according to the amount of electricity demands; when there is an increase in demand, the electricity cost goes up correspondingly, and vice versa [11]. The practice of TOU electricity tariffs not only provides significant opportunities for the industrial sector to enhance energy efficiency, but also avoids power rationing during on-peak periods, and improves the stability of the power grid [7].

Using low-energy equipment and improving the efficiency of production management are two important methods to save energy [12]. As a widely used production management method, scheduling can effectively control energy consumption [13], which brings a lower cost of operation. However, the studies about energy-saving scheduling are still limited [14]. Over recent years, energy-efficient scheduling problems have gradually aroused the attention of scholars. To achieve the goal of energy saving during the production process, some researchers have investigated the problems with various energy-efficient mechanisms to reduce electricity costs by minimizing overall energy consumption, such as speed-scaling [15–18] and power-down [19–21], while others have studied the problems from the perspective of TOU electricity tariffs, which has become a frontier issue in this field.

As for researching scheduling problems under TOU electricity tariffs, there has been a growing interest recently. Considering both production and energy efficiency, Luo et al. [22] proposed an ant colony optimization meta-heuristic algorithm for hybrid flow shop scheduling problems under TOU electricity tariffs. Zhang et al. [12] studied a flow shop scheduling problem with production throughput constraints to minimize electricity cost and the carbon footprint simultaneously. Sharma et al. [23] presented a so called "econological scheduling" model for a speed-scaling multi-machine scheduling problem aimed to minimize the electricity cost and environmental impact. Moon et al. [24] examined the unrelated parallel machine scheduling problem under TOU electricity tariffs to optimize the weighted sum of makespan and electricity cost. Ding et al. [7] and Che et al. [25] addressed a similar parallel machine scheduling problem under TOU electricity tariffs to minimize the total electricity cost. The former developed a time-interval-based mixed-integer linear programming (MILP) model and a column generation heuristic algorithm. The latter improved the former model by providing a linear programming relaxation and a two-stage heuristic algorithm.

Single machine scheduling problems are of great significance both in theory and practice. On one hand, there are many single machine scheduling problems in the real industrial environment. For example, a Computer Numerical Control (CNC for short) planer horizontal milling and boring machine can be regarded as a single machine. On the other hand, the research results and methods of single machine scheduling problems can provide reference for other scheduling problems, such as flow shop, job shop, and parallel machine scheduling problems. For single machine scheduling problems under TOU electricity tariffs, Wang et al. [26] investigated a single-machine batch scheduling problem to minimize the makespan and the total energy costs simultaneously. Considering the TOU electricity tariffs and the power-down mechanism, Shrouf et al. [27] proposed a model that enables the operations manager to determine the "turning on" time, "turning off" time, and idle time at machine level, leading to a significant reduction in electricity cost by avoiding on-peak periods. Gong et al. [28] developed a mixed integer linear programming model and a genetic algorithm for the same problem, reducing electricity cost and greenhouse gas emissions effectively during peak time periods. Without considering a power-down mechanism, Fang et al. [29] studied the single machine scheduling problem under TOU electricity tariffs systematically to minimize the total electricity costs. They divided the problem into the two cases of uniform-speed and speed-scalable, in which a preemptive version and a non-preemptive version were investigated respectively. For the uniform-speed case with non-preemptive assumption (Problem U-pyr), they demonstrated that the problem is strongly non-deterministic polynomial-time hard (NP-Hard). Note that the Problem U-pyr is the same as our problem. Based on Fang et al. [29], Che et al. [9] investigated a continuous-time MILP model for Problem U-pyr and developed a greedy insertion heuristic algorithm(GIH) that is the most classic method for this problem until now, according to our best knowledge. In their algorithm, the jobs are inserted into available periods with lower electricity prices in sequence, and the jobs with higher power consumption rates are mostly assigned to periods with lower electricity prices by traversing all non-full "forward blocks" and "backward blocks".

However, Fang et al. [29] did not establish a complete mathematical model for the single machine scheduling problem (Problem U-pyr) under TOU electricity tariffs, and their algorithm is only feasible in the condition that all the jobs have the same workload and the TOU tariffs follow a so-called pyramidal structure. Regrettably, the TOU electricity tariffs rarely follow a complete pyramidal structure in most provinces in China. To perfect the theory of Fang et al. [29], Che et al. [9] developed a new model and algorithm, but their algorithm requires that all the jobs must traverse all non-full "forward blocks" and "backward blocks", causing a strong high-time complexity. Especially when the processing times of the jobs are relatively short, it usually takes several jobs to fill one period and generates uneven remaining idle times, which leads to an increase in the number of forward and backward blocks required to be calculated. In addition, the generation process of "forward (backward) block" is limited to the job movements that do not incur additional electricity cost, which may cause some jobs to miss the optimum positions.

Thus, by focusing on the jobs with short processing times, this paper proposes a more efficient greedy insertion algorithm with a multi-stage filtering mechanism (GIH-F) based on the continuous-time MILP model to address these issues. The proposed algorithm mainly consists of two stages: coarse granularity filtering and fine granularity filtering. In the coarse granularity filtering stage, all the possible positions are first divided into three levels (i.e., three layers) according to the price of electricity, corresponding to high-price, medium-price, and low-price layers. Then, all the jobs are sorted in non-increasing order of their electricity consumption rates and assigned to the layer with a lower price successively. Based on the concentration and diffusion strategy, once the layer to which a job belongs is determined, the hunting zone of possible positions of a job is concentrated in a certain layer. To find the optimal position, the job to be inserted can search for its position in a relatively large space in the selected layer. Then, considering processing times, electricity consumption rates of the jobs and electricity prices, the electricity cost with respect to each possible position can be compared using characteristic polynomials. Based on these, several judgment conditions can be set up in for a fine granularity filtering stage to determine the position of each job to be inserted.

To summarize, the proposed algorithm can filter out impossible layers, and then judge each condition that belongs to the selected layer successively. Once the condition is satisfied, the job is just inserted into the corresponding position. Through coarse granularity filtering and fine granularity filtering, our algorithm does not have to traverse all possible positions, which leads to a great reduction in the time complexity. A real case study and two sets of randomly generated instances are used to test the performance of the proposed algorithm in this paper.

The rest of this paper is organized as follows. Section 2 presents a description of the single machine scheduling problem under TOU electricity tariffs. In Section 3, an efficient GIH-F is developed. Next, a real case study and two sets of experimental tests are provided in Section 4. Finally, the conclusions and prospects are presented in Section 5.

## 2. MILP Formulation for the Problem

This paper studies an energy-efficient single machine scheduling problem under TOU electricity tariffs. The problem can also be called a single machine scheduling problem with electricity costs (SMSEC). Consider a set of jobs $N = \{1, 2, \ldots, n\}$ that need to be processed on a single machine with the objective of minimizing the electricity cost. It is assumed that all the jobs must be processed at a uniform speed. Each job $i \in N$ has its unique processing time $t_i$ and power consumption per hour $p_i$. A machine can process, at most, one job at a time, and when it is processing a job, no preemption is allowed. Each job and the machine are available for processing at time instant 0. Machine breakdown and preventive maintenance are not considered in this paper.

The machine is mainly powered by electricity. The electricity price follows a TOU pricing scheme represented by a set of time periods $M = \{1, 2, \ldots, m\}$, with each period $k \in M$, having an electricity price $c_k$ and a starting time $b_k$. The interval of period $k$ is represented by $[b_k, b_{k+1}]$, $k \in M$, and $b_1 = 0$ is always established. It is assumed that the $C_{max}$ is the given makespan and $b_{m+1} \geq C_{max}$. This means that a feasible solution always exists.

The main work of this problem is to assign a set of jobs to periods with different electricity prices in the time horizon $[0, b_{m+1}]$ to minimize total electricity cost, and the main task is to determine to which period(s) a job is assigned and how long a job is processed in each period. Hence, two decision variables are given as follows. Note that the starting time of each job can be determined by the decision variables (i.e., $x_{i,k}$ and $y_{i,k}$).

$x_{i,k}$: assigned processing time of job $i$ in period $k$, $i \in N, k \in M$;

$$y_{i,k} = \begin{cases} 1, \text{if job } i \text{ or part of job } i \text{ is processed in period } k \\ 0, \text{othertwise} \end{cases}, i \in N, k \in M.$$

In addition, a job is called processed within a period if both its starting time and completion time are within the same period. Otherwise, it is called processed across periods [9]. Let $d_k$ and $X_k$, $k \in M$, represent the duration of period $k$ and the total already assigned processing times in period $k$, respectively. The difference between $d_k$ and $X_k$ is defined as the remaining idle time of period $k$ which is represented by $I_k$, $k \in M$. If $I_k = 0$, the period $k$ is called *full*.

The MILP model for the single machine scheduling problem can be presented as follows:

$$\text{Min } TEC = \sum_{i=1}^{n} \sum_{k=1}^{m} p_i x_{i,k} c_i \tag{1}$$

s.t.

$$\sum_{k=1}^{m} x_{i,k} = t_i, i \in N; \tag{2}$$

$$\sum_{i=1}^{n} x_{i,k} \leq b_{k+1} - b_k, k \in M; \tag{3}$$

$$x_{i,k} \leq t_i y_{i,k}, i \in N, k \in M; \tag{4}$$

$$\sum_{k=j+1}^{l-1} y_{i,k} \geq (l - j - 1)(y_{i,l} + y_{i,j} - 1), i \in N, 3 \leq l \leq m, 1 \leq j \leq l - 2; \tag{5}$$

$$x_{i,k} \geq (y_{i,k-1} + y_{i,k+1} - 1)(b_{k+1} - b_k), i \in N, 2 \leq k \leq m - 1; \tag{6}$$

$$y_{i,k} + y_{i,k+1} + y_{j,k} + y_{j,k+1} \leq 3, i \in N, j \in N, 1 \leq k \leq m - 1, i \neq j. \tag{7}$$

Equation (1), the objective is to minimize the total electricity cost (*TEC*). Constraints (2)–(4) are associated with the processing time assigned to periods. Constraints (5) and (6) are used to guarantee the non-preemptive assumption. Specifically, constraint (5) guarantees that if a job is processed across more than one period, it must occupy several continuous periods. Constraint (6) ensures that if a job is processed across three periods, the middle period must be fully occupied by the job. Constraint (7) ensures that, at most, one job is processed across any pair of adjacent periods. An illustration to the MILP model is given by an example of a 2-job single machine scheduling problem under a 3-period TOU scheme, as shown in Figure 1. Note that more detailed explanations of the formulas can be seen in Che et al. [9].

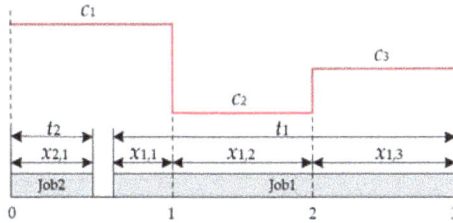

**Figure 1.** An example of the mixed-integer linear programming (MILP) model.

## 3. A Greedy Insertion Heuristic Algorithm with Multi-Stage Filtering Mechanism

### 3.1. The Characteristics of TOU Electricity Tariffs

In China, the TOU electricity tariffs can be mainly divided into two types according to the relative position of the off-peak period: (1) the off-peak period lies between an on-peak period and a mid-peak period; (2) the off-peak period lies between two mid-peak periods. In addition, there is only one off-peak period in a day, and the duration of the off-peak period is longest. This paper will investigate the single-machine scheduling problem under the first type of TOU electricity tariffs, which are being implemented in many places in China, such as Shanxi, Guangxi, and Jiangxi Provinces and so on. Next, let $A$, $B$, and $\Gamma$ represent the off-peak, mid-peak, and on-peak periods, respectively. That is, $A$, $B$, $\Gamma \subseteq M$, $A \cup B \cup \Gamma = M$, $A \cap B = \varnothing$, $A \cap \Gamma = \varnothing$, and $B \cap \Gamma = \varnothing$. An illustration is given in Figure 2. Accordingly, $M = \{1, 2, \ldots, 10\}$, $A = \{5, 10\}$, $B = \{1, 3, 6, 8\}$, and $\Gamma = \{2, 4, 7, 9\}$.

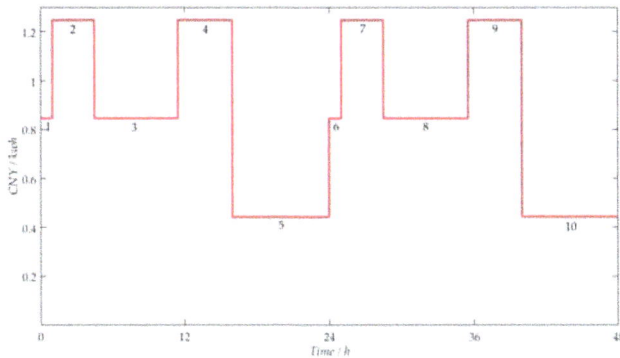

**Figure 2.** An illustration of the first type of time-of-use (TOU) electricity tariffs.

### 3.2. Multi-Stage Filtering Mechanism Design

The proposed algorithm is built on the concentration and diffusion strategy. Firstly, all the possible positions are divided into three layers based on the price of electricity. Specifically, in layer 1, all the jobs are processed within an off-peak period or processed across a pair of periods consisting of an off-peak period and a mid-peak period. In layer 2, the vast majority of the jobs are processed within a mid-peak period or processed across a set of adjacent periods consisting of an off-peak period, a mid-peak period, and an on-peak period. Obviously, the electricity prices corresponding to the positions in this layer are relatively higher than layer 1. In layer 3, the vast majority of the jobs are processed across a pair of periods consisting of a mid-peak period and an on-peak period or processed within an on-peak period. The electricity prices of the positions corresponding to 3rd layer are the highest.

Then, all the jobs are sorted in non-increasing order of their electricity consumption rates and assigned to the layer with a lower price successively to achieve the preliminary coarse granularity filtering target. Once the layer to which a job belongs is determined, the hunting zone of possible positions of a job is concentrated in a certain layer. In the fine granularity filtering stage, several judgment conditions are set up to determine the position of each job to be inserted using characteristic polynomials based on considering processing time, electricity consumption rate of the job, and electricity price. What is more, aiming to find the optimal position, the hunting zone of possible positions of a job is properly expanded in this stage.

Assuming that a job, say job $i$, whose processing time does not exceed the duration of the shortest on-peak period, is to be inserted, the idea of the proposed algorithm is given as follows.

Layer 1 includes conditions 1–2. If $\exists k \in A, t_i \leq \max_{k \in A}\{I_k\} + I_{k+1}$, job $i$ is assigned to layer 1. Obviously, at the very start, the off-peak periods are preferentially occupied by the jobs with high power consumption rates, since all the jobs are sorted in advance.

Condition 1: $\exists k \in A, t_i \leq I_k$.

If Condition 1 is satisfied, job $i$ can be directly processed within an off-peak period with the lowest electricity price. Note that each job processed within a period is placed to the leftmost side of the period. An illustration is given in Figure 3 (i.e., job 3 is a job to be inserted, and $t_3 < I_4$). It is noteworthy that the smaller the number of the job, the greater the power consumption rate.

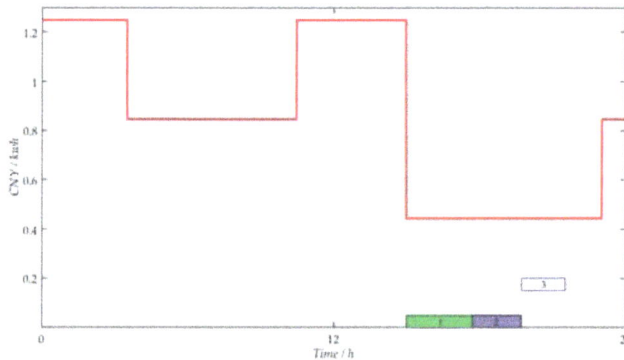

**Figure 3.** Illustration of Condition 1.

Condition 2: $t_i > \max_{k \in A}\{I_k\} > 0$ and $d_{k+1} > 0$.

When Condition 2 is satisfied, it means that job $i$ cannot be processed within an off-peak period. As a second-best choice, job $i$ can be processed across periods $k$ and $k + 1$ in such a condition. An illustration is given in Figure 4. Note that when the job 5 is inserted into the position, it should be adjacent to job 2.

Layer 2 includes Conditions 3–5. If $t_i > \max_{k \in A}\{I_k\} + I_{k+1}$ and $\exists k' \in B, t_i \leq I_{k'}$, job $i$ is assigned to layer 2.

Condition 3: $\max_{k \in A}\{I_k\} > 0$ and $d_{k+2} > 0$.

Since $d_{k+2} > 0$, it follows that period $k + 2$ always exists. To minimize the electricity cost for processing job $i$, the off-peak period with maximal remaining idle time (period $k$) should be given priority, compared with other off-peak, mid-peak, and on-peak periods. Thus, job $i$ is preferred to be processed across periods $k$, $k + 1$, and $k + 2$. Meanwhile, the corresponding electricity cost is named as *cost1*. However, if job $i$ is processed within a mid-peak period (i.e., the corresponding electricity cost is named as *cost2*), the electricity cost may be lower. Hence, two positions are considered and an illustration is given in Figure 5. Let $c_A$, $c_B$, and $c_\Gamma$ represent the electricity prices of off-peak, mid-peak, and on-peak periods, respectively. To select the optimal position, the key property 1 is given as follows.

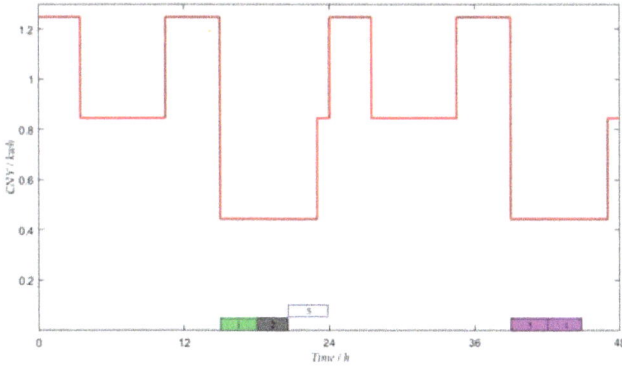

**Figure 4.** Illustration of Condition 2.

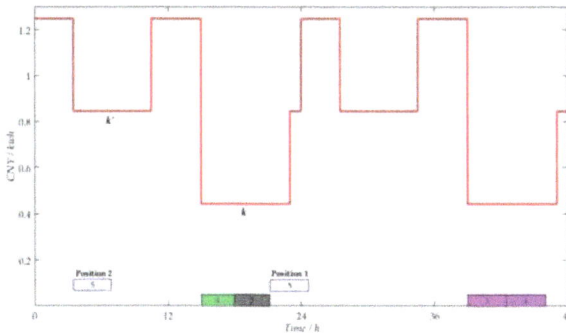

**Figure 5.** Illustration of Condition 3.

**Property 1.** *If Condition 3 is satisfied and $x_{i,k} \times (c_B - c_A) < x_{i,k+2} \times (c_\Gamma - c_B)$ holds, the best choice for job $i$ is to be processed within a mid-peak period.*

**Proof.** $cost1 = p_i \times (x_{i,k} \times c_A + x_{i,k+1} \times c_B + x_{i,k+2} \times c_\Gamma)$ and $cost2 = p_i \times t_i \times c_B$. It is assumed that $cost1 > cost2$, that is, $cost1 - cost2 = p_i \times (x_{i,k} \times c_A + x_{i,k+1} \times c_B + x_{i,k+2} \times c_\Gamma) - p_i \times t_i \times c_B > 0$. Since $t_i = x_{i,k} + x_{i,k+1} + x_{i,k+2}$, it follows that:

$$x_{i,k} \times (c_B - c_A) < x_{i,k+2} \times (c_\Gamma - c_B). \tag{8}$$

Therefore, when inequality (8) holds, job $i$ can be directly processed within a mid-peak period. Otherwise, job $i$ must be processed across periods $k$, $k + 1$, and $k + 2$. This ends the proof. □

Condition 4: $\max_{k \in A}\{I_k\} > 0$ and $d_{k+2} = 0$.

The only difference between Conditions 3 and 4 is whether $d_{k+2} = 0$ or not. This implies that period $k$ is the last off-peak period and period $k + 2$ does not exist. Hence, period $k$ must be in the last cycle (i.e., on the last day) and there are two scenarios that should be considered as follows (the zero points of the two scenarios are different).

Scenario 1: In this scenario, neither period $k + 1$ nor period $k + 2$ exists and four possible positions marked in Figure 6a are given. Similar to Condition 3, period $k$ is critical due to its lowest price and the longest remaining idle time. To insert job $i$ into period $k$, the other jobs processed within period $k$ should be moved right or left. Hence, two positions are analyzed first.

1. To occupy off-peak periods as much as possible, a set of already inserted jobs in period $k$ should be moved to the rightmost side, and then job $i$ can be processed across periods $k$ and $k - 1$.
2. All inserted jobs in period $k$ should be moved left so that job $i$ can be processed within period $k$.

Note that the movement of jobs may lead to a change in their electricity costs, defined as movement cost. The insertion cost of job $i$ is equal to the electricity cost for processing job $i$ plus the corresponding movement cost if some already inserted jobs must be moved to enlarge the idle time of period(s). If the movement cost is zero, then the insertion cost is equal to the electricity cost for processing the job.

Let $cost1$ and $cost2$, respectively, denote the insertion costs of the above corresponding positions. Since the power consumption rate of job $i$ is not higher than any one of the already inserted jobs, it follows that $cost1 \leq cost2$. That is, the Position 1 is the only one that should be considered when job $i$ is inserted into period $k$. Next, let $k^{sm}$ denote the off-peak period with submaximal remaining idle time. Positions 3 and 4 are given as follows.

3. Suppose that job $i$ is processed across periods $k^{sm}$, $k^{sm} + 1$, and $k^{sm} + 2$. If $I_k{}^{sm}$ is slightly smaller than $I_k$, $cost3$ may be less than $cost1$ as period $k^{sm} + 1$ is a mid-peak period. Hence, Position 3 needs to be considered.
4. Similar to Condition 3, job $i$ can be processed within a mid-peak period.

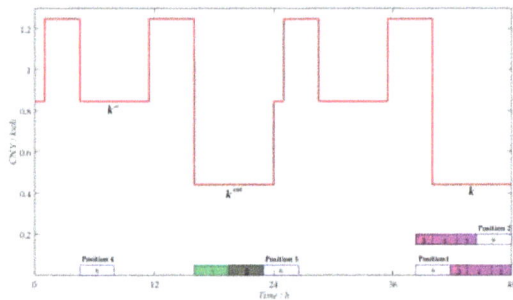

(a) Illustration of Scenario 1

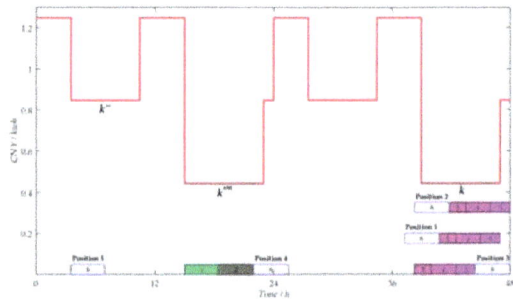

(b) Illustration of Scenario 2

**Figure 6.** Illustration of Condition 4.

To ensure that $cost1 = p_i \times (x_{i,k} \times c_A + x_i x_{i,k-1} \times c_\Gamma)$ always holds, Property 2 is given first.

**Property 2.** *When Condition 4 is satisfied and $d_{k+1} = 0$, job i can be directly inserted into Position 1 without moving any already inserted jobs in period $k - 1$.*

**Proof.** Since $\exists k \in A, I_k > 0$, there must be no jobs processed within period $k - 1$. It is only possible that a job, say job $j$, $j < i$, is processed across periods $k - 2$ and $k - 1$. Therefore, $x_{j,k-2}$ may be the value of the maximal remaining idle time of all mid-peak periods before inserting job $j$. Since $\exists k' \in B$, $t_i \leq I_{k'}$ and $j < i$, it is understandable that $t_i \leq I_{k'} \leq x_{j,k-2}$. Now, job $i$ is to be inserted, it follows that $t_i + x_{j,k-1} \leq x_{j,k-2} + x_{j,k-1} = t_j$. As mentioned earlier, the processing times of all the jobs do not exceed the duration of the shortest on-peak period, that is, $t_j \leq d_{k-1}$. Hence, $t_i + x_{j,k-1} \leq d_{k-1}$. If job $i$ is processed across periods $k$ and $k - 1$, then $x_{i,k-1} + x_{j,k-1} \leq x_{j,k-1} + x_{i,k} + x_{j,k-1} = t_i + x_{j,k-1} \leq d_{k+1}$. That is, $x_{i,k-1} + x_{j,k-1} \leq d_{k-1}$. Thus, $d_{k-1} - x_{j,k-1} - x_{i,k-1} = I_{k-1} \geq 0$. This suggests when job $i$ is inserted into Position 1, period $k - 1$ cannot be full. Hence, job $i$ can be directly inserted into Position 1 without moving any already inserted jobs in period $k - 1$. Note that this property applies to Scenario 2 as well. $\square$

According to Property 2, $cost1 = p_i \times (x_{i,k} \times c_A + x_{i,k-1} \times c_\Gamma)$ is always satisfied. In the following part, three formulas for calculating the insertion costs of Positions 1, 3, and 4 are given.

$$cost1 = p_i \times (x_{i,k} \times c_A + x_{i,k-1} \times c_\Gamma); \tag{9}$$

$$cost3 = p_i \times (x_{i,k^{sm}} \times c_A + x_{i,k^{sm}+1} \times c_B + x_{i,k^{sm}+2} \times c_\Gamma); \tag{10}$$

$$cost4 = p_i \times t_i \times c_B. \tag{11}$$

Since $cost1$ is always less than $cost2$, there is no need to compute $cost2$. Eventually, the insertion costs of all the possible positions that job $i$ can be inserted into can be easily and directly calculated by the above formulas, and then the position with minimum insertion cost will be chosen.

Scenario 2: It can be seen from Figure 6b that the Positions 1, 4, and 5 in Scenario 2 are the same as the Positions 1, 3, and 4 in Scenario 1. The only difference between Scenarios 1 and 2 is whether $d_{k+1} > 0$ or not (i.e., period $k + 1$ exists). Since period $k + 1$ is a mid-peak period, two additional positions need to be considered in comparison with Scenario 1.

1. A set of already inserted jobs in period $k$ are moved to the rightmost side of the period $k + 1$, and then job $i$ is processed across periods $k$ and $k - 1$.
2. A set of already inserted jobs in period $k$ should be moved to the left until job $i$ can be processed across periods $k$ and $k + 1$.

The size of the two insertion costs (i.e., the processing electricity cost of job 6 plus the movement costs of jobs 3, 4, and 5) corresponding to Positions 2 and 3 is uncertain, because the electricity cost for processing job $i$ is greater at Position 2 than at Position 3, while the movement costs are just the opposite. Eventually, it should calculate insertion costs of five possible positions, and then choose the position with minimum cost.

Condition 5: $\forall k \in A, I_k = 0$.

When Condition 5 is satisfied, this means all the off-peak periods are full and job $i$ can be directly processed within a mid-peak period.

Layer 3 includes Conditions 6 and 7. Most of the jobs are processed across a pair of periods consisting of a mid-peak period and an on-peak period or processed within an on-peak period in this layer.

Condition 6: $t_i > \max_{k' \in B}\{I_{k'}\} > 0$.

If Condition 6 is satisfied, it means job *i* cannot be processed within any one of the mid-peak periods, let alone off-peak periods. In other words, job *i* can only be used to fill the remaining idle time of a certain period with lower electricity price. There is an obvious regularity that the more the remaining idle time of the off-peak or mid-peak period job *i* occupy, the lower is its processing electricity cost. Figure 7 shows all the possible positions for a scenario. The analysis process of the possible positions that job *i* can be inserted into is similar to Condition 4, and, therefore, will not be repeated here.

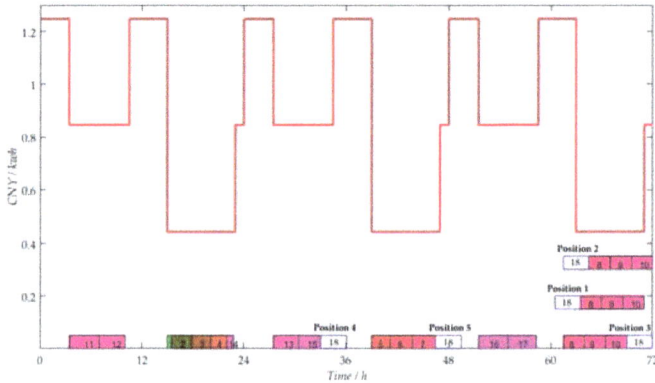

**Figure 7.** Illustration of Condition 6.

Condition 7: $\max_{k' \in B}\{I_{k'}\} = 0$.

Condition 7 implies that all the off-peak periods and mid-peak periods are full, and job *i* can only be inserted into on-peak periods. If $\max_{k_n \in \Gamma}\{I_{k_n}\} < t_i$, job *i* should be inserted into all non-full on-peak periods by moving a set of already inserted jobs, and then the position with the lowest insertion cost can be chosen. The movement method can refer to Che et al. [9].

The core component of the heuristic algorithm is described as a pseudo code, shown in Algorithm 1. Note that the argmin($I_k \geq t_i$) denotes the minimal index *k* for $I_k \geq t_i$.

**Theorem 1.** *The proposed heuristic algorithm runs in $O(n^2 m\,|\Gamma|)$ in the worst case.*

**Proof.** Step 1 runs in $O(n\log n)$ time for sorting *n* numbers and Step 2 requires $O(m)$ time. For each given job, Step C1 runs in $O(|A|)$ in the worst case and Step C2 requires $O(1)$. Steps C3 and C5 both require $O(|B|)$ operations in the worst case. Steps C4 and C6 demand $O(nm)$ to compute the movement cost when calculating the insertion cost. Step C7 includes steps C7.1 and C7.2, wherein the complexity of step C7.1 is $O(|\Gamma|)$, and the complexity of step C7.2 is $O(nm\,|\Gamma|)$. Therefore, step C7 requires $O(nm\,|\Gamma|)$ operations in the worst case. To summarize, the proposed algorithm runs in $O(n^2 m\,|\Gamma|)$ in the worst case. □

Now, assume that no jobs need to traverse all non-full on-peak periods, that is, the Step C7.2 has never been used. In this case, the time complexity of the proposed algorithm is $O(n^2 m)$. However, the classic greedy insertion algorithm proposed by Che et al. [9] requires $O(n^2 m^2)$ operations in the worst case when dealing with the same problem, because their algorithm requires all the jobs to traverse all non-full periods to find an optimum position.

---

**Algorithm 1:** Greedy insertion heuristic algorithm with multi-stage filtering mechanism

---

1. Sort all jobs in non-increasing order of their power consumption rates
2. Initialization: $I_k = b_{k+1} - b_k$, for all $1 \leq k \leq m$
3. **For** $i = 1$ to $n$ **do**
   3.1. **If** layer 1
      C1. **If** Condition 1 is satisfied
      Initial the period index $kk = \text{argmin}_{k \in A}(I_k \geq t_i)$
      //Job $i$ is processed within period $kk$.
      C2. **Else if** Condition 2 is satisfied
      //Job $i$ is processed across periods $k$ and $k + 1$.
   3.2. **Else if** layer 2
      C3. **If** Condition 3 is satisfied
         C3.1. **If** inequality (8) is not satisfied
      //Job $i$ is processed across periods $k$, $k + 1$, and $k + 2$.
         C3.2. **Else**
      Initial the period index $kk' = \text{argmin}_{k' \in B}(I_{k'} \geq t_i)$
      //Job $i$ is processed within period $kk'$.
      C4. **Else if** Condition 4 is satisfied
         C4.1. **If** $d_{k+1} = 0$
      //Calculate $cost1$, $cost3$, and $cost4$ and insert job $i$ into the position with minimal insertion cost.
         C4.2. **Else if** $d_{k+2} = 0$ and $d_{k+1} > 0$
      //Calculate $cost1$, $cost2$, $cost3$, $cost4$, and $cost5$ and insert job $i$ into the position with minimal
insertion cost.
      C5. **Else if** Condition 5 is satisfied
      Initial the period index $kk' = \text{argmin}_{k' \in B}(I_{k'} \geq t_i)$
      //Job $i$ is processed within period $kk'$.
   3.3. **Else if** layer 3
      C6. **If** Condition 6 is satisfied
      //Similarly to Condition 4, it needs to calculate the insertion cost of several possible positions and insert
job $i$ into the position with minimal insertion cost.
      C7. **Else if** Condition 7 is satisfied
         C7.1. **If** $\max_{k_{//} \in \Gamma}\left\{ I_{k_{//}} \right\} > t_i$
      Initial the period index $kk_{//} = \text{argmin}_{k_{//} \in \Gamma}\left( I_{k_{//}} \geq t_i \right)$
      //Job $i$ is processed within period $kk''$.
         C7.2. **Else**
      //Job $i$ traverses all non-full on-peak periods and insert job $i$ into the position with minimal
insertion cost.

---

## 4. Computational Results

In this section, a real-life instance from a machinery manufacturing company in China is provided to further illustrate the MILP model and the proposed algorithm. Then, two sets of contrasting experiments with randomly generated instances are conducted, aiming to show the good performance of the algorithm. The algorithm is coded in MATLAB R2015b and the experiments are run on an Intel(R) Core(TM) i7-4790 3.60 GHz processor with 16 GB of memory under the Windows 7 operating system. For benchmarking, the greedy insertion heuristic algorithm proposed by Che et al. [9] is adopted for our contrast tests.

The TOU electricity tariffs used for all instances are those implemented in Shanxi Province, China, as given in Table 1. It can be seen from Table 1 that the off-peak period is between an on-peak period and a mid-peak period, which means that the actual electricity prices meets the first type of TOU electricity tariffs. Assume that the time horizon starts at 8:00 a.m. of the first day.

**Table 1.** The TOU tariffs used for all instances.

| Period Type | Electricity Price (CNY/kwh) | Time Periods |
|---|---|---|
| On-peak | 1.2473 | 8:00–11:30 |
| | | 18:30–23:00 |
| Mid-peak | 0.8451 | 7:00–8:00 |
| | | 11:30–18:30 |
| Off-peak | 0.4430 | 23:00–7:00 |

$C_{max}$ and $C_B$ denote the given makespan and the total processing times of all the jobs, respectively. The parameter $e = C_{max}/C_B$ ($e \geq 1$) is used to measure the degree of time tightness. In these instances, $C_{max}$ can be obtained by the formula $C_{max} = e \times C_B$ as long as the parameter $e$ is set. Obviously, as $e$ increases, $C_{max}$ increases. Note that $b_{m+1}$ is calculated by $b_{m+1} = \lceil C_{max}/24 \rceil \times 24$. Let $TEC_F$ and $TEC_H$ be the total electricity cost calculated by our algorithm (GIH-F)and Che et al.'s algorithm (GIH), respectively. The runtimes of the two algorithms are represented by $CT_F$ and $CT_H$, respectively. The gaps between $TEC_F$ and $TEC_H$ are represented by $G$, $G = (TEC_F - TEC_H)/TEC_H \times 100\%$. The ratio of $CT_H/CT_F$ is represented by $R$.

### 4.1. A Real Case Study

The MILP model and the proposed algorithm are applied to a vertical machining center (SMTCL VMC 1600P) from a machinery manufacturing company shown in Figure 8. In this real-life instance, the company receives some orders of processing rectangular parts with three product models for continuous casting machines of the steel plants. Figure 9 is the picture of the rectangular part, and the complex cavity surfaces including the planes, camber surfaces, and spherical surfaces are to be processed on the VMC. The power consumption per hour required by the VMC is related to the material of the job, cutting depth, feed speed, and so on. To obtain the average power consumption rate of the machine, a power measurement is performed using a power meter (UNI-T UT232), and the measurement data is presented by Table 2. In addition, the order quantity of the three models is 15, 35, and 10 parts, respectively.

**Figure 8.** Vertical machining center.

**Figure 9.** The geometry of the rectangular part.

**Table 2.** The measurement data of the real-life instance.

| Product Model | Average Power Consumption Rate (kW) | Processing Time (h) | The Number of Parts |
|---|---|---|---|
| 40 | 4.4 | 2.4 | 15 |
| 70 | 4.7 | 2.6 | 35 |
| 100 | 5.3 | 3.1 | 10 |

Let us temporarily put aside the above real-life instance, and talk about solving a hypothetical instance given in Tables 3 and 4 based on the above data. Specifically, it is assumed that there are twelve parts which need to be processed within two days. That is, the number of periods is ten (i.e., $m = 10$). According to this example, we will demonstrate the MILP model and the main computational process of GIH-F.

**Table 3.** A hypothetical instance based on real-life data.

| Part (Job) | 1 | 2 | 3 | 4 | 5 | 6 | 7 | 8 | 9 | 10 | 11 | 12 |
|---|---|---|---|---|---|---|---|---|---|---|---|---|
| Processing time (h) | 2.4 | 2.4 | 2.4 | 2.4 | 2.6 | 2.6 | 3.1 | 3.1 | 3.1 | 3.1 | 3.1 | 3.1 |
| Power consumption rate (kW) | 4.4 | 4.4 | 4.4 | 4.4 | 4.7 | 4.7 | 5.3 | 5.3 | 5.3 | 5.3 | 5.3 | 5.3 |

**Table 4.** TOU tariffs for the hypothetical instance.

| Period | 1 | 2 | 3 | 4 | 5 | 6 | 7 | 8 | 9 | 10 |
|---|---|---|---|---|---|---|---|---|---|---|
| Duration (h) | 3.5 | 7 | 4.5 | 8 | 1 | 3.5 | 7 | 4.5 | 8 | 1 |
| Price (CNY/kwh) | 1.2473 | 0.8451 | 1.2473 | 0.443 | 0.8451 | 1.2473 | 0.8451 | 1.2473 | 0.443 | 0.8451 |

First, the twelve jobs are sorted in non-increasing order according to their power consumption rates, that is, job 7, job 8, job 9, job 10, job 11, job 12, job 5, job 6, job 1, job 2, job 3, and job 4. Second, the remaining idle time of all the periods are initialized. Obviously, $t_7 \leq \max_{k \in A}\{I_k\} + I_{k+1}$ and $t_7 \leq I_4$ (i.e., $k = 4$), hence job 7 can be inserted into the low-price layer 1 and is placed in the position corresponding to the Condition 1. The same applies to jobs 8, 9, and 10. In this way, the off-peak periods are fully utilized by the jobs with high power consumption rates, resulting in lower total electricity costs. At this stage, the remaining idle time of each period is as follows: $I_1 = 3.5$, $I_2 = 7$, $I_3 = 4.5$, $I_4 = 1.8$, $I_5 = 1$, $I_6 = 3.5$, $I_7 = 7$, $I_8 = 4.5$, $I_9 = 1.8$, $I_{10} = 1$. An illustration is given in Figure 10a.

Now, $t_{11} > \max_{k \in A}\{I_k\} + I_{k+1}$ and $\exists k_r \in B, t_{11} \leq I_{k_r}$ (i.e., $k = 4$ and $k' = 2$), hence job 11 is assigned to medium-price layer 2, and is placed in the position corresponding to the Condition 3 because $\max_{k \in A}\{I_k\} > 0$ and $d_{k+2} = 3.5 > 0$. Moreover, $x_{i,k} \times (c_B - c_A) > x_{i,k+2} \times (c_\Gamma - c_B)$ where $k = 4$ and $i = 11$, thus, job 11 is to be inserted into the position across periods 4–6. At this stage, the remaining idle time of each period is as follows: $I_1 = 3.5$, $I_2 = 7$, $I_3 = 4.5$, $I_4 = 0$, $I_5 = 0$, $I_6 = 3.2$, $I_7 = 7$, $I_8 = 4.5$, $I_9 = 1.8$, $I_{10} = 1$. An illustration is given in Figure 10b.

Let us continue, and it is not hard to check that the job 12 is to be inserted into the position corresponding to Condition 4. As mentioned earlier, there will be five positions waiting for selection at this moment. The insertion costs for these five positions are 12.8, 10.7, 10.7, Inf, and 13.9, respectively. Therefore, job 12 is to be inserted into the Position 2 which crosses periods 8 and 9. Note that $I_k^{sm} = 0$ (i.e., $I_4 = 0$), hence job 12 cannot be processed across periods $k^{sm}$, $k^{sm} + 1$, and $k^{sm} + 2$, and the corresponding insertion cost is infinity. At this stage, the remaining idle time of each period is as follows: $I_1 = 3.5$, $I_2 = 7$, $I_3 = 4.5$, $I_4 = 0$, $I_5 = 0$, $I_6 = 3.2$, $I_7 = 7$, $I_8 = 4.2$, $I_9 = 0$, $I_{10} = 0$. An illustration is given in Figure 10b.

Next, let us explain the insertion process of other jobs concisely. Jobs 5 and 6 are assigned to layer 2, and they satisfy Condition 5. Therefore, these two jobs are to be inserted into mid-peak period 2. Similarly, jobs 1 and 2 are to be inserted into period 7. At this stage, the remaining idle time of each period is as follows: $I_1 = 3.5$, $I_2 = 1.8$, $I_3 = 4.5$, $I_4 = 0$, $I_5 = 0$, $I_6 = 3.2$, $I_7 = 2.2$, $I_8 = 4.2$, $I_9 = 0$, $I_{10} = 0$. An illustration is given in Figure 10c.

Finally, jobs 3 and 4 are assigned to high-price layer 3, and they both satisfy Condition 6. A final complete scheduling diagram is given in Figure 10c.

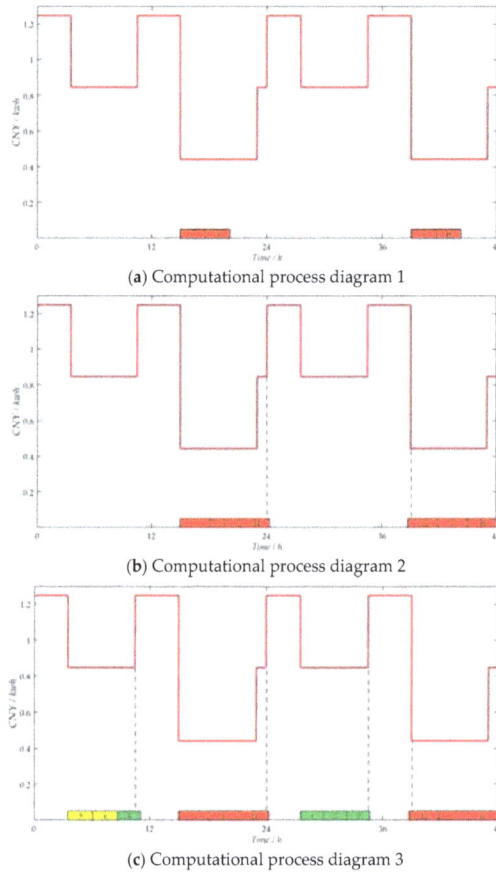

(a) Computational process diagram 1

(b) Computational process diagram 2

(c) Computational process diagram 3

**Figure 10.** Illustration of the computational process of GIH-F.

Returning to the real-life instance, let us now talk about the efficiency of the MILP model and the proposed algorithm. Currently, the company plans to produce 70-Model, 40-Model, and 100-Model rectangular parts in 7, 3, and 2 days respectively. That is, five parts are to be produced every day from 8:00 to 24:00. This suggests that it needs 12 days to process all the parts and the total electricity cost (*TEC*) can be computed as follows:

$$TEC_{70} = 4.7 \times (3.5 \times 1.2473 + 7 \times 0.8451 + 2.5 \times 1.2473) \times 7 = 440.8 \text{ CNY};$$

$$TEC_{40} = 4.4 \times (3.5 \times 1.2473 + 7 \times 0.8451 + 1.5 \times 1.2473) \times 3 = 160.4 \text{ CNY};$$

$$TEC_{100} = 5.3 \times (3.5 \times 1.2473 + 7 \times 0.8451 + 4.5 \times 1.2473 + 0.5 \times 0.4430) \times 2 = 170.8 \text{ CNY};$$

$$TEC = TEC_{70} + TEC_{40} + TEC_{100} = 772.0 \text{ CNY}.$$

Figure 11 is the scheduling result diagram of the real-life instance. It can be seen from Figure 11 that with our scheduling, the total electricity cost for processing all the parts is 447.9 CNY, which can be reduced by 42.0%.

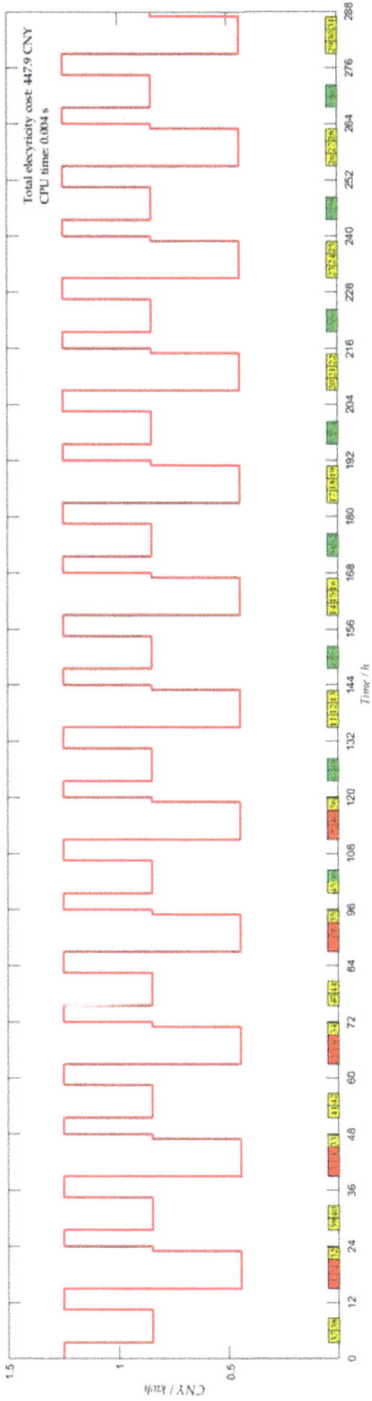

**Figure 11.** The scheduling result of the real-life instance.

*4.2. Randomly Generated Instances Studies*

In these tests, we utilize two sets of randomly generated instances to evaluate the performance of the proposed algorithm. For the first set of small-size instances, the range of jobs is [20, 100]. For the second set of large-size instances, the number of jobs is set as 500, 1000, 2000, 3000, 4000, and 5000. The processing time $t_i$ is randomly generated from a uniform distribution (30, 210) min and the power consumption per hour $p_i$ is randomly generated in (30, 100) kW. To measure the effect of the proposed algorithm, parameter $e$ is set as $e$ = 1.2, 1.5, 2, 3.

For each group of $n$ ($n \leq 2000$) and $e$, 10 random instances are generated, then the average values of 10 tests are calculated and recorded. When the number of jobs is set as 3000, 4000, and 5000, GIH has to run for more than 4 h (the longest is nearly two days) to find a feasible solution. Thus, considering the feasibility of the experiment, only 3 random instances are generated in such a group of tests. All the average values are recorded in Tables 5 and 6. Meanwhile, for the large-size instances, we add two rules to GIH to reduce the computation time without changing the computational accuracy. The improved algorithm is named GIH2.

**Table 5.** Computational results for the small-size instances.

| Instance | | | GIH | | | GIH-F | | |
|---|---|---|---|---|---|---|---|---|
| $n$ | $e$ | $m$ | $TEC_H$ | $CT_H$ (s) | $TEC_F$ | $CT_F$ (s) | $G$ (%) | $R$ |
| 20 | 1.2 | 12.0 | 1634.1 | 0.034 | 1632.5 | 0.002 | −0.10% | 17.0 |
| | 1.5 | 15.0 | 1370.1 | 0.037 | 1370.1 | 0.002 | 0.00% | 18.5 |
| | 2.0 | 19.0 | 1295.7 | 0.041 | 1295.1 | 0.002 | −0.05% | 20.5 |
| | 3.0 | 28.5 | 1168.0 | 0.056 | 1168.0 | 0.001 | 0.00% | 56.0 |
| 30 | 1.2 | 18.0 | 2414.6 | 0.064 | 2415.7 | 0.002 | 0.05% | 32.0 |
| | 1.5 | 20.0 | 2274.6 | 0.065 | 2274.1 | 0.002 | −0.02% | 32.5 |
| | 2.0 | 28.5 | 2005.0 | 0.083 | 2005.0 | 0.002 | 0.00% | 41.5 |
| | 3.0 | 39.5 | 1741.0 | 0.119 | 1741.0 | 0.002 | 0.00% | 59.5 |
| 40 | 1.2 | 23.0 | 3342.1 | 0.096 | 3342.0 | 0.004 | 0.00% | 24.0 |
| | 1.5 | 28.0 | 2900.1 | 0.109 | 2899.3 | 0.003 | −0.03% | 36.3 |
| | 2.0 | 36.0 | 2775.6 | 0.143 | 2775.0 | 0.003 | −0.02% | 47.7 |
| | 3.0 | 52.0 | 2380.3 | 0.194 | 2380.3 | 0.002 | 0.00% | 97.0 |
| 50 | 1.2 | 27.5 | 4242.5 | 0.137 | 4242.4 | 0.005 | 0.00% | 27.4 |
| | 1.5 | 34.0 | 3733.0 | 0.164 | 3732.6 | 0.003 | −0.01% | 54.7 |
| | 2.0 | 43.0 | 3243.8 | 0.212 | 3243.2 | 0.004 | −0.02% | 53.0 |
| | 3.0 | 64.5 | 2940.6 | 0.315 | 2940.6 | 0.003 | 0.00% | 105.0 |
| 60 | 1.2 | 34.0 | 4820.8 | 0.204 | 4819.7 | 0.006 | −0.02% | 34.0 |
| | 1.5 | 40.0 | 4536.5 | 0.224 | 4536.3 | 0.004 | 0.00% | 56.0 |
| | 2.0 | 52.0 | 4029.0 | 0.293 | 4028.9 | 0.004 | 0.00% | 73.3 |
| | 3.0 | 78.0 | 3544.1 | 0.464 | 3544.1 | 0.004 | 0.00% | 116.0 |
| 70 | 1.2 | 37.5 | 6133.5 | 0.249 | 6132.2 | 0.007 | −0.02% | 35.6 |
| | 1.5 | 46.0 | 5416.3 | 0.303 | 5416.2 | 0.007 | 0.00% | 43.3 |
| | 2.0 | 61.0 | 4676.0 | 0.413 | 4675.8 | 0.004 | 0.00% | 103.3 |
| | 3.0 | 90.0 | 4024.9 | 0.643 | 4024.9 | 0.005 | 0.00% | 128.6 |
| 80 | 1.2 | 43.0 | 7073.1 | 0.321 | 7072.9 | 0.009 | 0.00% | 35.7 |
| | 1.5 | 53.0 | 6049.6 | 0.401 | 6049.6 | 0.006 | 0.00% | 66.8 |
| | 2.0 | 68.5 | 5348.1 | 0.554 | 5348.1 | 0.007 | 0.00% | 79.1 |
| | 3.0 | 101.5 | 4514.4 | 0.868 | 4514.3 | 0.005 | 0.00% | 173.6 |
| 90 | 1.2 | 48.0 | 8128.5 | 0.399 | 8128.4 | 0.009 | 0.00% | 44.3 |
| | 1.5 | 58.0 | 6772.5 | 0.501 | 6772.4 | 0.011 | 0.00% | 45.5 |
| | 2.0 | 77.5 | 6172.7 | 0.697 | 6172.6 | 0.008 | 0.00% | 87.1 |
| | 3.0 | 104.1 | 5228.2 | 1.196 | 5228.2 | 0.009 | 0.00% | 132.9 |
| 100 | 1.2 | 53.5 | 8623.5 | 0.509 | 8622.9 | 0.017 | −0.01% | 29.9 |
| | 1.5 | 64.0 | 7607.1 | 0.614 | 7607.0 | 0.011 | 0.00% | 55.8 |
| | 2.0 | 86.5 | 6896.8 | 0.927 | 6896.8 | 0.014 | 0.00% | 66.2 |
| | 3.0 | 128.0 | 5815.2 | 1.482 | 5815.1 | 0.009 | 0.00% | 164.7 |

**Table 6.** Computational results for the large-size instances.

| Instance | | | GIH2 | | | GIH-F | |
|---|---|---|---|---|---|---|---|
| $n$ | $e$ | $m$ | $TEC_H$ | $CT_{H2}$ (s) | $TEC_F$ | $CT_F$ (s) | $R$ |
| 500 | 1.2 | 250.5 | 43,909.1 | 53.0 | 43,909.1 | 0.219 | 242.0 |
| | 1.5 | 315.0 | 38,637.8 | 52.2 | 38,637.9 | 0.187 | 279.1 |
| | 2.0 | 417.0 | 34,417.5 | 56.3 | 34,417.5 | 0.082 | 686.6 |
| | 3.0 | 628.5 | 28,948.1 | 56.9 | 28,948.0 | 0.093 | 611.8 |
| 1000 | 1.2 | 504.0 | 87,500.3 | 244.2 | 87,500.3 | 1.802 | 135.5 |
| | 1.5 | 628.5 | 77,598.3 | 230.3 | 77,597.0 | 0.873 | 263.8 |
| | 2.0 | 840.0 | 69,199.2 | 294.1 | 69,199.2 | 0.432 | 680.8 |
| | 3.0 | 1256.5 | 57,923.1 | 250.7 | 57,923.1 | 0.485 | 516.9 |
| 2000 | 1.2 | 1002.5 | 176,681.2 | 3910.8 | 176,680.7 | 15.701 | 249.1 |
| | 1.5 | 1255.5 | 155,205.3 | 3114.3 | 155,206.2 | 6.503 | 478.9 |
| | 2.0 | 1669.0 | 137,774.1 | 4316.2 | 137,774.1 | 3.346 | 1290.0 |
| | 3.0 | 2501.7 | 115,661.0 | 1785.8 | 115,661.0 | 3.574 | 499.7 |
| 3000 | 1.2 | 1511.7 | 263,511.1 | 19,136.9 | 263,511.6 | 46.551 | 411.1 |
| | 1.5 | 1880.0 | 231,954.6 | 14,429.7 | 231,954.9 | 25.560 | 564.5 |
| | 2.0 | 2483.3 | 205,368.8 | 19,759.8 | 205,368.8 | 11.219 | 1761.3 |
| | 3.0 | 3780.0 | 173,630.6 | 6571.9 | 173,630.6 | 12.432 | 528.6 |
| 4000 | 1.2 | 1991.7 | 352,975.8 | 59,016.2 | 352,977.0 | 107.610 | 548.4 |
| | 1.5 | 2511.7 | 306,983.3 | 43,971.5 | 306,983.3 | 66.669 | 659.5 |
| | 2.0 | 3335.0 | 275,694.8 | 60,539.8 | 275,694.8 | 26.281 | 2303.6 |
| | 3.0 | 4986.7 | 231,148.4 | 17,014.0 | 231,148.4 | 29.728 | 572.3 |
| 5000 | 1.2 | 2498.3 | 438,717.9 | 136,314.9 | 438,718.6 | 168.581 | 808.6 |
| | 1.5 | 3131.1 | 386,546.1 | 101,071.7 | 386,548.1 | 106.764 | 946.7 |
| | 2.0 | 4161.1 | 341,504.3 | 139,122.7 | 341,504.3 | 50.931 | 2731.6 |
| | 3.0 | 6257.8 | 291,685.7 | 51,471.0 | 291,685.7 | 58.821 | 875.0 |

Rule 1: If $t_i \leq \max_{k \in A}\{I_k\}$ and $kk = \operatorname{argmin}_{k \in A}\{t_i \leq I_k\}$, where $\operatorname{argmin}_{k \in A}\{t_i \leq I_k\}$ denotes the minimal index $k$ for $t_i \leq I_k$, then job $i$ can be directly inserted into the off-peak period $kk$ and the job no longer traverses all the non-full periods.

Rule 2: If $t_i \leq \min_{k'' \in \Gamma}\left\{I_{k''}\right\}$, $\max_{k \in A}\{I_k\} > 0$ or $\max_{k' \in B}\{I_{k'}\} > 0$, then job $i$ no longer traverses on-peak periods.

As mentioned above, when there are no jobs that need to traverse non-full on-peak periods, the time complexity of GIH-F is $O(n^2 m)$ and GIH is $O(n^2 m^2)$. This implies that GIH-F is $m$ times faster than GIH, theoretically, and the experimental data of the small-size instances in Table 5 can verify this conclusion. From Table 5, we can see that $R$ and $m$ are almost the same order of magnitude. In addition, all the small-size instances can be solved within 0.02 s using GIH-F. By and large, the computation time increases slightly with $n$, and parameter $e$ has no significant effect on the computation time, which indicates that the algorithm is quite stable. In addition, it can be seen from Table 5 that the smaller the parameter $e$ (i.e., the shorter the makespan), the higher the total electricity cost. Therefore, in a specific instance, the decision-makers can obtain a set of Pareto solutions by adjusting the makespan, and they can choose a solution according to actual needs. What is more, it is amazing to see that our algorithm not only greatly improves computation speed but also further improves the accuracy.

Table 6 shows that the number of periods increases quickly with the number of jobs. Since the time complexity of GIH is $O(n^2 m^2)$, it's runtime will rise sharply. To ensure the feasibility of the contrast tests, we add two rules (i.e., Rule 1 and Rule 2) to improve the computational speed of GIH without changing the computational accuracy.

Intuitively, the $CT_F$ is significantly less than $CT_{H2}$, which means that the designed filtering mechanism is efficient in dealing with large-scale instances. Specially, as $n = 5000$ and $e = 2.0$, our algorithm can solve a randomly generated instance within 1 min and maintain the same accuracy as GIH2, while GIH2 takes nearly 39 h, let alone GIH. Note that when $e$ is set as 3.0, the given makespan is very abundant and there is no job processed within an on-peak period in our experimental environments. Thus, according to Rule 2, all the jobs do not have to traverse on-peak periods, and then $CT_{H2}$ is greatly reduced. Conversely, when $e$ is set as 1.2, the number of periods decreases and the jobs are arranged very tightly. There will be many jobs inserted into the periods with higher electricity prices. Therefore, our algorithm should filter more positions with lower electricity prices and constantly judge whether the job needs to be moved. Obviously, all these operations may increase the computation time. Thus, when dealing with large-size instances and setting $e$ to 1.2 or 1.5, our algorithm runs longer, but the computation time is still far less than GIH2.

## 5. Conclusions and Prospects

This paper develops a new greedy insertion heuristic algorithm with a multi-stage filtering mechanism for single machine scheduling problems under TOU electricity tariffs. The algorithm can quickly filter out many impossible positions in the coarse granularity filtering stage and then each job to be inserted can search for its optimal position in a relatively large space in the fine granularity filtering stage. Compared with the classic greedy insertion algorithm, the greatest advantage of our algorithm is that it no longer needs to traverse all non-full periods, so the time complexity of the algorithm is quite low, and it can easily address the large-scale single machine scheduling problems under TOU electricity tariffs. The real case study demonstrates that with our scheduling, the total electricity cost for processing all the parts can be reduced by 42.0%. In addition, two sets of experimental instances are provided. The computational results demonstrate that the small-size instances can be solved within 0.02 s using our algorithm, and the accuracy of the algorithm is further improved. For the large-size instances, we add two rules to the classic greedy insertion algorithm, which reduces the computation time without changing the calculation precision, but the results show that our algorithm still outperforms it. Specifically, when addressing the large-scale instances with 5000 jobs, the computation speed of our algorithm improves by nearly 2700 times. Computational experiments also reveal that the smaller the parameter $e$, the more significant the filtering mechanism is.

This paper focuses on the single machine scheduling problems under the first type of TOU electricity tariffs. In our future research, we will continue to study the problem under the second type of TOU tariffs (i.e., the off-peak period lies between two mid-peak periods). In addition, we will also strive to improve our algorithm and extend it to other machine environments, such as parallel machines and flow shop.

**Acknowledgments:** This research is supported by the National Natural Science Foundation of China (Grant No. 71772002).

**Author Contributions:** Hongliang Zhang contributed to the overall idea, algorithm, and writing of the manuscript; Youcai Fang coded the algorithm in MATLAB and contributed to the detailed writing; Ruilin Pan contributed to the ideas and discussions on the scheduling problem under TOU electricity tariffs, as well as the revision, preparation, and publishing of the paper; Chuanming Ge analyzed the characteristics of the single machine scheduling problem under TOU electricity tariffs. All authors have read and approved the final manuscript.

**Conflicts of Interest:** The authors declare no conflict of interest.

## References

1. International Energy Agency. *World Energy Investment Outlook*; International Energy Agency (IEA): Paris, France, 2015.
2. Li, C.; Tang, Y.; Cui, L.; Li, P. A quantitative approach to analyze carbon emissions of CNC-based machining systems. *J. Intell. Manuf.* **2015**, *26*, 911–922. [CrossRef]
3. Jovane, F.; Yoshikawa, H.; Alting, L.; Boër, C.R.; Westkamper, E.; Williams, D.; Tseng, M.; Seliger, G.; Paci, A.M. The incoming global technological and industrial revolution towards competitive sustainable manufacturing. *CIRP Ann. Manuf. Technol.* **2008**, *57*, 641–659. [CrossRef]
4. Lu, C.; Gao, L.; Li, X.; Pan, Q.; Wang, Q. Energy-efficient permutation flow shop scheduling problem using a hybrid multi-objective backtracking search algorithm. *J. Clean. Prod.* **2017**, *144*, 228–238. [CrossRef]

5. Sun, Z.; Li, L. Opportunity estimation for real-time energy control of sustainable manufacturing systems. *IEEE Trans. Autom. Sci. Eng.* **2013**, *10*, 38–44. [CrossRef]

6. Park, C.W.; Kwon, K.S.; Kim, W.B.; Min, B.K.; Park, S.J.; Sung, I.H.; Yoon, Y.S.; Lee, K.S.; Lee, J.H.; Seok, J. Energy consumption reduction technology in manufacturing—A selective review of policies, standards, and research. *Int. J. Precis. Eng. Manuf.* **2009**, *10*, 151–173. [CrossRef]

7. Ding, J.Y.; Song, S.; Zhang, R.; Chiong, R.; Wu, C. Parallel machine scheduling under time-of-use electricity prices: New models and optimization approaches. *IEEE Trans. Autom. Sci. Eng.* **2016**, *13*, 1138–1154. [CrossRef]

8. Merkert, L.; Harjunkoski, I.; Isaksson, A.; Säynevirta, S.; Saarela, A.; Sand, G. Scheduling and energy—Industrial challenges and opportunities. *Comput. Chem. Eng.* **2015**, *72*, 183–198. [CrossRef]

9. Che, A.; Zeng, Y.; Ke, L. An efficient greedy insertion heuristic for energy-conscious single machine scheduling problem under time-of-use electricity tariffs. *J. Clean. Prod.* **2016**, *129*, 565–577. [CrossRef]

10. Longe, O.; Ouahada, K.; Rimer, S.; Harutyunyan, A.; Ferreira, H. Distributed demand side management with battery storage for smart home energy scheduling. *Sustainability* **2017**, *9*, 120. [CrossRef]

11. Shapiro, S.A.; Tomain, J.P. Rethinking reform of electricity markets. *Wake For. Law Rev.* **2005**, *40*, 497–543.

12. Zhang, H.; Zhao, F.; Fang, K.; Sutherland, J.W. Energy-conscious flow shop scheduling under time-of-use electricity tariffs. *CIRP Ann. Manuf. Technol.* **2014**, *63*, 37–40. [CrossRef]

13. Che, A.; Wu, X.; Peng, J.; Yan, P. Energy-efficient bi-objective single-machine scheduling with power-down mechanism. *Comput. Oper. Res.* **2017**, *85*, 172–183. [CrossRef]

14. He, F.; Shen, K.; Guan, L.; Jiang, M. Research on energy-saving scheduling of a forging stock charging furnace based on an improved SPEA2 algorithm. *Sustainability* **2017**, *9*, 2154. [CrossRef]

15. Pruhs, K.; Stee, R.V.; Uthaisombut, P. Speed scaling of tasks with precedence constraints. In Proceedings of the 3rd International Workshop on Approximation and Online Algorithms, Palma de Mallorca, Spain, 6–7 October 2005; Erlebach, T., Persiano, G., Eds.; Springer: Berlin, Germany, 2005; pp. 307–319.

16. Fang, K.; Uhan, N.; Zhao, F.; Sutherland, J.W. A new approach to scheduling in manufacturing for power consumption and carbon footprint reduction. *J. Manuf. Syst.* **2011**, *30*, 234–240. [CrossRef]

17. Dai, M.; Tang, D.; Giret, A.; Salido, M.A.; Li, W.D. Energy-efficient scheduling for a flexible flow shop using an improved genetic-simulated annealing algorithm. *Robot. Comput. Integr. Manuf.* **2013**, *29*, 418–429. [CrossRef]

18. Liu, C.H.; Huang, D.H. Reduction of power consumption and carbon footprints by applying multi-objective optimisation via genetic algorithms. *Int. J. Prod. Res.* **2014**, *52*, 337–352. [CrossRef]

19. Mouzon, G.; Yildirim, M.B.; Twomey, J. Operational methods for minimization of energy consumption of manufacturing equipment. *Int. J. Prod. Res.* **2007**, *45*, 4247–4271. [CrossRef]

20. Mouzon, G.; Yildirim, M. A framework to minimise total energy consumption and total tardiness on a single machine. *Int. J. Sustain. Eng.* **2008**, *1*, 105–116. [CrossRef]

21. Liu, C.; Yang, J.; Lian, J.; Li, W.; Evans, S.; Yin, Y. Sustainable performance oriented operational decision-making of single machine systems with deterministic product arrival time. *J. Clean. Prod.* **2014**, *85*, 318–330. [CrossRef]

22. Luo, H.; Du, B.; Huang, G.Q.; Chen, H.; Li, X. Hybrid flow shop scheduling considering machine electricity consumption cost. *Int. J. Prod. Econ.* **2013**, *146*, 423–439. [CrossRef]

23. Sharma, A.; Zhao, F.; Sutherland, J.W. Econological scheduling of a manufacturing enterprise operating under a time-of-use electricity tariff. *J. Clean. Prod.* **2015**, *108*, 256–270. [CrossRef]

24. Moon, J.-Y.; Shin, K.; Park, J. Optimization of production scheduling with time-dependent and machine-dependent electricity cost for industrial energy efficiency. *Int. J. Adv. Manuf. Technol.* **2013**, *68*, 523–535. [CrossRef]

25. Che, A.; Zhang, S.; Wu, X. Energy-conscious unrelated parallel machine scheduling under time-of-use electricity tariffs. *J. Clean. Prod.* **2017**, *156*, 688–697. [CrossRef]

26. Wang, S.; Liu, M.; Chu, F.; Chu, C. Bi-objective optimization of a single machine batch scheduling problem with energy cost consideration. *J. Clean. Prod.* **2016**, *137*, 1205–1215. [CrossRef]

27. Shrouf, F.; Ordieres-Meré, J.; García-Sánchez, A.; Ortega-Mier, M. Optimizing the production scheduling of a single machine to minimize total energy consumption costs. *J. Clean. Prod.* **2014**, *67*, 197–207. [CrossRef]

28. Gong, X.; Pessemier, T.D.; Joseph, W.; Martens, L. An energy-cost-aware scheduling methodology for sustainable manufacturing. In Proceedings of the 22nd CIRP Conference on Life Cycle Engineering (LCE) Univ New S Wales, Sydney, Australia, 7–9 April 2015; Kara, S., Ed.; Elsevier: Amsterdam, The Netherlands, 2015; pp. 185–190.
29. Fang, K.; Uhan, N.A.; Zhao, F.; Sutherland, J.W. Scheduling on a single machine under time-of-use electricity tariffs. *Ann. Oper. Res.* **2016**, *238*, 199–227. [CrossRef]

algorithms

MDPI

*Article*

# Single Machine Scheduling Problem with Interval Processing Times and Total Completion Time Objective

Yuri N. Sotskov * and Natalja G. Egorova

United Institute of Informatics Problems, National Academy of Sciences of Belarus, Surganova Street 6, Minsk 220012, Belarus; NataMog@yandex.by
* Correspondence: sotskov48@mail.ru; Tel.: +375-17-284-2120

Received: 2 March 2018; Accepted: 23 April 2018; Published: 7 May 2018

**Abstract:** We consider a single machine scheduling problem with uncertain durations of the given jobs. The objective function is minimizing the sum of the job completion times. We apply the stability approach to the considered uncertain scheduling problem using a relative perimeter of the optimality box as a stability measure of the optimal job permutation. We investigated properties of the optimality box and developed algorithms for constructing job permutations that have the largest relative perimeters of the optimality box. Computational results for constructing such permutations showed that they provided the average error less than 0.74% for the solved uncertain problems.

**Keywords:** scheduling; uncertain durations; single machine; total completion time

## 1. Introduction

Since real-life scheduling problems involve different forms of uncertainties, several approaches have been developed in the literature for dealing with uncertain scheduling problems. In a stochastic approach, job processing times are assumed to be random variables with the known probability distributions [1,2]. If one has no sufficient information to characterize the probability distribution of all random processing times, other approaches are needed [3–5]. In the approach of seeking a robust schedule [3,6], the decision-maker prefers a schedule that hedges against the worst-case scenario. A fuzzy approach [7–9] allows a scheduler to determine best schedules with respect to fuzzy processing times. A stability approach [10–12] is based on the stability analysis of the optimal schedules to possible variations of the numerical parameters. In this paper, we apply the stability approach to a single machine scheduling problem with uncertain processing times of the given jobs. In Section 2, we present the setting of the problem and the related results. In Section 3, we investigate properties of an optimality box of the permutation used for processing the given jobs. Efficient algorithms are derived for finding a job permutation with the largest relative perimeter of the optimality box. In Section 5, we develop an algorithm for finding an approximate solution for the uncertain scheduling problem. In Section 6, we report on the computational results for finding the approximate solutions for the tested instances. Section 7 includes the concluding remarks.

## 2. Problem Setting and the Related Results

There are given $n$ jobs $\mathcal{J} = \{J_1, J_2, ..., J_n\}$ to be processed on a single machine. The processing time $p_i$ of the job $J_i \in \mathcal{J}$ can take any real value from the given segment $[p_i^L, p_i^U]$, where $p_i^U \geq p_i^L > 0$. The exact value $p_i \in [p_i^L, p_i^U]$ of the job processing time remains unknown until completing the job $J_i \in \mathcal{J}$. Let $R_+^n$ denote a set of all non-negative $n$-dimensional real vectors. The set of all possible vectors $(p_1, p_2, ..., p_n) = p \in R_+^n$ of the job processing times is presented as the Cartesian product of the segments $[p_i^L, p_i^U]$: $T = \{p \in R_+^n \ : \ p_i^L \le p_i \le p_i^U, \ i \in \{1, 2, \ldots, n\}\} = [p_1^L, p_1^U] \times [p_2^L, p_2^U] \times \ldots \times [p_n^L, p_n^U] =: \times_{i=1}^n [p_i^L, p_i^U]$. Each vector $p \in T$ is called a **scenario**.

Let $S = \{\pi_1, \pi_2, \ldots, \pi_{n!}\}$ be a set of all permutations $\pi_k = (J_{k_1}, J_{k_2}, \ldots, J_{k_n})$ of the jobs $\mathcal{J}$. Given a scenario $p \in T$ and a permutation $\pi_k \in S$, let $C_i = C_i(\pi_k, p)$ denote the completion time of the job $J_i \in \mathcal{J}$ in the schedule determined by the permutation $\pi_k$. The criterion $\sum C_i$ denotes the minimization of the sum of job completion times: $\sum_{J_i \in \mathcal{J}} C_i(\pi_t, p) = \min_{\pi_k \in S} \left\{ \sum_{J_i \in \mathcal{J}} C_i(\pi_k, p) \right\}$, where the permutation $\pi_t = (J_{t_1}, J_{t_2}, \ldots, J_{t_n}) \in S$ is optimal for the criterion $\sum C_i$. This problem is denoted as $1|p_i^L \le p_i \le p_i^U| \sum C_i$ using the three-field notation $\alpha|\beta|\gamma$ [13], where $\gamma$ denotes the objective function. If scenario $p \in T$ is fixed before scheduling, i.e., $[p_i^L, p_i^U] = [p_i, p_i]$ for each job $J_i \in \mathcal{J}$, then the uncertain problem $1|p_i^L \le p_i \le p_i^U| \sum C_i$ is turned into the deterministic one $1|| \sum C_i$. We use the notation $1|p| \sum C_i$ to indicate an instance of the problem $1|| \sum C_i$ with the fixed scenario $p \in T$. Any instance $1|p| \sum C_i$ is solvable in $O(n \log n)$ time [14] since the following claim has been proven.

**Theorem 1.** *The job permutation $\pi_k = (J_{k_1}, J_{k_2}, \ldots, J_{k_n}) \in S$ is optimal for the instance $1|p| \sum C_i$ if and only if the following inequalities hold: $p_{k_1} \le p_{k_2} \le \cdots \le p_{k_n}$. If $p_{k_u} < p_{k_v}$, then job $J_{k_u}$ precedes job $J_{k_v}$ in any optimal permutation $\pi_k$.*

Since a scenario $p \in T$ is not fixed for the uncertain problem $1|p_i^L \le p_i \le p_i^U| \sum C_i$, the completion time $C_i$ of the job $J_i \in \mathcal{J}$ cannot be exactly determined for the permutation $\pi_k \in S$ before the completion of the job $J_i$. Therefore, the value of the objective function $\sum C_i$ for the permutation $\pi_k$ remains uncertain until jobs $\mathcal{J}$ have been completed.

**Definition 1.** *Job $J_v$ dominates job $J_w$ (with respect to T) if there is no optimal permutation $\pi_k \in S$ for the instance $1|p| \sum C_i$, $p \in T$, such that job $J_w$ precedes job $J_v$.*

The following criterion for the domination was proven in [15].

**Theorem 2.** *Job $J_v$ dominates job $J_w$ if and only if $p_v^U < p_w^L$.*

Since for the problem $\alpha|p_i^L \le p_i \le p_i^U|\gamma$, there does not usually exist a permutation of the jobs $\mathcal{J}$ being optimal for all scenarios $T$, additional objectives or agreements are often used in the literature. In particular, a robust schedule minimizing the worst-case regret to hedge against data uncertainty has been developed in [3,8,16–20]. For any permutation $\pi_k \in S$ and any scenario $p \in T$, the difference $\gamma_p^k - \gamma_p^t =: r(\pi_k, p)$ is called the regret for permutation $\pi_k$ with the objective function $\gamma$ equal to $\gamma_p^k$ under scenario $p$. The value $Z(\pi_k) = \max\{r(\pi_k, p) \ : \ p \in T\}$ is called the worst-case absolute regret. The value $Z^*(\pi_k) = \max\{\frac{r(\pi_k, p)}{\gamma_p^t} \ : \ p \in T\}$ is called the worst-case relative regret. While the deterministic problem $1|| \sum C_i$ is polynomially solvable [14], finding a permutation $\pi_t \in S$ minimizing the worst-case absolute regret $Z(\pi_k)$ or the relative regret $Z^*(\pi_k)$ for the problem $1|p_i^L \le p_i \le p_i^U| \sum C_i$ are binary NP-hard even for two scenarios [19,21]. In [6], a branch-and-bound algorithm was developed to find a permutation $\pi_k$ minimizing the absolute regret for the problem $1|p_i^L \le p_i \le p_i^U| \sum w_i C_i$, where jobs $J_i \in \mathcal{J}$ have weights $w_i > 0$. The computational experiments showed that the developed algorithm is able to find such a permutation $\pi_k$ for the instances with up to 40 jobs. The fuzzy scheduling technique was used in [7–9,22] to develop a fuzzy analogue of dispatching rules or to solve mathematical programming problems to determine a schedule that minimizes a fuzzy-valued objective function.

In [23], several heuristics were developed for the problem $1|p_i^L \le p_i \le p_i^U| \sum w_i C_i$. The computational experiments including different probability distributions of the processing times showed that the error of the best performing heuristic was about 1% of the optimal objective function value $\sum w_i C_i$ obtained after completing the jobs when their factual processing times became known.

The stability approach [5,11,12] was applied to the problem $1|p_i^L \le p_i \le p_i^U|\sum w_i C_i$ in [15], where hard instances were generated and solved by the developed Algorithm MAX-OPTBOX with the average error equal to 1.516783%. Algorithm MAX-OPTBOX constructs a job permutation with the optimality box having the largest perimeter.

In Sections 3–6, we continue the investigation of the optimality box for the problem $1|p_i^L \le p_i \le p_i^U|\sum C_i$. The proven properties of the optimality box allows us to develop Algorithm 2 for constructing a job permutation with the optimality box having the largest relative perimeter and Algorithm 3, which outperforms Algorithm MAX-OPTBOX for solving hard instances of the problem $1|p_i^L \le p_i \le p_i^U|\sum C_i$. Algorithm 3 constructs a job permutation $\pi_k$, whose optimality box provides the minimal value of the error function introduced in Section 5. Randomly generated instances of the problem $1|p_i^L \le p_i \le p_i^U|\sum C_i$ were solved by Algorithm 3 with the average error equal to 0.735154%.

## 3. The Optimality Box

Let $M$ denote a subset of the set $N = \{1, 2, \ldots, n\}$. We define an **optimality box** for the job permutation $\pi_k \in S$ for the problem $1|p_i^L \le p_i \le p_i^U|\sum C_i$ as follows.

**Definition 2.** *The maximal (with respect to the inclusion) rectangular box $\mathcal{OB}(\pi_k, T) = \times_{k_i \in M}[l_{k_i}^*, u_{k_i}^*] \subseteq T$ is called the optimality box for the permutation $\pi_k = (J_{k_1}, J_{k_2}, \ldots, J_{k_n}) \in S$ with respect to $T$, if the permutation $\pi_k$ being optimal for the instance $1|p|\sum C_i$ with the scenario $p = (p_1, p_2, \ldots, p_n) \in T$ remains optimal for the instance $1|p'|\sum C_i$ with any scenario $p' \in \mathcal{OB}(\pi_k, T) \times \{\times_{k_j \in N \setminus M}[p_{k_j}, p_{k_j}]\}$. If there does not exist a scenario $p \in T$ such that the permutation $\pi_k$ is optimal for the instance $1|p|\sum C_i$, then $\mathcal{OB}(\pi_k, T) = \varnothing$.*

Any variation $p'_{k_i}$ of the processing time $p_{k_i}$, $J_{k_i} \in \mathcal{J}$, within the maximal segment $[l_{k_i}^*, u_{k_i}^*]$ indicated in Definition 2 cannot violate the optimality of the permutation $\pi_k \in S$ provided that the inclusion $p'_{k_i} \in [l_{k_i}^*, u_{k_i}^*]$ holds. The non-empty maximal segment $[l_{k_i}^*, u_{k_i}^*]$ with the inequality $l_{k_i}^* \le u_{k_i}^*$ and the length $u_{k_i}^* - l_{k_i}^* \ge 0$ indicated in Definition 2 is called an **optimality segment** for the job $J_{k_i} \in \mathcal{J}$ in the permutation $\pi_k$. If the maximal segment $[l_{k_i}^*, u_{k_i}^*]$ is empty for job $J_{k_i} \in \mathcal{J}$, we say that job $J_{k_i}$ **has no optimality segment** in the permutation $\pi_k$. It is clear that if job $J_{k_i}$ has no optimality segment in the permutation $\pi_k$, then the inequality $l_{k_i}^* > u_{k_i}^*$ holds.

### 3.1. An Example of the Problem $1|p_i^L \le p_i \le p_i^U|\sum C_i$

Following to [15], the notion of a block for the jobs $\mathcal{J}$ may be introduced for the problem $1|p_i^L \le p_i \le p_i^U|\sum C_i$ as follows.

**Definition 3.** *A maximal set $B_r = \{J_{r_1}, J_{r_2}, \ldots, J_{r_{|D_r|}}\} \subset \mathcal{J}$ of the jobs, for which the inequality $\max_{J_{r_i} \in B_r} p_{r_i}^L \le \min_{J_{r_i} \in B_r} p_{r_i}^U$ holds, is called a **block**. The segment $[b_r^L, b_r^U]$ with $b_r^L = \max_{J_{r_i} \in B_r} p_{r_i}^L$ and $b_r^U = \min_{J_{r_i} \in B_r} p_{r_i}^U$ is called a **core** of the block $B_r$.*

If job $J_i \in \mathcal{J}$ belongs to only one block, we say that job $J_i$ is **fixed** (in this block). If job $J_k \in \mathcal{J}$ belongs to two or more blocks, we say that job $J_k$ is **non-fixed**. We say that the block $B_v$ is **virtual**, if there is no fixed job in the block $B_v$.

**Remark 1.** *Any permutation $\pi_k \in S$ determines a distribution of all non-fixed jobs to their blocks. Due to the fixings of the positions of the non-fixed jobs, some virtual blocks may be destroyed for the permutation $\pi_k$. Furthermore, the cores of some non-virtual blocks may be increased in the permutation $\pi_k$.*

We demonstrate the above notions on a small example with input data given in Table 1. The segments $[p_i^L, p_i^U]$ of the job processing times are presented in a coordinate system in Figure 1, where the abscissa axis is used for indicating the segments given for the job processing times and the ordinate axis for the jobs from the set $\mathcal{J}$. The cores of the blocks $B_1, B_2, B_3$ and $B_4$ are dashed in Figure 1.

**Table 1.** Input data for the problem $1|p_i^L \le p_i \le p_i^U|\sum C_i$.

| $i$ | 1 | 2 | 3 | 4 | 5 | 6 | 7 | 8 | 9 | 10 |
|---|---|---|---|---|---|---|---|---|---|---|
| $p_i^L$ | 6 | 7 | 6 | 1 | 8 | 17 | 15 | 24 | 25 | 26 |
| $p_i^U$ | 11 | 11 | 12 | 19 | 16 | 21 | 35 | 28 | 27 | 27 |

There are four blocks in this example as follows: $\{B_1, B_2, B_3, B_4\} =: B$. The jobs $J_1, J_2, J_3, J_4$ and $J_5$ belong to the block $B_1$. The jobs $J_4, J_5$ and $J_7$ are non-fixed. The remaining jobs $J_1, J_2, J_3, J_6, J_8, J_9$ and $J_{10}$ are fixed in their blocks. The block $B_2$ is virtual. The jobs $J_4, J_5$ and $J_7$ belong to the virtual block $B_2$. The jobs $J_4, J_6$, and $J_7$ belong to the block $B_3$. The jobs $J_7, J_8, J_9$ and $J_{10}$ belong to the block $B_4$.

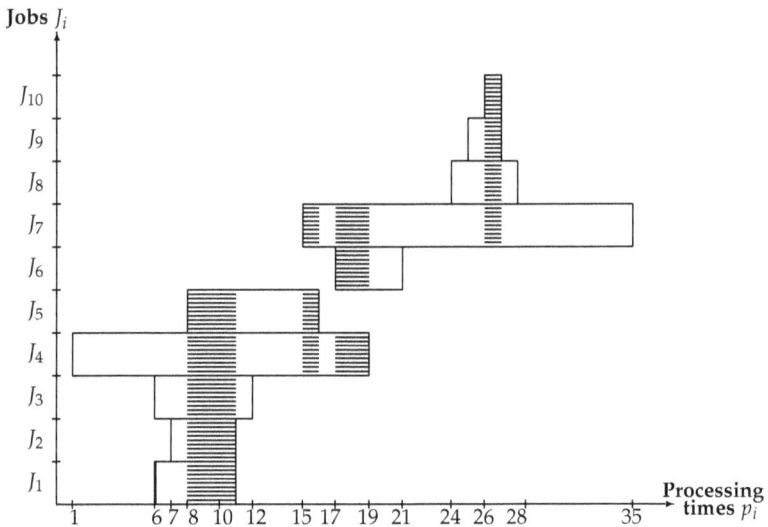

**Figure 1.** The segments $[p_i^L, p_i^U]$ given for the feasible processing times of the jobs $J_i \in \mathcal{J} = \{J_1, J_2, \dots, J_{10}\}$ (the cores of the blocks $B_1, B_2, B_3$ and $B_4$ are dashed).

*3.2. Properties of a Job Permutation Based on Blocks*

The proof of Lemma 1 is based on Procedure 1.

**Lemma 1.** *For the problem* $1|p_i^L \le p_i \le p_i^U|\sum C_i$, *the set* $B = \{B_1, B_2, \dots, B_m\}$ *of all blocks can be determined in* $O(n \log n)$ *time.*

**Proof.** The right bound $b_1^U$ of the core of the first block $B_1 \in B$ is determined as follows: $b_1^U = \min_{J_i \in \mathcal{J}} p_i^U$. Then, all jobs included in the block $B_1$ may be determined as follows: $B_1 = \{J_i \in \mathcal{J} : p_i^L \le b_1^U \le p_i^U\}$. The left bound $b_1^L$ of the core of the block $B_1$ is determined as follows: $b_1^L = \max_{J_i \in B_1} p_i^L$. Then, one can determine the second block $B_2 \in B$ via applying the above procedure to the subset of set $\mathcal{J}$ without jobs $J_i$, for which the equality $b_1^U = p_i^U$ holds. This process is continued until determining the last block $B_m \in B$. Thus, one can use the above procedure (we call it Procedure 1) for constructing the set $B = \{B_1, B_2, \dots, B_m\}$ of all blocks for the problem $1|p_i^L \le p_i \le p_i^U|\sum C_i$. Obviously, Procedure 1 has the complexity $O(n \log n)$. □

Any block from the set $B$ has the following useful property.

**Lemma 2.** *At most two jobs in the block $B_r \in B$ may have the non-empty optimality segments in any fixed permutation $\pi_k \in S$.*

**Proof.** Due to Definition 3, the inclusion $[b_r^L, b_r^U] \subseteq [p_i^L, p_i^U]$ holds for each job $J_i \in B_r$. Thus, due to Definition 2, only job $J_{k_i}$, which is the first one in the permutation $\pi_k$ among the jobs from the block $B_r$, and only job $J_{k_v} \in B_r$, which is the last one in the permutation $\pi_k$ among the jobs from the block $B_r$, may have the non-empty optimality segments. It is clear that these non-empty segments look as follows: $[p_{k_i}^L, u_{k_i}^*]$ and $[l_{k_v}^*, p_{k_v}^U]$, where $u_{k_i}^* \leq b_r^L$ and $b_r^U \leq l_{k_v}^*$. $\square$

**Lemma 3.** *If $\mathcal{OB}(\pi_k, T) \neq \varnothing$, then any two jobs $J_v \in B_r$ and $J_w \in B_s$, which are fixed in different blocks, $r < s$, must be ordered in the permutation $\pi_k \in S$ with the increasing of the left bounds (and the right bounds as well) of the cores of their blocks, i.e., the permutation $\pi_k$ looks as follows: $\pi_k = (\ldots, J_v, \ldots, J_w, \ldots)$, where $b_r^L < b_s^L$.*

**Proof.** For any two jobs $J_v \in B_r$ and $J_w \in B_s$, $r < s$, the condition $[p_v^L, p_v^U] \cap [p_w^L, p_w^U] = \varnothing$ holds. Therefore, the same permutation $\pi_k = (\ldots, J_v, \ldots, J_w, \ldots)$ is obtained if jobs $J_v \in B_r$ and $J_w \in B_s$ are ordered either in the increasing of the left bounds of the cores of their blocks or in the increasing of the right bounds of the cores of their blocks. We prove Lemma 3 by a contradiction. Let the condition $\mathcal{OB}(\pi_k, T) \neq \varnothing$ hold. However, we assume that there are fixed jobs $J_v \in B_r$ and $J_w \in B_s$, $r < s$, which are located in the permutation $\pi_k \in S$ in the decreasing order of the left bounds of their blocks $B_r$ and $B_s$. Note that blocks $B_r$ and $B_s$ cannot be virtual.

Due to our assumption, the permutation $\pi_k$ looks as follows: $\pi_k = (\ldots, J_w, \ldots, J_v, \ldots)$, where $b_r^L < b_s^L$. Using Definition 3, one can convince that the cores of any two blocks have no common point. Thus, the inequality $b_r^L < b_s^L$ implies the inequality $b_r^U < b_s^L$. The inequalities $p_v^L \leq p_v \leq b_r^U < b_s^L \leq p_w^L \leq p_w$ hold for any feasible processing time $p_v$ and any feasible processing time $p_w$, where $p \in T$. Hence, the inequality $p_v < p_w$ holds as well, and due to Theorem 1, the permutation $\pi_k \in S$ cannot be optimal for any fixed scenario $p \in T$. Thus, the equality $\mathcal{OB}(\pi_k, T) = \varnothing$ holds due to Definition 2. This contradiction with our assumption completes the proof. $\square$

Next, we assume that blocks in the set $B = \{B_1, B_2, \ldots, B_m\}$ are numbered according to the increasing of the left bounds of their cores, i.e., the inequality $b_v^L < b_u^L$ implies $v < u$. Due to Definition 3, each block $B_r = \{J_{r_1}, J_{r_2}, \ldots, J_{r_{|B_r|}}\}$ may include jobs of the following four types. If $p_{r_i}^L = b_r^L$ and $p_{r_i}^U = b_r^U$, we say that job $J_{r_i}$ is a **core** job in the block $B_r$. Let $B_r^*$ be a set of all core jobs in the block $B_r$. If $p_{r_i}^L < b_r^L$, we say that job $J_{r_i}$ is a **left** job in the block $B_r$. If $p_{r_i}^U > b_r^U$, we say that job $J_{r_i}$ is a **right** job in the block $B_r$. Let $B_r^-$ ($B_r^+$) be a set of all left (right) jobs in the block $B_r$. Note that some job $J_{r_i} \in B_r$ may be **left-right** job in the block $B_r$, since it is possible that $B \setminus \{B_r^* \sqcup B_r^- \cup B_r^+\} \neq \varnothing$.

Two jobs $J_v$ and $J_w$ are **identical** if both equalities $p_v^L = p_w^L$ and $p_v^U = p_w^U$ hold. Obviously, if the set $B_r \in B$ is a singleton, $|B_r| = 1$, then the equality $B_r = B_r^*$ holds. Furthermore, the latter equality cannot hold for a virtual block $B_t \in B$ since any trivial block must contain at least two non-fixed jobs.

**Theorem 3.** *For the problem $1|p_i^L \leq p_i \leq p_i^U| \sum C_i$, any permutation $\pi_k \in S$ has an empty optimality box $\mathcal{OB}(\pi_k, T) = \varnothing$, if and only if for each block $B_r \in B$, either condition $|B_r| = |B_r^*| \geq 2$ holds or $B_r = B_r^- \cup B_r^+$, all jobs in the set $B_r^-$ (in the set $B_r^+$) are identical and the following inequalities hold: $|B_r^-| \geq 2$ and $|B_r^+| \geq 2$.*

**Proof.** *Sufficiency.* It is easy to prove that there is no virtual block in the considered set $B$. First, we assume that for each block $B_r \in B$, the condition $|B_r| = |B_r^*| \geq 2$ holds.

Due to Lemma 3, in the permutation $\pi_k \in S$ with the non-empty optimality box $\mathcal{OB}(\pi_k, T) \neq \varnothing$, all jobs must be located in the increasing order of the left bounds of the cores of their blocks. However, in the permutation $\pi_k \in S$, the following two equalities hold for each block $B_r \in B$: $\min_{J_i \in B_r} p_i^U = \max_{J_i \in B_r} p_i^U$ and $\max_{J_i \in B_r} p_i^L = \min_{J_i \in B_r} p_i^L$. Since $|B_r| \geq 2$, there is no job $J_v \in B_r$, which has an optimality segment in any fixed permutation $\pi_k \in S$ due to Lemma 2. Hence, the equality $\mathcal{OB}(\pi_k, T) = \emptyset$ must hold, if the condition $|B_r| = |B_r^*| \geq 2$ holds for each block $B_r \in B$.

We assume that there exists a block $B_r \in B$ such that all jobs are identical in the set $B_r^-$ with $|B_r^-| \geq 2$ and all jobs are identical in the set $B_r^+$ with $|B_r^+| \geq 2$. Thus, for the permutation $\pi_k \in S$, the equalities $\min_{J_i \in B_r^-} p_i^U = \max_{J_i \in B_r^-} p_i^U$ and $\max_{J_i \in B_r^-} p_i^L = \min_{J_i \in B_r^-} p_i^L$ hold. Since $|B_r^-| \geq 2$ and all jobs are identical in the set $B_r^-$, there is no job $J_v \in B_r^-$, which has the optimality segment in any fixed permutation $\pi_k \in S$.

Similarly, one can prove that there is no job $J_v \in B_r^+$, which has the optimality segment in any fixed permutation $\pi_k \in S$. Thus, we conclude that if the condition of Theorem 3 holds for each block $B_r \in B$, then each job $J_i \in \mathcal{J}$ has no optimality segment in any fixed permutation $\pi_k \in S$. Thus, if the condition of Theorem 3 holds, the equality $\mathcal{OB}(\pi_k, T) = \emptyset$ holds for any permutation $\pi_k \in S$.

*Necessity.* We prove the necessity by a contradiction. We assume that any permutation $\pi_k \in S$ has an empty optimality box $\mathcal{OB}(\pi_k, T) = \emptyset$. Let the condition $|B_r| = |B_r^*| \geq 2$ do not hold. If $|B_r| = |B_r^*| = 1$, the set $B_r$ is a singleton and so job $J_{r_1} \in B_r$ has the optimality segment $[p_{r_1}^L, p_{r_1}^U]$ with the length $p_{r_1}^U - p_{r_1}^L \geq 0$ in the permutation $\pi_{k^*} \in S$, where all jobs are ordered according to the increasing of the left bounds of the cores of their blocks. Let the following condition hold:

$$|B_r| > |B_r^*| \geq 2. \tag{1}$$

The condition (1) implies that there exists at least one left job or right job or left-right job $J_{r_i} \in B_r$. If job $J_{r_i}$ is a left job or a left-right job (a right job) in the block $B_r$, then job $J_{r_i}$ must be located on the first place (on the last place, respectively) in the above permutation $\pi_{k^*} \in S$ among all jobs from the block $B_r$. All jobs from the set $B_r \setminus \{B_r^* \cup \{J_{r_i}\}\}$ must be located between job $J_{r_v} \in B_r^*$ and job $J_{r_w} \in B_r^*$. Therefore, the left job $J_{r_i}$ or the left-right job $J_{r_i}$ (the right job $J_{r_i}$) has the following optimality segment $[p_{r_i}^L, b_r^L]$ (segment $[b_r^U, p_{r_i}^U]$, respectively) with the length $b_r^L - p_{r_i}^L > 0$ (the length $p_{r_i}^U - b_r^U > 0$, respectively) in the permutation $\pi_{k^*}$. Let the equality $B_r = B_r^- \cup B_r^+$ do not hold. If there exist a job $J_{r_i} \in B_r^*$, this job must be located between the jobs $J_{r_v} \in B_r^-$ and $J_{r_w} \in B_r^-$. It is clear that the job $J_{r_i}$ has the optimality segment $[b_r^L, b_r^U]$ with the length $b_r^U - b_r^L > 0$ in the permutation $\pi_{k^*}$.

Let the following conditions hold: $B_r = B_r^- \cup B_r^+$, $|B_r^-| \geq 2$, $|B_r^+| \geq 2$. However, we assume that jobs from the set $B_r^-$ are not identical. Then the job $J_{r_u} \in B_r^-$ with the largest segment $[p_{r_u}^L, p_{r_u}^U]$ among the jobs from the set $B_r^-$ must be located in the permutation $\pi_{k^*}$ before other jobs from the set $B_r^-$. It is clear that the job $J_{r_u}$ has the optimality segment in the permutation $\pi_{k^*}$.

Similarly, we can construct the permutation $\pi_{k^*}$ with the non-empty optimality box, if the jobs from the set $B_r^+$ are not identical.

Let the equality $B_r = B_r^- \cup B_r^+$ hold. Let all jobs in the set $B_r^-$ (and all jobs in the set $B_r^+$) be identical. However, we assume that the equality $|B_r^-| = 1$ holds. The job $J_{r_i} \in B_r^-$ has the optimality segment in the permutation $\pi_{k^*}$, if the job $J_{r_i}$ is located before other jobs in the set $B_r$.

Similarly, one can construct the permutation $\pi_{k^*}$ such that job $J_{r_i} \in B_r^+$ has the optimality segment in the permutation $\pi_{k^*}$, if the inequality $|B_r^+| \geq 2$ is violated. Thus, we conclude that in all cases of the violating of the condition of Theorem 3, we can construct the permutation $\pi_{k^*} \in S$ with the non-empty optimality box $\mathcal{OB}(\pi_{k^*}, T) \neq \emptyset$. This conclusion contradicts to our assumption that any permutation $\pi_k \in S$ has an empty optimality box $\mathcal{OB}(\pi_k, T) = \emptyset$. $\square$

Next, we present Algorithm 1 for constructing the optimality box $\mathcal{OB}(\pi_k, T)$ for the fixed permutation $\pi_k \in S$. In steps 1–4 of Algorithm 1, the problem $1|\hat{p}_i^L \leq p_i \leq \hat{p}_i^U|\sum C_i$ with the reduced segments of the job processing times is constructed. It is clear that the optimality box for the permutation $\pi_k \in S$ for the problem $1|p_i^L \leq p_i \leq p_i^U|\sum C_i$ coincides with the optimality box for the same permutation for the problem $1|\hat{p}_i^L \leq p_i \leq \hat{p}_i^U|\sum C_i$, where given segments of the job processing times are reduced. In steps 5 and 6, the optimality box for the permutation $\pi_k$ for the problem $1|\hat{p}_i^L \leq p_i \leq \hat{p}_i^U|\sum C_i$ is constructed. It takes $O(n)$ time for constructing the optimality box $\mathcal{OB}(\pi_k, T)$ for any fixed permutation $\pi_k \in S$ using Algorithm 1.

---

**Algorithm 1**

---

**Input**: Segments $[p_i^L, p_i^U]$ for the jobs $J_i \in \mathcal{J}$. The permutation $\pi_k = (J_{k_1}, J_{k_2}, \dots, J_{k_n})$.
**Output**: The optimality box $\mathcal{OB}(\pi_k, T)$ for the permutation $\pi_k \in S$.
    *Step 1*: **FOR** $i = 1$ to $n$ **DO** set $\hat{p}_i^L = p_i^L$, $\hat{p}_i^U = p_i^U$ **END FOR**
        set $t_L = \hat{p}_1^L$, $t_U = \hat{p}_n^U$
    *Step 2*: **FOR** $i = 2$ to $n$ **DO**
        **IF** $\hat{p}_i^L < t_L$ **THEN** set $\hat{p}_i^L = t_L$ **ELSE** set $t_L = \hat{p}_i^L$ **END FOR**
    *Step 3*: **FOR** $i = n - 1$ to $1$ **STEP** $-1$ **DO**
        **IF** $\hat{p}_i^U > t_U$ **THEN** set $\hat{p}_i^U = t_U$ **ELSE** set $t_U = \hat{p}_i^U$ **END FOR**
    *Step 4*: Set $\hat{p}_0^U = \hat{p}_1^L$, $\hat{p}_{n+1}^L = \hat{p}_n^U$.
    *Step 5*: **FOR** $i = 1$ to $n$ **DO** set $\hat{d}_{k_i}^- = \max \left\{ \hat{p}_{k_i}^L, \hat{p}_{k_{i-1}}^U \right\}$, $\hat{d}_{k_i}^+ = \min \left\{ \hat{p}_{k_i}^U, \hat{p}_{k_{i+1}}^L \right\}$
    **END FOR**
    *Step 6*: Set $\mathcal{OB}(\pi_k, T) = \times_{\hat{d}_i^- \le \hat{d}_i^+} \left[ \hat{d}_{k_i}^-, \hat{d}_{k_i}^+ \right]$ **STOP**.

---

*3.3. The Largest Relative Perimeter of the Optimality Box*

If the permutation $\pi_k \in S$ has the non-empty optimality box $\mathcal{OB}(\pi_k, T) \ne \varnothing$, then one can calculate the length of the relative perimeter of this box as follows:

$$Per\mathcal{OB}(\pi_k, T) = \sum_{J_{k_i} \in \mathcal{J}(\pi_k)} \frac{u_{k_i}^* - l_{k_i}^*}{p_{k_i}^U - p_{k_i}^L}, \tag{2}$$

where $\mathcal{J}(\pi_k)$ denotes the set of all jobs $J_i \in \mathcal{J}$ having optimality segments $[l_{k_i}^*, u_{k_i}^*]$ with the positive lengths, $l_{k_i}^* < u_{k_i}^*$, in the permutation $\pi_k$. It is clear that the inequality $l_{k_i}^* < u_{k_i}^*$ may hold only if the inequality $p_{k_i}^L < p_{k_i}^U$ holds. Theorem 3 gives the sufficient and necessary condition for the smallest value of $Per\mathcal{OB}(\pi_k, T)$, i.e., the equalities $\mathcal{J}(\pi_k) = \varnothing$ and $Per\mathcal{OB}(\pi_k, T) = 0$ hold for each permutation $\pi_k \in S$. A necessary and sufficient condition for the largest value of $Per\mathcal{OB}(\pi_k, T) = n$ is given in Theorem 4. The sufficiency proof of Theorem 4 is based on Procedure 2.

**Theorem 4.** *For the problem $1|p_i^L \le p_i \le p_i^U| \sum C_i$, there exists a permutation $\pi_k \in S$, for which the equality $Per\mathcal{OB}(\pi_k, T) = n$ holds, if and only if for each block $B_r \in B$, either $|B_r| = |B_r^*| = 1$ or $B_r = B_r^- \cup B_r^+$ with $|B_r| = 2$.*

**Proof.** *Sufficiency.* Let the equalities $|B_r| = |B_r^*| = 1$ hold for each block $B_r \in B$. Therefore, both equalities $B_r = B_r^*$ and $|B| = n$ hold. Due to Theorem 2 and Lemma 3, all jobs must be ordered with the increasing of the left bounds of the cores of their blocks in each permutation $\pi_k \in S$ such that $\mathcal{OB}(\pi_k, T) \ne \varnothing$. Each job $J_{k_i} \in \mathcal{J}$ in the permutation $\pi_k = (J_{k_1}, J_{k_2}, \dots, J_{k_n})$ has the optimality segments $[l_{k_i}^*, u_{k_i}^*]$ with the maximal possible length

$$u_{k_i}^* - l_{k_i}^* = p_{k_i}^U - p_{k_i}^L. \tag{3}$$

Hence, the desired equalities

$$Per\mathcal{OB}(\pi_k, T) = \sum_{J_{k_i} \in \mathcal{J}(\pi_k)} \frac{p_{k_i}^U - p_{k_i}^L}{p_{k_i}^U - p_{k_i}^L} = n \tag{4}$$

hold, if the equalities $|B_r| = |B_r^*| = 1$ hold for each block $B_r \in B$.

Let there exist a block $B_r \in B$ such that the equalities $B_r = B_r^- \cup B_r^+$ and $|B_r| = 2$ hold. It is clear that the equalities $|B_r^-| = |B_r^+| = 1$ hold as well, and job $J_{k_i}$ from the set $B_r^-$ (from the set $B_r^+$) in the permutation $\pi_k$ has the optimality segments $[l_{k_i}^*, u_{k_i}^*]$ with the maximal possible length determined

27

in ($3$). Hence, the equalities ($4$) hold, if there exist blocks $B_r \in B$ with the equalities $B_r = B_r^- \cup B_r^+$ and $|B_r| = 2$. This sufficiency proof contains the description of Procedure 2 with the complexity $O(n)$.

*Necessary.* If there exists at least one block $B_r \in B$ such that neither condition $|B_r| = |B_r^*| = 1$ nor condition $B_r = B_r^- \cup B_r^+$, $|B_r| = 2$ hold, then the equality ($3$) is not possible for at least one job $J_{k_i} \in B_r$. Therefore, the inequality $Per\mathcal{OB}(\pi_k, T) < n$ holds for any permutation $\pi_k \in S$. $\square$

The length of the relative perimeter $Per\mathcal{OB}(\pi_k, T)$ may be used as a stability measure for the optimal permutation $\pi_k$. If the permutation $\pi_k \in S$ has the non-empty optimality box $\mathcal{OB}(\pi_k, T)$ with a larger length of the relative perimeter than the optimality box $\mathcal{OB}(\pi_t, T)$ has, $\pi_k \neq \pi_t \in S$, then the permutation $\pi_k \in S$ may provide a smaller value of the objective function $\sum C_i$ than the permutation $\pi_t$. One can expect that the inequality $\sum_{J_i \in \mathcal{J}} C_i(\pi_k, p) \leq \sum_{J_i \in \mathcal{J}} C_i(\pi_t, p)$ holds for more scenarios $p$ from the set $T$ than the opposite inequality $\sum_{J_i \in \mathcal{J}} C_i(\pi_k, p) > \sum_{J_i \in \mathcal{J}} C_i(\pi_t, p)$. Next, we show how to construct a permutation $\pi_k \in S$ with the largest value of the relative perimeter $Per\mathcal{OB}(\pi_k, T)$. The proof of Theorem $4$ is based on Procedure 3.

**Theorem 5.** *If the equality $B_1 = \mathcal{J}$ holds, then it takes $O(n)$ time to find the permutation $\pi_k \in S$ and the optimality box $\mathcal{OB}(\pi_k, T)$ with the largest length of the relative perimeter $Per\mathcal{OB}(\pi_k, T)$ among all permutations $S$.*

**Proof.** Since the equality $B_1 = \mathcal{J}$ holds, each permutation $\pi_k \in S$ looks as follows: $\pi_k = (J_{k_1}, J_{k_2}, \dots, J_{k_n})$, where $\{J_{k_1}, J_{k_2}, \dots, J_{k_n}\} = B_1$. Due to Lemma $2$, only job $J_{k_1}$ may have the optimality segment (this segment looks as follows: $[p_{k_1}^L, u_{k_1}^*]$) and only job $J_n$ may have the following optimality segment $[l_{k_n}^*, p_{k_n}^U]$. The length $u_{k_1}^* - p_{k_1}^L$ of the optimality segment $[p_{k_1}^L, u_{k_1}^*]$ for the job $J_{k_1}$ is determined by the second job $J_{k_2}$ in the permutation $\pi_k$. Similarly, the length $p_{k_n}^U - l_{k_n}^*$ of the optimality segment $[l_{k_n}^*, p_{k_n}^U]$ for the job $J_{k_n}$ is determined by the second-to-last job $J_{k_{n-1}}$ in the permutation $\pi_k$. Hence, to find the optimality box $\mathcal{OB}(\pi_k, T)$ with the largest value of $Per\mathcal{OB}(\pi_k, T)$, it is needed to test only four jobs $J_{k_1}, J_{k_2}, J_{k_{n-1}}$ and $J_{k_n}$ from the block $B_1 = \mathcal{J}$, which provide the largest relative optimality segments for the jobs $J_{k_1}$ and $J_{k_n}$.

To this end, we should choose the job $J_{k_i}$ such that the following equality holds: $p_{k_i}^L = \min_{J_{k_s} \in B_1} \frac{b_1^U - p_{k_s}^L}{p_{k_s}^U - p_{k_s}^L}$. Then, if $B_1 \setminus \{J_{k_i}\} \neq \varnothing$, we should choose the job $J_{k_v}$ such that the equality $p_{k_v}^U = \max_{J_{k_s} \in B_1 \setminus \{J_{k_i}\}} \frac{p_{k_s}^U - b_1^L}{p_{k_s}^U - p_{k_s}^L}$ holds. Then, if $B_1 \setminus \{J_{k_i}, J_{k_v}\} \neq \varnothing$, we should choose the job $J_{k_{i+1}}$ such that the equality $p_{k_{i+1}}^L = \min_{J_{k_s} \in B_1 \setminus \{J_{k_i}, J_{k_v}\}} p_{k_s}^L$ holds. Then, if $B_1 \setminus \{J_{k_i}, J_{k_{i+1}}, J_{k_v}\} \neq \varnothing$, we should choose the job $J_{k_{v-1}}$ such that the equality $p_{k_{v-1}}^U = \max_{J_{k_s} \in B_1 \setminus \{J_{k_i}, J_{k_{i+1}}, J_{k_v}\}} p_{k_s}^U$ holds. Thus, to determine the largest value of $Per\mathcal{OB}(\pi_k, T)$, one has either to test a single job or to test two jobs or three jobs or to choose and test four jobs from the block $B_1$, where $|B_1| \geq 4$, independently of how large the cardinality $|B_1|$ of the set $B_1$ is. In the worst case, one has to test at most $4! = 24$ permutations of the jobs chosen from the set $B_1$. Due to the direct testing of the chosen jobs, one selects a permutation of $|B_1|$ jobs, which provides the largest length of the relative perimeter $Per\mathcal{OB}(\pi_k, T)$ for all $|B_1|!$ permutations $\pi_k$. If $B_1 = \mathcal{J}$, the above algorithm (we call it Procedure 3) for finding the permutation $\pi_k$ with the largest value of $Per\mathcal{OB}(\pi_k, T)$ takes $O(n)$ time. Theorem $5$ has been proven. $\square$

**Remark 2.** *If $J_i \notin B_r \in B$, then the job $J_i$ must be located either on the left side from the core of the block $B_r$ or on the right side from this core in any permutation $\pi_k$ having the largest relative perimeter of the optimality box $\mathcal{OB}(\pi_k, T)$. Hence, job $J_i$ must precede (or succeed, respectively) each job $J_v \in B_k$. The permutation $\pi_t$ of the jobs from the block $B_r$ obtained using Procedure 3 described in the proof of Theorem $5$, where jobs $B_r$ are sequenced, remains the same if job $J_i \notin B_r \in B$ is added to the permutation $\pi_t$.*

**Lemma 4.** *Let there exist two adjacent blocks $B_r \in B$ and $B_{r+1} \in B$, such that the equality $B_r \cap B_{r+1} = \varnothing$ holds. Then the problem $1|p_i^L \leq p_i \leq p_i^U|\sum C_i$ can be decomposed into the subproblem $P_1$ with the set of jobs $\mathcal{J}_1 := \bigcup_{k=1}^r B_k$ and the subproblem $P_2$ with the set of jobs $\mathcal{J}_2 := \bigcup_{k=r+1}^m B_k = \mathcal{J} \setminus \mathcal{J}_1$. The optimality box*

$\mathcal{OB}(\pi_k, T)$ with the largest length of the relative perimeter may be obtained as the Cartesian product of the optimality boxes constructed for the subproblems $P_1$ and $P_2$.

**Proof.** Since the blocks $B_r$ and $B_{r+1}$ have no common jobs, the inequality $p_u^U < p_v^L$ holds for each pair of jobs $J_u \in \bigcup_{k=1}^r B_k$ and $J_v \in \bigcup_{k=r+1}^m B_k$. Due to Theorem 2, any job $J_u \in \bigcup_{k=1}^r B_k$ dominates any job $J_v \in \bigcup_{k=r+1}^m B_k$. Thus, due to Definition 1, there is no optimal permutation $\pi_k \in S$ for the instance $1|p| \sum C_i$ with a scenario $p \in T$ such that job $J_v$ precedes job $J_u$ in the permutation $\pi_k$. Due to Definition 2, a non-empty optimality box $\mathcal{OB}(\pi_k, T)$ is possible only if job $J_u$ is located before job $J_v$ in the permutation $\pi_r$. The optimality box $\mathcal{OB}(\pi_k, T)$ with the largest relative perimeter for the problem $1|p_i^L \le p_i \le p_i^U| \sum C_i$ may be obtained as the Cartesian product of the optimality boxes with the largest length of the relative perimeters constructed for the subproblems $P_1$ and $P_2$, where the set of jobs $\bigcup_{k=1}^r B_k$ and the set of jobs $\bigcup_{k=r+1}^m B_k$ are considered separately one from another. $\square$

Let $B^*$ denote a subset of all blocks of the set $B$, which contain only fixed jobs. Let $\mathcal{J}^*$ denote a set of all non-fixed jobs in the set $\mathcal{J}$. Let the set $B(J_u) \subseteq B^*$ denote a set of all blocks containing the non-fixed job $J_u \in \mathcal{J}^*$. Theorem 5, Lemma 6 and the constructive proof of the following claim allows us to develop an $O(n \log n)$-algorithm for constructing the permutation $\pi_k$ with the largest value of $PerOB(\pi_k, T)$ for the special case of the problem $1|p_i^L \le p_i \le p_i^U| \sum C_i$. The proof of Theorem 6 is based on Procedure 4.

**Theorem 6.** *Let each block $B_r \in B$ contain at most one non-fixed job. The permutation $\pi_k \in S$ with the largest value of the relative perimeter $PerOB(\pi_k, T)$ is constructed in $O(n \log n)$ time.*

**Proof.** There is no virtual block in the set $B$, since any virtual block contains at least two non-fixed jobs while each block $B_r \in B$ contains at most one non-fixed job for the considered problem $1|p_i^L \le p_i \le p_i^U| \sum C_i$. Using $O(n \log n)$-Procedure 1 described in the proof of Lemma 1, we construct the set $B = \{B_1, B_2, \dots, B_m\}$ of all blocks. Using Lemma 3, we order the jobs in the set $\mathcal{J} \setminus \mathcal{J}^*$ according to the increasing of the left bounds of the cores of their blocks. As a result, all $n - |\mathcal{J}^*|$ jobs are linearly ordered since all jobs $\mathcal{J} \setminus \mathcal{J}^*$ are fixed in their blocks. Such an ordering takes $O(n \log n)$ time due to Lemma 1.

Since each block $B_r \in B$ contain at most one non-fixed job, the equality $B(J_u) \cap B(J_v) = \varnothing$ holds for each pair of jobs $J_u \in B(J_u) \subseteq B \setminus B^*$ and $J_v \in B(J_v) \subseteq B \setminus B^*$, $u \ne v$. Furthermore, the problem $1|p_i^L \le p_i \le p_i^U| \sum C_i$ may be decomposed into $h = |\mathcal{J}^*| + |B^*|$ subproblems $P_1, P_2, \dots, P_{|\mathcal{J}^*|}, P_{|\mathcal{J}^*|+1}, \dots, P_h$ such that the condition of Lemma 4 holds for each pair $P_r$ and $P_{r+1}$ of the adjacent subproblems, where $1 \le r \le h - 1$. Using Lemma 4, we decompose the problem $1|p_i^L \le p_i \le p_i^U| \sum C_i$ into $h$ subproblems $\{P_1, P_2, \dots, P_{|\mathcal{J}^*|}, P_{|\mathcal{J}^*|+1}, \dots, P_h\} = \mathcal{P}$. The set $\mathcal{P}$ is partitioned into two subsets, $\mathcal{P} = \mathcal{P}^1 \cup \mathcal{P}^2$, where the subset $\mathcal{P}^1 = \{P_1, P_2, \dots, P_{|\mathcal{J}^*|}\}$ contains all subproblems $P_f$ containing all blocks from the set $B(J_{i_f})$, where $J_{i_f} \in \mathcal{J}^*$ and $|\mathcal{P}^1| = |\mathcal{J}^*|$.

The subset $\mathcal{P}^2 = \{P_{|\mathcal{J}^*|+1}, \dots, P_h\}$ contains all subproblems $P_d$ containing one block $B_{r_d}$ from the set $B^*$, $|\mathcal{P}^2| = |B^*|$. If subproblem $P_d$ belongs to the set $\mathcal{P}^2$, then using $O(n)$-Procedure 3 described in the proof of Theorem 5, we construct the optimality box $\mathcal{OB}(\pi^{(d)}, T^{(d)})$ with the largest length of the relative perimeter for the subproblem $P_d$, where $\pi^{(d)}$ denotes the permutation of the jobs $B_{r_d} \subset B^*$ and $T^{(d)} \subset T$. It takes $O(|B_{r_d}|)$ time to construct a such optimality box $\mathcal{OB}(\pi^{(d)}, T^{(d)})$.

If subproblem $P_f$ belongs to the set $\mathcal{P}^1$, it is necessary to consider all $|B(J_{i_f})|$ fixings of the job $J_{i_f}$ in the block $B_{f_j}$ from the set $B(J_{i_f}) = \{B_{f_1}, B_{f_2}, \dots, B_{f_{|B(J_{i_f})|}}\}$. Thus, we have to solve $|B(J_{i_f})|$ subproblems $P_f^g$, where job $J_{i_f}$ is fixed in the block $B_{f_g} \in B(J_{i_f})$ and job $J_{i_f}$ is deleted from all other blocks $B(J_{i_f}) \setminus \{B_{f_g}\}$. We apply Procedure 3 for constructing the optimality box $\mathcal{OB}(\pi^{(g)}, T^{(f)})$ with the largest length of the relative perimeter $PerOB(\pi^{(g)}, T^{(f)})$ for each subproblem $P_f^g$, where $g \in \{1, 2, \dots, |B(J_{i_f})|\}$. Then, we have to choose the largest length of the relative perimeter among $|B(J_{i_f})|$ constructed ones: $PerOB(\pi^{(f^*)}, T^{(f)}) = \max_{g \in \{1,2,\dots,|B(J_{i_f})|\}} PerOB(\pi^{(g)}, T^{(f)})$.

Due to Lemma 4, the optimality box $\mathcal{OB}(\pi_k, T)$ with the largest relative perimeter for the original problem $1|p_i^L \le p_i \le p_i^U|\sum C_i$ is determined as the Cartesian product of the optimality boxes $\mathcal{OB}(\pi^{(d)}, T^{(d)})$ and $\mathcal{OB}(\pi^{(f^*)}, T^{(f)})$ constructed for the subproblems $P_d \in \mathcal{P}^2$ and $P_f \in \mathcal{P}^1$, respectively. The following equality holds: $\mathcal{OB}(\pi_k, T) = \times_{P_f \in \mathcal{P}^1} \mathcal{OB}(\pi^{(f^*)}, T^{(f)}) \times (\times_{P_d \in \mathcal{P}^2} \mathcal{OB}(\pi^{(d)}, T^{(d)}))$. The permutation $\pi_k$ with the largest length of the relative perimeter of the optimality box $\mathcal{OB}(\pi_k, T)$ for the problem $1|p_i^L \le p_i \le p_i^U|\sum C_i$ is determined as the concatenation of the corresponding permutations $\pi^{(f^*)}$ and $\pi^{(d)}$. Using the complexities of the Procedures 1 and 3, we conclude that the total complexity of the described algorithm (we call it Procedure 4) can be estimated by $O(n \log n)$. □

**Lemma 5.** *Within constructing a permutation $\pi_k$ with the largest relative perimeter of the optimality box* $\mathcal{OB}(\pi_k, T)$, *any job $J_i$ may be moved only within the blocks $B(J_i)$.*

**Proof.** Let job $J_i$ be located in the block $B_r$ in the permutation $\pi_k$ such that $J_i \notin B_r$. Then, either the inequality $p_v^L > p_i^U$ or the inequality $p_v^U < p_i^L$ holds for each job $J_v \in B_r$. If $p_v^L > p_i^U$, job $J_u$ dominates job $J_i$ (due to Theorem 2). If $p_v^U < p_i^L$, job $J_i$ dominates job $J_u$. Hence, if job $J_i$ is located in the permutation $\pi_k$ between jobs $J_v \in B_r$ and $J_w \in B_r$, then $\mathcal{OB}(\pi_k, T) = \varnothing$ due to Definition 2. □

Due to Lemma 5, if job $J_i$ is fixed in the block $B_k \in B$ (or is non-fixed but distributed to the block $B_k \in B$), then job $J_i$ is located within the jobs from the block $B_k$ in any permutation $\pi_k$ with the largest relative perimeter of the optimality box $\mathcal{OB}(\pi_k, T)$.

## 4. An Algorithm for Constructing a Job Permutation with the Largest Relative Perimeter of the Optimality Box

Based on the properties of the optimality box, we next develop Algorithm 2 for constructing the permutation $\pi_k$ for the problem $1|p_i^L \le p_i \le p_i^U|\sum C_i$, whose optimality box $\mathcal{OB}(\pi_k, T)$ has the largest relative perimeter among all permutations in the set $S$.

---

**Algorithm 2**

---

**Input**: Segments $[p_i^L, p_i^U]$ for the jobs $J_i \in \mathcal{J}$.
**Output**: The permutation $\pi_k \in S$ with the largest relative perimeter $Per\mathcal{OB}(\pi_k, T)$.
  *Step 1*: **IF** the condition of Theorem 3 holds
        **THEN** $\mathcal{OB}(\pi_k, T) = \varnothing$ for any permutation $\pi_k \in S$ **STOP**.
  *Step 2*: **IF** the condition of Theorem 4 holds
        **THEN** construct the permutation $\pi_k \in S$ such that $Per\mathcal{OB}(\pi_k, T) = n$
        using Procedure 2 described in the proof of Theorem 4 **STOP**.
  *Step 3*: **ELSE** determine the set $B$ of all blocks using the
        $O(n \log n)$-Procedure 1 described in the proof of Lemma 1
  *Step 4*: Index the blocks $B = \{B_1, B_2, \ldots, B_m\}$ according to increasing
        left bounds of their cores (Lemma 3)
  *Step 5*: **IF** $\mathcal{J} = B_1$ **THEN** problem $1|p_i^L \le p_i \le p_i^U|\sum C_i$ is called problem $P_1$
        (Theorem 5) set $i = 0$ **GOTO** step 8 **ELSE** set $i = 1$
  *Step 6*: **IF** there exist two adjacent blocks $B_{r^*} \in B$ and $B_{r^*+1} \in B$ such
        that $B_{r^*} \cap B_{r^*+1} = \varnothing$; let $r$ denote the minimum of the above index $r^*$
        in the set $\{1, 2, \ldots, m\}$ **THEN** decompose the problem $P$ into
        subproblem $P_1$ with the set of jobs $\mathcal{J}_1 = \cup_{k=1}^{r} B_k$ and subproblem $P_2$
        with the set of jobs $\mathcal{J}_2 = \cup_{k=r+1}^{m} B_k$ using Lemma 4;
        set $P = P_1$, $\mathcal{J} = \mathcal{J}_1$, $B = \{B_1, B_2, \ldots, B_r\}$ **GOTO** step 7 **ELSE**
  *Step 7*: **IF** $B \ne \{B_1\}$ **THEN GOTO** step 9 **ELSE**
  *Step 8*: Construct the permutation $\pi^{s(i)}$ with the largest relative perimeter
        $Per\mathcal{OB}(\pi^{s(i)}, T)$ using Procedure 3 described in the proof of
        Theorem 5 **IF** $i = 0$ or $\mathcal{J}_2 = B_m$ **GOTO** step 12 **ELSE**

---

---

**Algorithm 2**

*Step 9:* **IF** there exists a block in the set $B$ containing more than one non-fixed jobs **THEN** construct the permutation $\pi^{s(i)}$ with the largest relative perimeter $Per\mathcal{OB}(\pi^{s(i)}, T)$ for the problem $P_1$ using Procedure 5 described in Section 4.1 **GOTO** step 11

*Step 10:* **ELSE** construct the permutation $\pi^{s(i)}$ with the largest relative perimeter $Per\mathcal{OB}(\pi^{s(i)}, T)$ for the problem $P_1$ using $O(n \log n)$-Procedure 4 described in the proof of Theorem 6

*Step 11:* Construct the optimality box $\mathcal{OB}(\pi^{s(i)}, T)$ for the permutation $\pi^{s(i)}$ using Algorithm 1 **IF** $\mathcal{J}_2 \neq B_m$ and $i \geq 1$ **THEN** set $i := i + 1$, $P = P_2$, $\mathcal{J} = \mathcal{J}_2$, $B = \{B_{r+1}, B_{r+2}, \ldots, B_m\}$ **GOTO** step 6 **ELSE IF** $\mathcal{J}_2 = B_m$ **THEN GOTO** step 8

*Step 12:* **IF** $i > 0$ **THEN** set $v = i$, determine the permutation $\pi_k = (\pi^{s(1)}, \pi^{s(2)}, \ldots, \pi^{s(v)})$ and the optimality box $\mathcal{OB}(\pi_k, T) = \times_{i \in \{1,2,\ldots,v\}} \mathcal{OB}(\pi^{s(i)}, T)$ **GOTO** step 13 **ELSE** $\mathcal{OB}(\pi_k, T) = \mathcal{OB}(\pi^{s(0)}, T)$

*Step 13:* The optimality box $\mathcal{OB}(\pi_k, T)$ has the largest value of $Per\mathcal{OB}(\pi_k, T)$ **STOP**.

---

*4.1. Procedure 5 for the Problem $1 | p_i^L \leq p_i \leq p_i^U | \sum C_i$ with Blocks Including More Than One Non-Fixed Jobs*

For solving the problem $P_1$ at step 9 of the Algorithm 2, we use Procedure 5 based on dynamic programming. Procedure 5 allows us to construct the permutation $\pi_k \in S$ for the problem $1 | p_i^L \leq p_i \leq p_i^U | \sum C_i$ with the largest value of $Per\mathcal{OB}(\pi_k, T)$, where the set $B$ consists of more than one block, $m \geq 2$, the condition of Lemma 4 does not hold for the jobs $\mathcal{J} = \{J_1, J_2, \ldots, J_n\} =: \mathcal{J}(\mathcal{B}_m^0)$, where $\mathcal{B}_m^0$ denotes the following set of blocks: $\{B_1, B_2, \ldots, B_m\} =: \mathcal{B}_m^0$. Moreover, the condition of Theorem 6 does hold for the set $B = \mathcal{B}_m^0$ of the blocks, i.e., there is a block $B_r \in \mathcal{B}_m^0$ containing more than one non-fixed jobs. For the problem $1 | p_i^L \leq p_i \leq p_i^U | \sum C_i$ with $\mathcal{J} = \{J_1, J_2, \ldots, J_n\} = \mathcal{J}(\mathcal{B}_m^0)$, one can calculate the following tight upper bound $Per_{max}$ on the length of the relative perimeter $Per\mathcal{OB}(\pi_k, T)$ of the optimality box $\mathcal{OB}(\pi_k, T)$:

$$Per_{max} = 2 \cdot |B \setminus \underline{B}| + |\underline{B}| \geq Per\mathcal{OB}(\pi_k, T), \tag{5}$$

where $\underline{B}$ denotes the set of all blocks $B_r \in B$ which are singletons, $|B_r| = 1$. The upper bound (5) on the relative perimeter $Per\mathcal{OB}(\pi_k, T)$ holds, since the relative optimality segment $\frac{u_{k_j}^* - l_{k_j}^*}{p_{k_j}^U - p_{k_j}^L}$ for any job $J_i \in \mathcal{J}$ is not greater than one. Thus, the sum of the relative optimality segments for all jobs $J_i \in \mathcal{J}$ cannot be greater than $2m$.

Instead of describing Procedure 5 in a formal way, we next describe the first two iterations of Procedure 5 along with the application of Procedure 5 to a small example with four blocks and three non-fixed jobs (see Section 4.2). Let $\mathcal{T} = (V, E)$ denote the solution tree constructed by Procedure 5 at the last iteration, where $V$ is a set of the vertexes presenting states of the solution process and $E$ is a set of the edges presenting transformations of the states to another ones. A subgraph of the solution tree $\mathcal{T} = (V, E)$ constructed at the iteration $h$ is denoted by $\mathcal{T}_h = (V_h, E_h)$. All vertexes $i \in V$ of the solution tree have their ranks from the set $\{0, 1, \ldots, m = |B|\}$. The vertex 0 in the solution tree $\mathcal{T}_h = (V_h, E_h)$ has a zero rank. The vertex 0 is characterized by a partial job permutation $\pi^0(\varnothing; \varnothing)$, where the non-fixed jobs are not distributed to their blocks.

All vertexes of the solution tree $\mathcal{T}_h$ having the first rank are generated at iteration 1 from vertex 0 via distributing the non-fixed jobs $\mathcal{J}[B_1]$ of the block $B_1$, where $\mathcal{J}[B_1] \subseteq B_1$. Each job $J_v \in \mathcal{J}[B_1]$ must be distributed either to the block $B_1$ or to another block $B_j \in B$ with the inclusion $J_v \in \mathcal{J}[B_j]$. Let $B_k^t$ denote a set of all non-fixed jobs $J_i \in B_k$, which are not distributed to their blocks at the iterations with the numbers less than $t$. A partial permutation of the jobs is characterized by the notation $\pi^u(B_k; \mathcal{J}_{[u]})$, where $u$ denotes the vertex $u \in V_t$ in the constructed solution tree $\mathcal{T}_t = (V_t, E_t)$ and $\mathcal{J}_{[u]}$ denotes the

31

non-fixed jobs from the set $\mathcal{J}[B_k]$, which are distributed to the block $B_k$ in the vertex $j \in V_t$ of the solution tree such that the arc $(j, u)$ belongs to the set $E_t$.

At the first iteration of Procedure 5, the set $\mathcal{P}(\mathcal{J}[B_1^1]) = \{\mathcal{J}[B_1^1], \ldots, \varnothing\}$ of all subsets of the set $\mathcal{J}[B_1^1]$ is constructed and $2^{|\mathcal{J}[B_1^1]|}$ partial permutations are generated for all subsets $\mathcal{P}(\mathcal{J}[B_1^1])$ of the non-fixed jobs $\mathcal{J}[B_1^1]$. The constructed solution tree $\mathcal{T}_1 = (V_1, E_1)$ consists of the vertexes $0, 1, \ldots, 2^{|\mathcal{J}[B_1^1]|}$ and $2^{|\mathcal{J}[B_1^1]|}$ arcs connecting vertex 0 with the other vertexes in the tree $\mathcal{T}_1$. For each generated permutation $\pi^u(B_k; \mathcal{J}_{[u]})$, where $\mathcal{J}_{[u]} \in \mathcal{P}(\mathcal{J}[B_k^l])$, the penalty $\phi_u$ determined in (6) is calculated, which is equal to the difference between the lengths of the maximal possible relative perimeter $Per_{max}$ of the optimality box $\mathcal{OB}(\pi_k, T)$ and the relative perimeter of the optimality box $\mathcal{OB}(\pi^u(B_k; \mathcal{J}_{[u]}), T)$, which may be constructed for the permutation $\pi^u(B_k; \mathcal{J}_{[u]})$:

$$PerOB(\pi^u(B_k; \mathcal{J}_{[u]}), T) = \sum_{i \leq k} Per_{max}^u - \phi_u, \tag{6}$$

where $Per_{max}^u$ denotes the maximal length of the relative perimeter of the permutation $\pi^u(B_k; \mathcal{J}_{[u]}), T)$. The penalty $\phi_u$ is calculated using $O(n)$-Procedure 3 described in the proof of Theorem 5. The complete permutation $\pi^{0 \to u}(B_k; \mathcal{J}_{[u]})$ with the end $\pi^u(B_k; \mathcal{J}_{[u]})$ is determined based on the permutations $\pi^s(B_c; \mathcal{J}_{[s]})$, where each vertex $s$ belongs to the chain $(0 \to u)$ between vertexes 0 and $u$ in the solution tree $\mathcal{T}_t = (V_t, E_t)$ and each block $B_c$ belongs to the set $B^u = \{B_1, B_2, \ldots, B_k\}$. The permutation $\pi^{0 \to u}(B_k; \mathcal{J}_{[u]})$ includes all jobs, which are fixed in the blocks $B_c \in B^u$ or distributed to their blocks $B_c \in B^u$ in the optimal order for the penalty $\phi_u$.

The aim of Procedure 5 is to construct a complete job permutation $\pi^f(B_m; \mathcal{J}_{[f]}) = \pi_k \in S$ such that the penalty $\phi_f$ is minimal for all job permutations from the set $S$. At the next iteration, a partial permutation $\pi^s(B_2; \mathcal{J}_{[s]})$ is chosen from the constructed solution tree such that the penalty $\phi_s$ is minimal among the permutations corresponding to the leafs of the constructed solution tree.

At the second iteration, the set $\mathcal{P}(\mathcal{J}[B_2^2]) = \{\mathcal{J}[B_2^2], \ldots, \varnothing\}$ of all subsets of the set $\mathcal{J}[B_2^2]$ is generated and $2^{|\mathcal{J}[B_2^2]|}$ permutations for all subsets $\mathcal{P}(\mathcal{J}[B_2^2])$ of the jobs from the block $B_2$ are constructed. For each generated permutation $\pi^v(B_2^2; \mathcal{J}_{[v]})$, where $\mathcal{J}_{[v]} \in \mathcal{P}(\mathcal{J}[B_2^2])$, the penalty $\phi_v$ is calculated using the equality $\phi_v = \phi_l + \Delta\phi_k$, where the edge $[l, v]$ belongs to the solution tree $\mathcal{T}_2 = (V_2, E_2)$ and $\Delta\phi_k$ denotes the penalty reached for the optimal permutation $\pi^{0 \to v}(B_2; \mathcal{J}_{[v]})$ constructed from the permutation $\pi^v(B_1; \mathcal{J}_{[v]})$ using $O(n)$-Procedure 3. For the consideration at the next iteration, one chooses the partial permutation $\pi^d(B_c; \mathcal{J}_{[d]})$ with the minimal value of the penalty for the partial permutations in the leaves of the constructed solution tree.

The whole solution tree is constructed similarly until there is a partial permutation with a smaller value of the penalty $\phi_t$ in the constructed solution tree.

### 4.2. The Application of Procedure 5 to the Small Example

Table 1 presents input data for the example of the problem $1|p_i^L \leq p_i \leq p_i^U|\sum C_i$ described in Section 3.1. The jobs $J_4$, $J_5$ and $J_7$ are non-fixed: $\mathcal{J}[B_1] = \{J_4, J_5\}$, $\mathcal{J}[B_2] = \{J_4, J_5, J_7\}$, $\mathcal{J}[B_3] = \{J_4, J_6, J_7\}$, $\mathcal{J}[B_4] = \{J_7\}$. The job $J_4$ must be distributed either to the block $B_1$, $B_2$, or to the block $B_3$. The job $J_5$ must be distributed either to the block $B_1$, or to the block $B_2$. The job $J_7$ must be distributed either to the block $B_2$, $B_3$, or to the block $B_4$. The relative perimeter of the optimality box $\mathcal{OB}(\pi_k, T)$ for any job permutation $\pi_k \in S$ cannot be greater than $Per_{max} = 4 \times 2 = 8$ due to the upper bound on the relative perimeter $PerOB(\pi_k, T)$ given in (5).

At the first iteration of Procedure 5, the set $\mathcal{P}(\mathcal{J}[B_1^1]) = \{\varnothing, \{J_4\}, \{J_5\}, \{J_4, J_5\}\}$ is constructed and permutations $\pi^1(B_1^1; \varnothing)$, $\pi^2(B_1^1; J_4)$, $\pi^3(B_1^1; J_5)$ and $\pi^4(B_1^1; J_4, J_5)$ are generated. For each element of the set $\mathcal{P}(\mathcal{J}[B_1^1])$, we construct a permutation with the maximal length of the relative perimeter of the optimality box and calculate the penalty. We obtain the following permutations with their penalties: $\pi^{0 \to 1}(B_1^1; \varnothing) = (J_1, J_2, J_3)$, $\phi_1 = 1\frac{19}{30} \approx 1.633333$; $\pi^{0 \to 2}(B_1^1; J_4) = (J_4, J_2, J_1, J_3)$, $\phi_2 = 1, 5$; $\pi^{0 \to 3}(B_1^1; J_5) = (J_1, J_5, J_2, J_3)$, $\phi_3 = 1\frac{13}{30} \approx 1.433333$; $\pi^{0 \to 4}(B_1^1; J_4, J_5) = (J_4, J_5, J_1, J_2, J_3)$, $\phi_4 = 1\frac{13}{30} \approx 1.433333$.

At the second iteration, the block $B_2$ is destroyed, and we construct the permutations $\pi^5(B_3^2;\varnothing)$ and $\pi^6(B_3^2;J_7)$ from the permutation $\pi^4(B_1^1;J_4,J_5)$. We obtain the permutation $\pi^{0\to5}(B_3^2;\varnothing) = (J_4,J_5,J_1,J_2,J_3,J_6)$ with the penalty $\phi_5 = 4\frac{13}{30} \approx 4.433333$ and the permutation $\pi^{0\to6}(B_3^2;J_7) = (J_4,J_5,J_1,J_2,J_3,J_7,J_6)$ with the penalty $\phi_6 = 5\frac{1}{3} \approx 5.333333$. Since all non-fixed jobs are distributed in the permutation $\pi^{0\to6}(B_1^1;J_3,J_8)$, we obtain the complete permutation $J_4,J_5,J_1,J_2,J_3,J_6,J_8,J_{10},J_9,J_7) \in S$ with the final penalty $\phi_6^* = 6\frac{7}{12} \approx 6.583333$. From the permutation $\pi^{0\to3}(B_1^1;J_5)$, we obtain the permutations with their penalties: $\pi^{0\to7}(B_3^3;\varnothing) = (J_1,J_5,J_3,J_2,J_4,J_6)$, $\phi_7 = 5\frac{29}{90} \approx 5.322222$ and $\pi^{0\to8}(B_3^3;J_7) = (J_1,J_5,J_3,J_2,J_4,J_7,J_6)$, $\phi_8 = 51$. We obtain the complete permutation $\pi = (J_1,J_5,J_3,J_2,J_4,J_7,J_6,J_9,J_{10},J_8)$ with the penalty $\phi_8^* = 6.35$.

We obtain the following permutations with their penalties: $\pi^{0\to9}(B_2^4;\varnothing) = (J_4,J_2,J_3,J_1,J_5)$, $\phi_9 = 3\frac{1}{24} \approx 3.041667$ and $\pi^{0\to10}(B_2^4;J_7) = (J_4,J_2,J_1,J_3,J_7,J_5)$, $\phi_{10} = 3.5$ from the permutation $\pi^{0\to2}(B_1^1;J_4)$. We obtain the following permutations with their penalties: $\pi^{0\to11}(B_2^5;\varnothing) = (J_1,J_2,J_3,J_5,)$, $\phi_{11} = 3.175$, $\pi^{0\to12}(B_2^5;J_4) = (J_1,J_2,J_3,J_4,J_5)$, $\phi_{12} = 3.3$, $\pi^{0\to13}(B_2^5;J_7) = (J_1,J_2,J_3,J_7,J_5)$, $\phi_{13} = 3\frac{19}{30} \approx 3.633333$, $\pi^{0\to14}(B_2^5;J_4,J_7) = (J_1,J_2,J_3,J_4,J_7,J_5)$, $\phi_{14} = 3\frac{19}{30} \approx 3.633333$ from $\pi^{0\to1}(B_1^1;\varnothing)$. We obtain the complete permutation $(J_1,J_2,J_3,J_4,J_7,J_5,J_6,J_9,J_{10},J_8) \in S$ with the final penalty $\phi_{14}^* = 5\frac{53}{60} \approx 5.883333$. We obtain the following permutations with their penalties: $\pi^{0\to15}(B_3^6;\varnothing) = (J_4,J_2,J_3,J_1,J_5,J_6)$, $\phi_{15} = 4\frac{1}{24} \approx 4.041667$, $\pi^{0\to16}(B_3^6;J_7) = (J_4,J_2,J_3,J_1,J_5,J_7,J_6)$, $\phi_{16} = 5\frac{7}{60} \approx 5.116667$ from the permutation $\pi^{0\to9}(B_2^4;\varnothing)$. We obtain the complete permutation $(J_4,J_2,J_3,J_1,J_5,J_7,J_6,J_9,J_{10},J_8) \in S$ with the final penalty $\phi_{16}^* = 6\frac{11}{30} \approx 6.366667$.

We obtain the following permutations with their penalties: $\pi^{0\to17}(B_3^7;\varnothing) = (J_1,J_2,J_3,J_5,J_4,J_6)$, $\phi_{17} = 5\frac{11}{45} \approx 5.244444$, $\pi^{0\to18}(B_3^7;J_7) = (J_1,J_2,J_3,J_5,J_7,J_4,J_6)$, $\phi_{18} = 4.925$ from the permutation $\pi^{0\to11}(B_2^5;\varnothing)$. We obtain the complete permutation $(J_1,J_2,J_3,J_5,J_7,J_4,J_6,J_9,J_{10},J_8) \in S$ with the final penalty $\phi_{18}^* = 6.175$.

We obtain the following permutations with their penalties: $\pi^{0\to19}(B_3^8;\varnothing) = (J_1,J_2,J_3,J_4,J_5,J_6)$, $\phi_{19} = 4.3$, $\pi^{0\to20}(B_3^8;J_7) = (J_1,J_2,J_3,J_4,J_5,J_7,J_6)$, $\phi_{20} = 5.7$ from the permutation $\pi^{0\to12}(B_2^5;J_4)$. We obtain the complete permutation $(J_1,J_2,J_3,J_4,J_5,J_7,J_6,J_9,J_{10},J_8) \in S$ with the final penalty $\phi_{20}^* = 6.95$.

From the permutation $\pi^{0\to10}(B_2^4;J_7)$, we obtain the permutation with its penalty as follows: $\pi^{0\to21}(B_3^9;J_4) = (J_4,J_2,J_1,J_3,J_7,J_5,J_4,J_6)$, $\phi_{21} = 5.05$. We obtain the complete permutation $J_4,J_2,J_1,J_3,J_7,J_5,J_4,J_6,J_9,J_{10},J_8) \in S$ with the final penalty $\phi_{21}^* = 6.3$. We obtain the following permutation with their penalties: $\pi^{0\to22}(B_3^{10};J_4) = (J_1,J_2,J_3,J_7,J_5,J_4,J_6)$, $\phi_{22} = 5\frac{1}{12} \approx 5.083333$ from the permutation $\pi^{0\to13}(B_2^5;J_7)$. We obtain the complete permutation $(J_1,J_2,J_3,J_7,J_5,J_4,J_6,J_9,J_{10},J_8) \in S$ with the final penalty $\phi_{22}^* = 6\frac{1}{3} \approx 6.333333$. We obtain the complete permutation $\pi^{0\to23}(B_4^{11};J_4) = (J_4,J_2,J_3,J_1,J_5,J_6,J_8,J_{10},J_9,J_7)$, $\phi_{23} = 5\frac{17}{120} \approx 5.141667$ from the permutation $\pi^{0\to15}(B_3^6;\varnothing)$. We obtain the complete permutation $\pi^{0\to24}(B_4^{12};J_4) = (J_1,J_2,J_3,J_4,J_5,J_6,J_8,J_{10},J_9,J_7)$, $\phi_{22} = 5.4$ from the permutation $\pi^{0\to19}(B_3^8;\varnothing)$. We obtain the complete permutation $\pi^{0\to25}(B_4^{12};J_4) = (J_4,J_5,J_1,J_2,J_3,J_6,J_8,J_{10},J_9,J_7)$, $\phi_{25} = 5\frac{8}{15} \approx 5.533333$ from the permutation $\pi^{0\to5}(B_3^2;\varnothing)$.

Using Procedure 5, we obtain the following permutation with the largest relative perimeter of the optimality box: $\pi^{0\to23}(B_4^{11};J_4)$. The maximal relative perimeter $Per(\mathcal{OB}(\pi_k,T))$ of the optimality box is equal to $2\frac{103}{120} \approx 2.858333$, where $Per_{max} = 22$ and the minimal penalty obtained for the permutation $\pi_k$ is equal to $5\frac{17}{120} \approx 5.141667$.

Since all non-fixed jobs are distributed in the permutation $\pi^{0\to23}(B_4^{11};J_4)$, we obtain the complete permutation $(J_4,J_2,J_3,J_1,J_5,J_6,J_8,J_{10},J_9,J_7) \in S$ with the final penalty $\phi_{23}^*$ equal to $5\frac{17}{120} \approx 5.141667$.

## 5. An Approximate Solution to the Problem $1|p_i^L \le p_i \le p_i^U|\sum C_i$

The relative perimeter of the optimality box $\mathcal{OB}(\pi_k,T)$ characterizes the probability for the permutation $\pi_k$ to be optimal for the instances $1|p|\sum C_i$, where $p \in T$. It is clear that this probability may be close to zero for the problem $1|p_i^L \le p_i \le p_i^U|\sum C_i$ with a high uncertainty of the job processing times (if the set $T$ has a large volume and perimeter).

If the uncertainty of the input data is high for the problem $1|p_i^L \le p_i \le p_i^U|\sum C_i$, one can estimate how the value of $\sum_{i=1}^{n} C_i(\pi_k, p)$ with a vector $p \in T$ may be close to the optimal value of $\sum_{i=1}^{n} C_i(\pi_t, p^*)$, where the permutation $\pi_t \in S$ is optimal for the factual scenario $p^* \in T$ of the job processing times. We call the scenario $p^* = (p_1^*, p_2^*, \ldots, p_n^*) \in T$ **factual**, if $p_i$ is equal to the exact time needed for processing the job $J_i \in \mathcal{J}$. The factual processing time $p_i^*$ becomes known after completing the job $J_i \in \mathcal{J}$.

If the permutation $\pi_k = (\pi_{k_1}, \pi_{k_2}, \ldots, \pi_{k_n}) \in S$ has the non-empty optimality box $\mathcal{OB}(\pi_k, T) \neq \emptyset$, then one can calculate the following error function:

$$F(\pi_k, T) = \sum_{i=1}^{n} \left(1 - \frac{u_{k_i}^* - l_{k_i}^*}{p_{k_i}^U - p_{k_i}^L}\right)(n - i + 1). \tag{7}$$

A value of the error function $F(\pi_k, T)$ characterizes how the objective function value of $\sum_{i=1}^{n} C_i(\pi_k, p^*)$ for the permutation $\pi_k$ may be close to the objective function value of $\sum_{i=1}^{n} C_i(\pi_t, p^*)$ for the permutation $\pi_t \in S$, which is optimal for the factual scenario $p^* \in T$ of the job processing times. The function $F(\pi_k, T)$ characterizes the following difference

$$\sum_{i=1}^{n} C_i(\pi_k, p^*) - \sum_{i=1}^{n} C_i(\pi_t, p^*), \tag{8}$$

which has to be minimized for a good approximate solution $\pi_k$ to the uncertain problem $1|p_i^L \leq p_i \leq p_i^U| \sum C_i$. The better approximate solution $\pi_k$ to the uncertain problem $1|p_i^L \leq p_i \leq p_i^U| \sum C_i$ will have a smaller value of the error function $F(\pi_k, T)$.

The formula (7) is based on the cumulative property of the objective function $\sum_{i=1}^{n} C_i(\pi_k, T)$, namely, each relative error $\left(1 - \frac{u_{k_i}^* - l_{k_i}^*}{p_{k_i}^U - p_{k_i}^L}\right) > 0$ obtained due to the wrong position of the job $J_{k_i}$ in the permutation $\pi_k$ is repeated $(n - i)$ times for all jobs, which succeed the job $J_{k_i}$ in the permutation $\pi_k$. Therefore, a value of the error function $F(\pi_k, T)$ must be minimized for a good approximation of the value of $\sum_{i=1}^{n} C_i(\pi_t, T)$ for the permutation $\pi_k$.

If the equality $Per\mathcal{OB}(\pi_k, T) = 0$ holds, the error function $F(\pi_k, T)$ has the maximal possible value determined as follows: $F(\pi_k, T) = \sum_{i=1}^{n}\left(1 - \frac{u_{k_i}^* - l_{k_i}^*}{p_{k_i}^U - p_{k_i}^L}\right)(n - i + 1) = \sum_{i=1}^{n}(n - i + 1) = \frac{n(n+1)}{2}$.

The necessary and sufficient condition for the largest value of $F(\pi_k, T) = \frac{n(n+1)}{2}$ is given in Theorem 3.

If the equality $Per\mathcal{OB}(\pi_k, T) = n$ holds, the error function $F(\pi_k, T)$ has the smallest value equal to 0: $F(\pi_k, T) = \sum_{i=1}^{n}\left(1 - \frac{u_{k_i}^* - l_{k_i}^*}{p_{k_i}^U - p_{k_i}^L}\right)(n - i + 1) = \sum_{i=1}^{n}(1 - 1)(n - i + 1) = 0$. The necessary and sufficient condition for the smallest value of $F(\pi_k, T) = 0$ is given in Theorem 4, where the permutation $\pi_k \in S$ is optimal for any factual scenario $p^* \in T$ of the job processing times.

Due to evident modifications of the proofs given in Section 3.3, it is easy to prove that Theorems 5 and 6 remain correct in the following forms.

**Theorem 7.** *If the equality $B_1 = \mathcal{J}$ holds, then it takes $O(n)$ time to find the permutation $\pi_k \in S$ and optimality box $\mathcal{OB}(\pi_k, T)$ with the smallest value of the error function $F(\pi_k, T)$ among all permutations $S$.*

**Theorem 8.** *Let each block $B_r \in B$ contain at most one non-fixed job. The permutation $\pi_k \in S$ with the smallest value of the error function $F(\pi_k, T)$ among all permutations $S$ is constructed in $O(n \log n)$ time.*

Using Theorems 7 and 8, we developed the modifications of Procedures 3 and 4 for the minimization of the values of $F(\pi_k, T)$ instead of the maximization of the values of $Per\mathcal{OB}(\pi_k, T)$. The derived modifications of Procedures 3 and 4 are called Procedures $3 + F(\pi_k, T)$ and $4 + F(\pi_k, T)$, respectively. We modify also Procedures 2 and 5 for minimization of the values of $F(\pi_k, T)$ instead of the maximization of the values of $Per\mathcal{OB}(\pi_k, T)$. The derived modifications of Procedures 2 and 5 are called Procedures $2 + F(\pi_k, T)$ and $5 + F(\pi_k, T)$, respectively. Based on the proven properties of the optimality box (Section 3), we propose the following Algorithm 3 for constructing the job permutation $\pi_k \in S$ for the problem $1|p_i^L \le p_i \le p_i^U| \sum C_i$, whose optimality box $\mathcal{OB}(\pi_k, T)$ provides the minimal value of the error function $F(\pi_k, T)$ among all permutations $S$.

---

**Algorithm 3**

---

**Input**: Segments $[p_i^L, p_i^U]$ for the jobs $J_i \in \mathcal{J}$.
**Output**: The permutation $\pi_k \in S$ and optimality box $\mathcal{OB}(\pi_k, T)$, which provide the minimal value of the error function $F(\pi_k, T)$.
*Step 1*: **IF** the condition of Theorem 3 holds
   **THEN** $\mathcal{OB}(\pi_k, T) = \varnothing$ for any permutation $\pi_k \in S$
   and the equality $F(\pi_k, T) = \frac{n(n+1)}{2}$ holds **STOP**.
*Step 2*: **IF** the condition of Theorem 4 holds
   **THEN** using Procedure 2+$F(\pi_k, T)$ construct
   the permutation $\pi_k \in S$ such that both equalities
   $Per\mathcal{OB}(\pi_k, T) = n$ and $F(\pi_k, T) = 0$ hold **STOP**.
*Step 3*: **ELSE** determine the set $B$ of all blocks using the
   $O(n \log n)$-Procedure 1 described in the proof of Lemma 1
*Step 4*: Index the blocks $B = \{B_1.B_2, \ldots, B_m\}$ according to increasing
   left bounds of their cores (Lemma 3)
*Step 5*: **IF** $\mathcal{J} = B_1$ **THEN** problem $1|p_i^L \le p_i \le p_i^U| \sum C_i$ is called problem $P_1$
   (Theorem 5) set $i = 0$ **GOTO** step 8 **ELSE** set $i = 1$
*Step 6*: **IF** there exist two adjacent blocks $B_{r^*} \in B$ and $B_{r^*+1} \in B$ such
   that $B_{r^*} \cap B_{r^*+1} = \varnothing$; let $r$ denote the minimum of the above index $r^*$
   in the set $\{1, 2, \ldots, m\}$ **THEN** decompose the problem $P$ into
   subproblem $P_1$ with the set of jobs $\mathcal{J}_1 = \cup_{k=1}^{r} B_k$ and subproblem $P_2$
   with the set of jobs $\mathcal{J}_2 = \cup_{k=r+1}^{m} B_k$ using Lemma 4;
   set $P = P_1$, $\mathcal{J} = \mathcal{J}_1$, $B = \{B_1, B_2, \ldots, B_r\}$ **GOTO** step 7 **ELSE**
*Step 7*: **IF** $B \ne \{B_1\}$ **THEN GOTO** step 9 **ELSE**
*Step 8*: Construct the permutation $\pi^{s(i)}$ with the minimal value of
   the error function $F(\pi_k, T)$ using Procedure 3+$F(\pi_k, t)$
   **IF** $i = 0$ or $\mathcal{J}_2 = B_m$ **GOTO** step 12 **ELSE**
*Step 9*: **IF** there exists a block in the set $B$ containing more than one
   non-fixed jobs **THEN** construct the permutation $\pi^{s(i)}$ with
   the minimal value of the error function $F(\pi_k, T)$ for the problem $P_1$
   using Procedure 5+$F(\pi_k, t)$ **GOTO** step 11
*Step 10*: **ELSE** construct the permutation $\pi^{s(i)}$ with the minimal value of the
   error function $F(\pi_k, T)$ for the problem $P_1$ using Procedure 3+$F(\pi_k, t)$
*Step 11*: Construct the optimality box $\mathcal{OB}(\pi^{s(i)}, T)$ for the permutation $\pi^{s(i)}$
   using Algorithm 1 **IF** $\mathcal{J}_2 \ne B_m$ and $i > 1$ **THEN**
   set $i := i + 1$, $P = P_2$, $\mathcal{J} = \mathcal{J}_2$, $B = \{B_{r+1}, B_{r+2}, \ldots, B_m\}$
   **GOTO** step 6 **ELSE IF** $\mathcal{J}_2 = B_m$ **THEN GOTO** step 8
*Step 12*: **IF** $i > 0$ **THEN** set $v = i$, determine the permutation
   $\pi_k = (\pi^{s(1)}, \pi^{s(2)}, \ldots, \pi^{s(v)})$ and the optimality box:
   $\mathcal{OB}(\pi_k, T) = \times_{i \in \{1, 2, \ldots, v\}} \mathcal{OB}(\pi^{s(i)}, T)$ **GOTO** step 13
   **ELSE** $\mathcal{OB}(\pi_k, T) = \mathcal{OB}(\pi^{s(0)}, T)$
*Step 13*: The optimality box $\mathcal{OB}(\pi_k, T)$ has the minimal value
   of the error function $F(\pi_k, T)$ **STOP**.

---

In Section 6, we describe the results of the computational experiments on applying Algorithm 3 to the randomly generated problems $1|p_i^L \le p_i \le p_i^U| \sum C_i$.

## 6. Computational Results

For the benchmark instances $1|p_i^L \le p_i \le p_i^U| \sum C_i$, where there are no properties of the randomly generated jobs $\mathcal{J}$, which make the problem harder, the mid-point permutation $\pi_{mid-p} = \pi_e \in S$, where all jobs $J_i \in \mathcal{J}$ are ordered according to the increasing of the mid-points $\frac{p_i^U + p_i^L}{2}$ of their segments $[p_i^L, p_i^U]$, is often close to the optimal permutation. In our computational experiments, we tested seven classes of harder problems, where the job permutation $\pi_k \in S$ constructed by Algorithm 3 outperforms the mid-point permutation $\pi_{mid-p}$ and the permutation $\pi_{max}$ constructed by Algorithm MAX-OPTBOX derived in [15]. Algorithm MAX-OPTBOX and Algorithm 3 were coded in C# and tested on a PC with Intel Core (TM) 2 Quad, 2.5 GHz, 4.00 GB RAM. For all tested instances, inequalities $p_i^L < p_i^U$ hold for all jobs $J_i \in \mathcal{J}$. Table 2 presents computational results for randomly generated instances of the problem $1|p_i^L \le p_i \le p_i^U| \sum C_i$ with $n \in \{100, 500, 1000, 5000, 10,000\}$. The segments of the possible processing times have been randomly generated similar to that in [15].

An integer center $C$ of the segment $[p_i^L, p_i^U]$ was generated using a uniform distribution in the range $[1, 100]$. The lower bound $p_i^L$ of the possible processing time $p_i$ was determined using the equality $p_i^L = C \cdot (1 - \frac{\delta}{100})$. Hereafter, $\delta$ denotes the maximal relative error of the processing times $p_i$ due to the given segments $[p_i^L, p_i^U]$. The upper bound $p_i^U$ was determined using the equality $p_i^U = C \cdot (1 + \frac{\delta}{100})$. Then, for each job $J_i \in \mathcal{J}$, the point $\underline{p}_i$ was generated using a uniform distribution in the range $[p_i^L, p_i^U]$. In order to generate instances, where all jobs $\mathcal{J}$ belong to a single block, the segments $[p_i^L, p_i^U]$ of the possible processing times were modified as follows: $[\widetilde{p}_i^L, \widetilde{p}_i^U] = [p_i^L + \underline{p} - \underline{p}_i, \ p_i^U + \underline{p} - \underline{p}_i]$, where $\underline{p} = \max_{i=1}^n \underline{p}_i$. Since the inclusion $\underline{p} \in [\widetilde{p}_i^L, \widetilde{p}_i^U]$ holds, each constructed instance contains a single block, $|B| = 1$. The maximum absolute error of the uncertain processing times $p_i$, $J_i \in \mathcal{J}$, is equal to $\max_{i=1}^n (p_i^U - p_i^L)$, and the maximum relative error of the uncertain processing times $p_i$, $J_i \in \mathcal{J}$, is not greater than $2\delta\%$. We say that these instances belong to class 1.

Similarly as in [15], three distribution laws were used in our experiments to determine the factual processing times of the jobs. (We remind that if inequality $p_i^L < p_i^U$ holds, then the factual processing time of the job $J_i$ becomes known after completing the job $J_i$.) We call the uniform distribution as the distribution law with number 1, the gamma distribution with the parameters $\alpha = 9$ and $\beta = 2$ as the distribution law with number 2 and the gamma distribution with the parameters $\alpha = 4$ and $\beta = 2$ as the distribution law with number 3. In each instance of class 1, for generating the factual processing times for different jobs of the set $\mathcal{J}$, the number of the distribution law was randomly chosen from the set $\{1, 2, 3\}$. We solved 15 series of the randomly generated instances from class 1. Each series contains 10 instances with the same combination of $n$ and $\delta$.

In the conducted experiments, we answered the question of how large the obtained relative error $\Delta = \frac{\gamma_{p*}^k - \gamma_{p*}^t}{\gamma_{p*}^t} \cdot 100\%$ of the value $\gamma_{p*}^k$ of the objective function $\gamma = \sum_{i=1}^n C_i$ was for the permutation $\pi_k$ with the minimal value of $F(\pi_k, T)$ with respect to the actually optimal objective function value $\gamma_{p*}^t$ calculated for the factual processing times $p^* = (p_1^*, p_2^*, \ldots, p_n^*) \in T$. We also answered the question of how small the obtained relative error $\Delta$ of the value $\gamma_{p*}^k$ of the objective function $\sum C_i$ was for the permutation $\pi_k$ with the minimal value of $F(\pi_k, T)$. We compared the relative error $\Delta$ with the relative error $\Delta_{mid-p}$ of the value $\gamma_{p*}^m$ of the objective function $\sum C_i$ for the permutation $\pi_{mid-p}$ obtained for determining the job processing times using the mid-points of the given segments. We compared the relative error $\Delta$ with the relative error $\Delta_{max}$ of the value of the objective function $\sum C_i$ for the permutation $\pi_{max}$ constructed by Algorithm MAX-OPTBOX derived in [15].

The number $n$ of jobs in the instance is given in column 1 in Table 2. The half of the maximum possible errors $\delta$ of the random processing times (in percentage) is given in column 2. Column 3 gives the average error $\Delta$ for the permutation $\pi_k$ with the minimal value of $F(\pi_k, T)$. Column 4 presents the average error $\Delta_{mid-p}$ obtained for the mid-point permutations $\pi_{mid-p}$, where all jobs are ordered according to increasing mid-points of their segments. Column 5 presents the average relative perimeter of the optimality box $\mathcal{OB}(\pi_k, T)$ for the permutation $\pi_k$ with the minimal value of $F(\pi_k, T)$. Column 6 presents the relation $\Delta_{mid-p}/\Delta$. Column 7 presents the relation $\Delta_{max}/\Delta$. Column 8 presents the average CPU-time of Algorithm 3 for the solved instances in seconds.

The computational experiments showed that for all solved examples of class 1, the permutations $\pi_k$ with the minimal values of $F(\pi_k, T)$ for their optimality boxes generated good objective function values $\gamma_{p*}^k$, which are smaller than those obtained for the permutations $\pi_{mid-p}$ and for the permutations

$\pi_{max}$. The smallest errors, average errors, largest errors for the tested series of the instances are presented in the last rows of Table 2.

**Table 2.** Computational results for randomly generated instances with a single block (class 1).

| $n$ | $\delta$ (%) | $\Delta$ (%) | $\Delta_{mid-p}$ (%) | $Per\mathcal{OB}(\pi_k, T)$ | $\Delta_{mid-p}/\Delta$ | $\Delta_{max}/\Delta$ | CPU-Time (s) |
|---|---|---|---|---|---|---|---|
| 1 | 2 | 3 | 4 | 5 | 6 | 7 | 8 |
| 100 | 1 | 0.08715 | 0.197231 | 1.6 | 2.263114 | 1.27022 | 0.046798 |
| 100 | 5 | 0.305088 | 0.317777 | 1.856768 | 1.041589 | 1.014261 | 0.031587 |
| 100 | 10 | 0.498286 | 0.500731 | 1.916064 | 1.001077 | 1.000278 | 0.033953 |
| 500 | 1 | 0.095548 | 0.208343 | 1.6 | 2.18052 | 1.0385 | 0.218393 |
| 500 | 5 | 0.273933 | 0.319028 | 1.909091 | 1.164623 | 1.017235 | 0.2146 |
| 500 | 10 | 0.469146 | 0.486097 | 1.948988 | 1.036133 | 1.006977 | 0.206222 |
| 1000 | 1 | 0.093147 | 0.21632 | 1.666667 | 2.322344 | 1.090832 | 0.542316 |
| 1000 | 5 | 0.264971 | 0.315261 | 1.909091 | 1.189795 | 1.030789 | 0.542938 |
| 1000 | 10 | 0.472471 | 0.494142 | 1.952143 | 1.045866 | 1.000832 | 0.544089 |
| 5000 | 1 | 0.095824 | 0.217874 | 1.666667 | 2.273683 | 1.006018 | 7.162931 |
| 5000 | 5 | 0.264395 | 0.319645 | 1.909091 | 1.208965 | 1.002336 | 7.132647 |
| 5000 | 10 | 0.451069 | 0.481421 | 1.952381 | 1.06729 | 1.00641 | 7.137556 |
| 10,000 | 1 | 0.095715 | 0.217456 | 1.666667 | 2.271905 | 1.003433 | 25.52557 |
| 10,000 | 5 | 0.26198 | 0.316855 | 1.909091 | 1.209463 | 1.003251 | 25.5448 |
| 10,000 | 10 | 0.454655 | 0.486105 | 1.952381 | 1.069175 | 1.003809 | 25.50313 |
| Minimum | | 0.08715 | 0.197231 | 1.6 | 1.001077 | 1.000278 | 0.031587 |
| Average | | 0.278892 | 0.339619 | 1.827673 | 1.489703 | 1.033012 | 6.692502 |
| Maximum | | 0.498286 | 0.500731 | 1.952381 | 2.322344 | 1,27022 | 25.5448 |

In the second part of our experiments, Algorithm 3 was applied to the randomly generated instances from other hard classes 2–7 of the problem $1|p_i^L \le p_i \le p_i^U| \sum C_i$. We randomly generated non-fixed jobs $J_1, J_2, \ldots, J_s$, which belong to all blocks $B_1, B_2, \ldots, B_m$ of the randomly generated $n - s$ fixed jobs. The lower bound $p_i^L$ and the upper bound $p_i^U$ on the feasible values of $p_i \in R_+^1$ of the processing times of the fixed jobs, $p_i \in [p_i^L, p_i^U]$, were generated as follows. We determined a bound of blocks $[\tilde{b}_i^L, \tilde{b}_i^U]$ for generating the cores of the blocks $[b_i^L, b_i^U] \subseteq [\tilde{b}_i^L, \tilde{b}_i^U]$ and for generating the segments $[p_i^L, p_i^U]$ for the processing times of $|B_i|$ jobs from all blocks $B_i$, $i \in \{1, 2, 3\}$, $[b_i^L, b_i^U] \subseteq [p_i^L, p_i^U] \subseteq [\tilde{b}_i^L, \tilde{b}_i^U]$. Each instance in class 2 has fixed jobs $J_i \in \mathcal{J}$ with rather closed centers $(p_i^L + p_i^U)/2$ and large difference between segment lengths $p_i^U - p_i^L$.

Each instance in class 3 or in class 4 has a single non-fixed job $J_v$, whose bounds are determined as follows: $p_{J_v}^L \le \tilde{b}_1^L \le \tilde{b}_1^U < \tilde{b}_2^L \le \tilde{b}_2^U < \tilde{b}_3^L \le \tilde{b}_3^U \le p_{J_v}^U$. Classes 3 and 4 of the solved instances differ one from another by the numbers of non-fixed jobs and the distribution laws used for choosing the factual processing times of the jobs $\mathcal{J}$. Each instance from classes 5 and 6 has two non-fixed jobs. In each instance from classes 2, 3, 5, 6 and 7, for generating the factual processing times for the jobs $\mathcal{J}$, the numbers of the distribution law were randomly chosen from the set $\{1, 2, 3\}$, and they are indicated in column 4 in Table 3. In the instances of class 7, the cores of the blocks were determined in order to generate different numbers of non-fixed jobs in different instances. The numbers of non-fixed jobs were randomly chosen from the set $\{2, 3, \ldots, 8\}$. Numbers $n$ of the jobs are presented in column 1. In Table 3, column 2 represents the number $|B|$ of blocks in the solved instance and column 3 the number of non-fixed jobs. The distribution laws used for determining the factual job processing times are indicated in column 4 in Table 3. Each solved series contained 10 instances with the same combination of $n$ and the other parameters. The obtained smallest, average and largest values of $\Delta$, $\Delta_{mid-p}$, $\frac{\Delta_{mid-p}}{\Delta}$ and $\frac{\Delta_{max}}{\Delta}$ for each series of the tested instances are presented in columns 5, 6, 8 and 9 in Table 3 at the end of series. Column 7 presents the average relative perimeter of the optimality box $\mathcal{OB}(\pi_k, T)$ for the permutation $\pi_k$ with the minimal value of $F(\pi_k, T)$. Column 10 presents the average CPU-time of Algorithm 3 for the solved instances in seconds.

Table 3. Computational results for randomly generated instances from classes 2–7.

| $n$ | $\lvert B \rvert$ | N-Fix Jobs | Laws | $\Delta$ (%) | $\Delta_{mid-p}$ (%) | Per $\mathcal{OB}(\pi_k, T)$ | $\Delta_{mid-p}/\Delta$ | $\Delta_{max}/\Delta$ | CPU-Time (s) |
|---|---|---|---|---|---|---|---|---|---|
| 1 | 2 | 3 | 4 | 5 | 6 | 7 | 8 | 9 | 10 |
| | | | | | Class 2 | | | | |
| 50 | 1 | 0 | 1.2.3 | 1.023598 | 2.401925 | 1.027 | 2.346551 | 1,395708 | 0,020781 |
| 100 | 1 | 0 | 1.2.3 | 0.608379 | 0.995588 | 0.9948 | 1.636461 | 1.618133 | 0.047795 |
| 500 | 1 | 0 | 1.2.3 | 0.265169 | 0.482631 | 0.9947 | 1.820092 | 1.630094 | 0.215172 |
| 1000 | 1 | 0 | 1.2.3 | 0.176092 | 0.252525 | 0.9952 | 1.434053 | 1.427069 | 0.535256 |
| 5000 | 1 | 0 | 1.2.3 | 0.111418 | 0.14907 | 0.9952 | 1.33793 | 1.089663 | 7.096339 |
| 10,000 | 1 | 0 | 1.2.3 | 0.117165 | 0.13794 | 0.9948 | 1.177313 | 1.004612 | 25.28328 |
| | Minimum | | | 0.111418 | 0.13794 | 0.9947 | 1.177313 | 1.004612 | 0.020781 |
| | Average | | | 0.383637 | 0.736613 | 0.99494 | 1.6254 | 1.399944 | 5.533104 |
| | Maximum | | | 1.023598 | 2.401925 | 0.9952 | 2.346551 | 1.630094 | 25.28328 |
| | | | | | Class 3 | | | | |
| 50 | 3 | 1 | 1.2.3 | 0.636163 | 0.657619 | 1.171429 | 1.033727 | 1.004246 | 0.047428 |
| 100 | 3 | 1 | 1.2.3 | 1.705078 | 1.789222 | 1.240238 | 1.049349 | 1.009568 | 0.066329 |
| 500 | 3 | 1 | 1.2.3 | 0.332547 | 0.382898 | 1.205952 | 1.151412 | 1.138869 | 0.249044 |
| 1000 | 3 | 1 | 1.2.3 | 0.286863 | 0.373247 | 1.400833 | 1.301132 | 1.101748 | 0.421837 |
| 5000 | 3 | 1 | 1.2.3 | 0.246609 | 0.323508 | 1.380833 | 1.311825 | 1.140728 | 2.51218 |
| 10,000 | 3 | 1 | 1.2.3 | 0.26048 | 0.338709 | 1.098572 | 1.300324 | 1.095812 | 5.46782 |
| | Minimum | | | 0.246609 | 0.323508 | 1.098572 | 1.033727 | 1.004246 | 0.047428 |
| | Average | | | 0.577957 | 0.644201 | 1.249643 | 1.191295 | 1.0818286 | 1.460773 |
| | Maximum | | | 1.705078 | 1.789222 | 1.400833 | 1.311825 | 1.140728 | 5.46782 |
| | | | | | Class 4 | | | | |
| 50 | 3 | 1 | 1 | 0.467885 | 0.497391 | 1.17369 | 1.063064 | 1.035412 | 0.043454 |
| 100 | 3 | 1 | 1 | 0.215869 | 0.226697 | 1.317222 | 1.05016 | 1.031564 | 0.067427 |
| 500 | 3 | 1 | 1 | 0.128445 | 0.15453 | 1.424444 | 1.203083 | 1.17912 | 0.256617 |
| 1000 | 3 | 1 | 1 | 0.111304 | 0.118882 | 1.307738 | 1.068077 | 1.042852 | 0.50344 |
| 5000 | 3 | 1 | 1 | 0.076917 | 0.085504 | 1.399048 | 1.111631 | 1.046061 | 2.612428 |
| 10,000 | 3 | 1 | 1 | 0.067836 | 0.076221 | 1.591905 | 1.123606 | 1.114005 | 4.407236 |
| | Minimum | | | 0.067836 | 0.076221 | 1.17369 | 1.05016 | 1.031564 | 0.043454 |
| | Average | | | 0.178043 | 0.193204 | 1.369008 | 1.10327 | 1.074836 | 1.3151 |
| | Maximum | | | 0.467885 | 0.497391 | 1.591905 | 1.203083 | 1.17912 | 4.407236 |
| | | | | | Class 5 | | | | |
| 50 | 3 | 2 | 1.2.3 | 1.341619 | 1.508828 | 1.296195 | 1.124632 | 1.035182 | 0.049344 |
| 100 | 3 | 2 | 1.2.3 | 0.700955 | 0.867886 | 1.271976 | 1.238149 | 1.037472 | 0.070402 |
| 500 | 3 | 2 | 1.2.3 | 0.182378 | 0.241735 | 1.029 | 1.32546 | 1.296414 | 0.255463 |
| 1000 | 3 | 2 | 1.2.3 | 0.098077 | 0.11073 | 1.473451 | 1.129006 | 1.104537 | 0.509969 |
| 5000 | 3 | 2 | 1.2.3 | 0.074599 | 0.084418 | 1.204435 | 1.131624 | 1.056254 | 2.577595 |
| 10,000 | 3 | 2 | 1.2.3 | 0.064226 | 0.074749 | 1.359181 | 1.163846 | 1.042676 | 5.684847 |
| | Minimum | | | 0.064226 | 0.074749 | 1.029 | 1.124632 | 1.035182 | 0.049344 |
| | Average | | | 0.410309 | 0.481391 | 1.272373 | 1.185453 | 1.095422 | 1.524603 |
| | Maximum | | | 1.341619 | 1.508828 | 1.473451 | 1.32546 | 1.296414 | 5.684847 |
| | | | | | Class 6 | | | | |
| 50 | 4 | 2 | 1.2.3 | 0.254023 | 0.399514 | 1.818905 | 1.57275 | 1.553395 | 0.058388 |
| 100 | 4 | 2 | 1.2.3 | 0.216541 | 0.260434 | 1.868278 | 1.202704 | 1.03868 | 0.091854 |
| 500 | 4 | 2 | 1.2.3 | 0.081932 | 0.098457 | 1.998516 | 1.201691 | 1.1292 | 0.365865 |
| 1000 | 4 | 2 | 1.2.3 | 0.06145 | 0.067879 | 1.933984 | 1.104622 | 1.061866 | 0.713708 |
| 5000 | 4 | 2 | 1.2.3 | 0.050967 | 0.060394 | 1.936453 | 1.184953 | 1.048753 | 3.602502 |
| 10,000 | 4 | 2 | 1.2.3 | 0.045303 | 0.05378 | 2.332008 | 1.187101 | 1.038561 | 7.426986 |
| | Minimum | | | 0.045303 | 0.05378 | 1.818905 | 1.104622 | 1.038561 | 0.058388 |
| | Average | | | 0.118369 | 0.156743 | 1.981357 | 1.242304 | 1.1450756 | 2.043217 |
| | Maximum | | | 0.254023 | 0.399514 | 2.332008 | 1.57275 | 1.553395 | 7.426986 |
| | | | | | Class 7 | | | | |
| 50 | 2 | 2–4 | 1.2.3 | 4.773618 | 6.755918 | 0.262946 | 1.415262 | 1.308045 | 0.039027 |
| 100 | 2 | 2–4 | 1.2.3 | 3.926612 | 4.991843 | 0.224877 | 1.271285 | 1.160723 | 0.059726 |
| 500 | 2 | 2–6 | 1.2.3 | 3.811794 | 4.600017 | 0.259161 | 1.206785 | 1.132353 | 0.185564 |
| 1000 | 2 | 2–8 | 1.2.3 | 3.59457 | 4.459855 | 0.337968 | 1.24072 | 1.08992 | 0.474514 |
| 5000 | 2 | 2–8 | 1.2.3 | 3.585219 | 4.297968 | 0.261002 | 1.198802 | 1.031319 | 2.778732 |
| 10,000 | 2 | 2–8 | 1.2.3 | 3.607767 | 4.275581 | 0.299311 | 1.185105 | 1.013096 | 5.431212 |
| | Minimum | | | 3.585219 | 4.275581 | 0.224877 | 1.185105 | 1.013096 | 0.039027 |
| | Average | | | 3.883263 | 4.896864 | 0.274211 | 1.252993 | 1.122576 | 1.494796 |
| | Maximum | | | 4.773618 | 6.755918 | 0.337968 | 1.415262 | 1.308045 | 5.431212 |

## 7. Concluding Remarks

The uncertain problem $1|p_i^L \le p_i \le p_i^U|\sum C_i$ continues to attract the attention of the OR researchers since the problem is widely applicable in real-life scheduling and is commonly used in many multiple-resource scheduling systems, where only one of the machines is the bottleneck and uncertain. The right scheduling decisions allow the plant to reduce the costs of productions due to better utilization of the machines. A shorter delivery time is archived with increasing customer satisfaction. In Sections 2–6, we used the notion of an optimality box of a job permutation $\pi_k$ and proved useful properties of the optimality box $\mathcal{OB}(\pi_k, T)$. We investigated permutation $\pi_k$ with the largest relative perimeter of the optimality box. Using these properties, we derived efficient algorithms for constructing the optimality box for a job permutation $\pi_k$ with the largest relative perimeter of the box $\mathcal{OB}(\pi_k, T)$.

From the computational experiments, it follows that the permutation $\pi_k$ with the smallest values of the error function $F(\pi_k, t)$ for the optimality box $\mathcal{OB}(\pi_k, T)$ is close to the optimal permutation, which can be determined after completing the jobs when their processing times became known. In our computational experiments, we tested classes 1–7 of hard problems $1|p_i^L \le p_i \le p_i^U|\sum C_i$, where the permutation constructed by Algorithm 3 outperforms the mid-point permutation, which is often used in the published algorithms applied to the problem $1|p_i^L \le p_i \le p_i^U|\sum C_i$. The minimal, average and maximal errors $\Delta$ of the objective function values were 0.045303%, 0.735154% and 4.773618%, respectively, for the permutations with smallest values of the error function $F(\pi_k, t)$ for the optimality boxes. The minimal, average and maximal errors $\Delta_{mid-p}$ of the objective function values were 0.05378%, 0.936243% and 6.755918%, respectively, for the mid-point permutations. The minimal, average and maximal errors $\Delta_{max}$ of the objective function values were 0.04705%, 0.82761% and 6.2441066%, respectively. Thus, Algorithm 3 solved all hard instances with a smaller error $\Delta$ than other tested algorithms. The average relation $\frac{\Delta_{mid-p}}{\Delta}$ for the obtained errors for all instances of classes 1–7 was equal to 1.33235. The average relation $\frac{\Delta_{max}}{\Delta}$ for the obtained errors for all instances of classes 1–7 was equal to 1.1133116.

**Author Contributions:** Y.N. proved theoretical results; Y.N. and N.E. jointly conceived and designed the algorithms; N.E. performed the experiments; Y.N. and N.E. analyzed the data; Y.N. wrote the paper.

**Conflicts of Interest:** The authors declare no conflict of interest.

## References

1. Davis, W.J.; Jones, A.T. A real-time production scheduler for a stochastic manufacturing environment. *Int. J. Prod. Res.* **1988**, *1*, 101–112. [CrossRef]
2. Pinedo, M. *Scheduling: Theory, Algorithms, and Systems*; Prentice-Hall: Englewood Cliffs, NJ, USA, 2002.
3. Daniels, R.L.; Kouvelis, P. Robust scheduling to hedge against processing time uncertainty in single stage production. *Manag. Sci.* **1995**, *41*, 363–376. [CrossRef]
4. Sabuncuoglu, I.; Goren, S. Hedging production schedules against uncertainty in manufacturing environment with a review of robustness and stability research. *Int. J. Comput. Integr. Manuf.* **2009**, *22*, 138–157. [CrossRef]
5. Sotskov, Y.N.; Werner, F. *Sequencing and Scheduling with Inaccurate Data*; Nova Science Publishers: Hauppauge, NY, USA, 2014.
6. Pereira, J. The robust (minmax regret) single machine scheduling with interval processing times and total weighted completion time objective. *Comput. Oper. Res.* **2016**, *66*, 141–152. [CrossRef]
7. Grabot, B.; Geneste, L. Dispatching rules in scheduling: A fuzzy approach. *Int. J. Prod. Res.* **1994**, *32*, 903–915. [CrossRef]
8. Kasperski, A.; Zielinski, P. Possibilistic minmax regret sequencing problems with fuzzy parameteres. *IEEE Trans. Fuzzy Syst.* **2011**, *19*, 1072–1082. [CrossRef]
9. Özelkan, E.C.; Duckstein, L. Optimal fuzzy counterparts of scheduling rules. *Eur. J. Oper. Res.* **1999**, *113*, 593–609. [CrossRef]

10. Braun, O.; Lai, T.-C.; Schmidt, G.; Sotskov, Y.N. Stability of Johnson's schedule with respect to limited machine availability. *Int. J. Prod. Res.* **2002**, *40*, 4381–4400. [CrossRef]
11. Sotskov, Y.N.; Egorova, N.M.; Lai, T.-C. Minimizing total weighted flow time of a set of jobs with interval processing times. *Math. Comput. Model.* **2009**, *50*, 556–573. [CrossRef]
12. Sotskov, Y.N.; Lai, T.-C. Minimizing total weighted flow time under uncertainty using dominance and a stability box. *Comput. Oper. Res.* **2012**, *39*, 1271–1289. [CrossRef]
13. Graham, R.L.; Lawler, E.L.; Lenstra, J.K.; Kan, A.H.G.R. Optimization and Approximation in Deterministic Sequencing and Scheduling. *Ann. Discret. Appl. Math.* **1979**, *5*, 287–326.
14. Smith, W.E. Various optimizers for single-stage production. *Nav. Res. Logist. Q.* **1956**, *3*, 59–66. [CrossRef]
15. Lai, T.-C.; Sotskov, Y.N.; Egorova, N.G.; Werner, F. The optimality box in uncertain data for minimising the sum of the weighted job completion times. *Int. J. Prod. Res.* **2017**. [CrossRef]
16. Burdett, R.L.; Kozan, E. Techniques to effectively buffer schedules in the face of uncertainties. *Comput. Ind. Eng.* **2015**, *87*, 16–29. [CrossRef]
17. Goren, S.; Sabuncuoglu, I. Robustness and stability measures for scheduling: Single-machine environment. *IIE Trans.* **2008**, *40*, 66–83. [CrossRef]
18. Kasperski, A.; Zielinski, P. A 2-approximation algorithm for interval data minmax regret sequencing problems with total flow time criterion. *Oper. Res. Lett.* **2008**, *36*, 343–344. [CrossRef]
19. Kouvelis, P.; Yu, G. *Robust Discrete Optimization and Its Application*; Kluwer Academic Publishers: Boston, MA, USA, 1997.
20. Lu, C.-C.; Lin, S.-W.; Ying, K.-C. Robust scheduling on a single machine total flow time. *Comput. Oper. Res.* **2012**, *39*, 1682–1691. [CrossRef]
21. Yang, J.; Yu, G. On the robust single machine scheduling problem. *J. Comb. Optim.* **2002**, *6*, 17–33. [CrossRef]
22. Harikrishnan, K.K.; Ishii, H. Single machine batch scheduling problem with resource dependent setup and processing time in the presence of fuzzy due date. *Fuzzy Optim. Decis. Mak.* **2005**, *4*, 141–147. [CrossRef]
23. Allahverdi, A.; Aydilek, H.; Aydilek, A. Single machine scheduling problem with interval processing times to minimize mean weighted completion times. *Comput. Oper. Res.* **2014**, *51*, 200–207. [CrossRef]

**MDPI**

*Article*

# Scheduling a Single Machine with Primary and Secondary Objectives

## Nodari Vakhania [iD]

Centro de Investigación en Ciencias, Universidad Autonoma del Estado de Morelos, 62210 Cuernavaca, Mexico;
nodari@uaem.mx

Received: 27 February 2018; Accepted: 31 May 2018; Published: 5 June 2018

**Abstract:** We study a scheduling problem in which jobs with release times and due dates are to be processed on a single machine. With the primary objective to minimize the maximum job lateness, the problem is strongly NP-hard. We describe a general algorithmic scheme to minimize the maximum job lateness, with the secondary objective to minimize the maximum job completion time. The problem of finding the Pareto-optimal set of feasible solutions with these two objective criteria is strongly NP-hard. We give the dominance properties and conditions when the Pareto-optimal set can be formed in polynomial time. These properties, together with our general framework, provide the theoretical background, so that the basic framework can be expanded to (exponential-time) implicit enumeration algorithms and polynomial-time approximation algorithms (generating the Pareto sub-optimal frontier with a fair balance between the two objectives). Some available in the literature experimental results confirm the practical efficiency of the proposed framework.

**Keywords:** scheduling single machine; release time; lateness; makespan; bi-criteria scheduling; Pareto-optimal solution

---

## 1. Introduction

We study a scheduling problem in which jobs with release times and due dates are to be processed by a single machine. The problem can be described as follows. There are $n$ jobs $j$ ($j = 1, ..., n$) and a single machine available from Time 0. Every job $j$ becomes available at its release time $r_j$, needs continuous processing time $p_j$ on the machine and is also characterized by the due date $d_j$, which is the desirable time moment for the completion of that job. These are are non-negative integral numbers. A feasible schedule $S$ assigns to each job $j$ the starting time $t_j(S)$, such that $t_j(S) \geq r_j$ and $t_j(S) \geq t_k(S) + p_k$, for any job $k$ included earlier in $S$; the first inequality says that a job cannot be started before its release time, and the second one reflects the restriction that the machine can handle at most one job at a time. If job $j$ completes behind its due date, i.e., $c_j = t_j(S) + p_j > d_j$, then it has a positive lateness $L_j = c_j - d_j$, otherwise its lateness is non-positive. Our primary objective is to find a feasible schedule in which the maximum job lateness is the minimal possible among all feasible schedules. However, we also consider a secondary objective to minimize the maximum job completion time or the makespan.

The problems involving both above objective criteria are strongly NP-hard, and the two objectives are, in general, contradictory; i.e., both of them cannot simultaneously be reached. Then, one may look for a subset of feasible solutions that are somehow acceptable with respect to both objective criteria. A Pareto-optimal frontier is constituted by a subset of all feasible solutions that are not dominated by some other feasible solution, in the sense that no other solution can surpass the former one with respect to both objective values. Finding Pareto-optimal frontier often remains NP-hard, as it is in our case. In particular, for the problems of finding among all feasible schedules with a given maximum job lateness one with the minimum makespan, and vice versa, among all feasible schedules with a given makespan, finding one with the minimum maximum job lateness is strongly NP-hard (see Section 2).

According to the conventional three-field notation introduced by Graham et al., our primary problem with the objective to minimize maximum job lateness (the secondary one to minimize the makespan, respectively) is abbreviated as $1|r_j|L_{max}$ ($1|r_j|C_{max}$, respectively); here in the first field, the single-machine environment is indicated; the second field specifies the job parameters; and in the third field, the objective criterion is given. The problem $1|r_j|L_{max}$ is known to be strongly NP-hard (Garey and Johnson [1]), whereas $1|r_j|C_{max}$ is easily solvable by a greedy algorithm that iteratively includes any earliest released job at its release time or the completion time of the latest so far assigned job, whichever magnitude is larger. Although being strongly NP-hard, the former problem can be efficiently approximated. A venerable $O(n \log n)$ two-approximation algorithm that is commonly used for problem $1|r_j|L_{max}$ was originally proposed by Jackson [2] for the version of the problem without release times and then was extended by Schrage [3] to take into account job release times. This heuristics, that is also referred to as the ED (Earliest Due date) heuristics, iteratively, at each scheduling time $t$ (given by job release or completion time), among the jobs released by time $t$, schedules the one with the smallest due date. Let us note here that, in terms of the minimization of the makespan, the ED heuristics has an important advantage that it creates no machine-idle time that can be evaded. Potts [4] showed that by a repeated application of the heuristics $O(n)$ times, the performance ratio can be improved to $3/2$ resulting in an $O(n^2 \log n)$ time performance. Later, Hall and Shmoys [5] illustrated that the application of the ED heuristics to the original and a specially-defined reversed problem may lead to a further improved approximation of $4/3$. Nowicki and Smutnicki [6] have shown that the approximation ratio $3/2$ can also be achieved in time $O(n \log n)$. In practice, the ED heuristics turned out to be the most efficient fast solution method for problem $1|r_j|L_{max}$, far better than one would suggest based on the above theoretical worst-case performance ratio (see Larson et al. [7], Kise et al. [8] and Vakhania et al. [9]). We shall reveal the benefit of such performance for our bi-criteria scheduling problem later.

As for the exact solution methods for problem $1|r_j|L_{max}$, the branch-and-bound algorithm with good practical behavior was proposed in McMahon and Florian [10] and Carlier [11] (the practical behavior of these two algorithms was compared also more recently by Sadykov and Lazarev [12]). Two more exact implicit enumerative algorithms were proposed, in Grabowski et al. [13] and Larson et al. [14]. More recently, Pan and Shi [15] and Liu [16] presented other branch-and-bound algorithms (in the latter reference, the version with precedence constraints was studied). The preemptive setting $1|r_j, pmtn|L_{max}$ is easily solved by the preemptive version of Jackson's extended heuristics, as it always interrupts non-urgent jobs in favor of a newly-released more urgent one. In fact, the preemptive ED heuristics is also useful for the solution of the non-preemptive version, as it gives a strong lower bound for the latter problem (see, for example, Gharbi and Labidi [17] for a recent study on this subject).

The version in which a feasible schedule may not include any idle time interval, abbreviated $1|r_j, nmit|L_{max}$, is strongly NP-hard according to Lenstra et al. [18] (minimization of the makespan and that of the length of idle time intervals are closely related objectives). The problem admits the same approximation as the unrestricted version; as Kacem and Kellerer [19] have observed, job release times can be adopted so that the ED heuristics and Potts's above-mentioned extension maintain the same approximation ratio for problem $1|r_j, nmit|L_{max}$ as for $1|r_j|L_{max}$. Hoogeveen has considered the no machine idle time version in the bi-criteria setting. Instead of minimizing the lateness, he has introduced the so-called target start time $s_j$ of job $j$; $s_j$ is the desirable starting time for job $j$, similarly to the due date $d_j$ being the desirable completion time for job $j$. Then, besides the minimization of the maximum job lateness, the maximum job promptness (the difference between the target and real start times of that job) can also be minimized. The above paper considers the corresponding bi-criteria scheduling problem and finds the Pareto-optimal set of feasible solutions. Lazarev [20] and Lazarev et al. [21] have proposed a polynomial time solution finding the Pareto-optimal set for two special cases of our bi-criteria problem with specially-ordered job parameters and equal-length jobs, respectively. An exact enumerative algorithm for the no idle time problem with the objective to minimize maximum job lateness was proposed by Carlier et al. [22], whereas the practical importance of the solutions with no idle time intervals was emphasized by Chrétienne [23].

In general, in an optimal solution to problem $1|r_j|L_{\max}$, some idle time intervals might be unavoidable (a non-urgent job has to wait until a more urgent one gets released; hence, the corresponding idle time interval arises): a problem instance might easily be constructed possessing an optimal solution to $1|r_j|L_{\max}$, but with no feasible solution for $1|r_j, nmit|L_{\max}$ (see Section 2).

In this paper, based on the study of the properties of the schedules generated by the ED heuristics and the ties of the scheduling problem $1|r_j|L_{\max}$ with a version of the bin packing problem, we propose a general solution method that is oriented toward both objective criteria, the maximum job lateness and the maximum job completion time (the makespan). Considering a single criterion problem, we first observe that different feasible solutions with different makespans may posses the same maximum job lateness. Though, as we will see, finding one with the minimum makespan is NP-hard. Viewing the problem in light of the bin packing problem with different bin capacities, we show how the known heuristic algorithms with reasonably good approximations can be adopted for the solution of the bi-criteria problem.

We establish the conditions when the Pareto-optimal frontier with the two objective functions (the maximum lateness and the makespan) can be formed in polynomial time. This, together with a general algorithmic framework that we present, provides a method for the construction of (exponential-time) implicit enumeration algorithms and polynomial-time approximation algorithms (the latter approximation algorithms generate a Pareto-sub-optimal frontier with a "fair balance" between the two objectives).

Our framework applies the partition of the scheduling horizon into two types of segments containing urgent and non-urgent jobs, respectively. Intuitively, segments with urgent jobs are to occupy quite restricted time intervals, whereas the second type of segments can be filled out, in some optimal fashion, with non-urgent jobs. The urgency of jobs is determined based on the first objective criterion, i.e., such jobs may potentially realize the maximum lateness in the optimal schedule $S^{opt}$ for problem $1|r_j|L_{\max}$. We determine the time intervals within which a sequence of urgent jobs is to be scheduled. We refer to such sequences as kernels and to the intervals in between them as bins. Our scheme, iteratively, fills in the bin intervals by the non-urgent jobs; it determines the corresponding kernel segments before that. The less gaps are left within the bin intervals, the less is the resultant maximum job completion time, whereas the way the bin intervals are filled out is reflected also by the maximum job lateness.

We exploit the observation that the total idle time interval in a bin of a given feasible schedule can be reduced without increasing the maximum job lateness of that schedule. This kind of "separate" scheduling turns out to be helpful in the search for a compromise between the two objectives. In practice, analyzing the computational experiments for the problem $1|r_j|L_{\max}$ reported in [9], we may assert that, by incorporating the ED heuristics into our scheme, a balanced distribution of the jobs within the bins imposing almost neglectable delays for the kernel jobs can be obtained. For the vast majority of the tested instances in [9], the maximum job lateness is very close to the optimum. At the same time, since the ED heuristics creates no avoidable machine idle time, the minimum makespan is achieved. We discuss this issue in more detail in Section 5.

In Section 2, we study our two objective functions and establish the time complexity of related Pareto-optimal problems. In Section 3, we give necessary preliminaries to our algorithmic scheme. In Section 4, we describe our method for partitioning the whole scheduling horizon into urgent and non-urgent zones. In Section 5, we introduce the binary search procedure that is used to verify the existence of a feasible solution with some threshold lateness and describe the rest of the proposed framework. In Section 6, we give final remarks.

## 2. Complexity of the Bi-Objective Problem

In this section, we will see why finding the Pareto-optimal frontier of feasible solutions for our bi-criteria scheduling problem is NP-hard. In the Introduction, we gave an intuitive description of the Pareto-optimality concept (for a detailed guideline on multi-objective optimization, the reader may have a look at, for example, Ehrgott [24]). More formally, in a bi-criteria optimization problem, two objective functions $f_1$ and $f_2$ over the set $\mathcal{F}$ of feasible solutions is given. These functions might be mutually contradictory; hence, there may exist no feasible solution minimizing both objective functions simultaneously (considering the minimization version). More precisely, if $F_i^*$ is the optimal value of a single-criterion problem with the objective to minimize function $f_i$, then there may exist no feasible solution to the problem that attains simultaneously both optimal values $F_1^*$ and $F_2^*$. In this situation, it is reasonable to look for feasible solutions that are somehow acceptable for both the objective functions. A commonly-used dominance relation is defined as follows.

Solution $\sigma_1 \in \mathcal{F}$ dominates solution $\sigma_2 \in \mathcal{F}$ if $f_i\sigma_1 < f_i\sigma_2$, for $i = 1, 2$; in fact, we allow $\leq$, instead of $<$ for either $i = 1$ or $i = 2$, and require having at least one strict inequality.

Now, we refer to $\sigma \in \mathcal{F}$ as a Pareto-optimal solution if no other solution from set $\mathcal{F}$ dominates it, and we call the set of all such solutions the Pareto-optimal frontier of feasible solutions. Forming a Pareto-optimal frontier of feasible solutions may not be easy; for instance, the corresponding single-objective problems may already be intractable (clearly, the solution of a bi-objective setting requires the solution of the corresponding single-objective settings).

Consider now the two basic objective functions dealt with in this paper, the maximum job lateness and the makespan. The two corresponding single-objective problems are $1|r_j|L_{\max}$ and $1|r_j|C_{\max}$.

The two objective criteria are contradictory, i.e., minimizing the maximum job lateness $L_{\max}$ does not necessarily minimize the makespan $C_{\max}$, and vice versa: it can be easily seen that an optimal schedule minimizing $L_{\max}$ may contain avoidable gaps and hence may not minimize $C_{\max}$. For example, consider an instance of $1|r_j|L_{\max}$ with two jobs with $r_1 = 0, r_2 = 3, p_1 = p_2 = 5, d_1 = 20, d_2 = 9$. In the schedule minimizing $L_{\max}$, there is a gap $[0, 3)$; Job 2 is included at its release time and is followed by Job 1, whereas in the schedule minimizing $C_{\max}$, Job 1 starts at its release time and is followed by Job 2. The maximum job lateness and the makespan in the first and the second schedules are $-1$ and 13 (respectively) and 1 and 10 (respectively). Below, we state related Pareto-optimality problems.

**Problem 1.** *Among all feasible schedules with a given makespan, find one with the minimum maximum job lateness.*

**Problem 2.** *Among all feasible schedules with a given maximum job lateness, find one with the minimum makespan.*

Clearly, Problem 1 is strongly NP-hard since, as earlier mentioned, problem $1|r_j|L_{\max}$ is strongly NP-hard [1]. On the other hand, problem $1|r_j|C_{\max}$ can easily be solved by the following linear-time list scheduling algorithm, which leaves no idle time interval that can be avoided: at every scheduling time, stating from the minimum job release time, include any released job, update the current scheduling time as the maximum between the completion time of that job and the next minimum job release time of an unscheduled job, and repeat the procedure until all jobs are scheduled (a modification of this greedy algorithm that at each scheduling time, among the released jobs, includes one with the minimum due date is the ED heuristics mentioned in the Introduction). This greedy solution also provides a polynomial-time solution for the decision version of problem $1|r_j|C_{\max}$ in which one wishes to know if there exists a feasible schedule with some fixed makespan $M$. Indeed, if $M$ is less than the optimum, then clearly, the answer is "no". If $M$ is greater than or equal to the minimum makespan $M^*$, then the answer is "yes": we can introduce in a schedule with the makespan $M^*$ gaps with a total length $M - M^*$, appropriately, so that the resultant makespan will be exactly $M$.

To show that Problem 2 is strongly NP-hard, we prove that its decision version is strongly NP-complete. Let $M$ be some threshold value for the makespan. Without loss of generality, we assume that this threshold is attainable, i.e., there is a feasible schedule with the objective value $M$. Then, the decision version of Problem 2 can be formulated as follows:

Problem 2-D. For an instance of problem $1|r_j|L_{max}$ (the input), among all feasible schedules with a given maximum job lateness, does there exist one with the makespan not exceeding magnitude $M$ (here, the output is a "yes" or a "no" answer)?

**Theorem 1.** *Problem 2-D is strongly NP-complete.*

**Proof.** To begin with, it is easy to see that the problem is in class NP. First, we observe that the size of an instance of problem $1|r_j|L_{max}$ with $n$ jobs is $O(n)$. Indeed, every job $j$ has three parameters, $r_j, p_j$ and $d_j$; let $M$ be the maximum such magnitude, i.e., the maximum among all job parameters. Then, the number of bits to represent our problem instance is bounded above by $3n \log M$. Since constant $M$ fits within a computer word, the size of the problem is $O(n)$. Now, given a feasible schedule with some maximum job lateness, we can clearly verify its makespan in no more than $n$ steps, which is a polynomial in the size of the input. Hence, Problem 2-D belongs to class NP.

Next, we use the reduction from a strongly NP-complete three-PARTITION problem to show the NP-hardness of our decision problem. In three-PARTITION, we are given a set $A$ of $3m$ elements, their sizes $C = \{c_1, c_2, \dots, c_{3m}\}$ and an integer number $B$ with $\sum_{i=1}^{3m} c_i = mB$, whereas the size of every element from set $A$ is between $B/4$ and $B/2$. We are asked if there exists the partition of set $A$ into $m$ (disjoint) sets $A_1, \dots, A_m$ such that the size of the elements in every subset sums up to $B$.

Given an arbitrary instance of three-PARTITION, we construct our scheduling instance with $3m + m$ jobs with the total length of $\sum_{i=1}^{n} c_i + m$ as follows. We have $3m$ partition jobs $1, \dots, 3m$ with $p_i = c_i, r_i = 0$ and $d_i = 2mB + m$, for $i = 1, \dots, 3m$. Besides, we have $m$ separator jobs $j_1, \dots, j_m$ with $p_{j_i} = 1, r_{j_i} = Bi + i - 1, d_{j_i} = Bi + i$. Note that this transformation creating a scheduling instance is polynomial as the number of jobs is bounded by the polynomial in $m$, and all magnitudes can be represented in binary encoding in $O(m)$ bits.

First, we easily observe that there exist feasible schedules in which the maximum job lateness is zero: we can just schedule the separator jobs at their release times and include the partition jobs arbitrarily without pushing any separator job. Now, we wish to know whether among the feasible schedules with the maximum job lateness of zero there is one with makespan $mB + m$. We claim that this decision problem has a "yes" answer iff there exists a solution to three-PARTITION.

In one direction, suppose three-PARTITION has a solution. Then, we schedule the partition jobs corresponding to the elements from set $A_1$ within the interval $[0, B]$ and partition jobs from set $A_i$ within the interval $[(i - 1)B + i - 1, iB + i - 1]$, for $i = 2, \dots, m$ (we schedule the jobs from each group in a non-delay fashion, using, for instance, the ED heuristics). We schedule the separator job $j_i$ within the interval $[iB + i - 1, iB + i]$, for $i = 1, \dots, m$. Therefore, the latest separator job will be completed at time $mB + m$, and the makespan in the resultant schedule is $mB + m$.

In the other direction, suppose there exists a feasible schedule $S$ with makespan $mB + m$. Since the total sum of job processing times is $mB + m$, there may exist no gap in that schedule. However, then, the time interval of length $B$ before every separator job must be filled in completely with the partition jobs, which obviously gives a solution to three-PARTITION. □

## 3. Basic Definitions and Properties

This section contains necessary preliminaries to the proposed framework including some useful properties of the feasible schedules created by the ED heuristics (we call them ED-schedules). Our approach is based on the analysis of the properties of ED-schedules. This analysis is initially carried out for the decision version of single-criterion problem $1|r_j|L_{max}$: we shall particularly be interested in ED-schedules with the maximum job lateness not exceeding a given threshold (so our analysis is carried out in terms of the minimization of the maximum job lateness). First, we give a detailed description of the ED heuristics.

The heuristics distinguishes $n$ scheduling times, the time moments at which a job is assigned to the machine. Initially, the earliest scheduling time is set to the minimum job release time. Among all jobs released by that time, a job with the minimum due date is assigned to the machine (ties being broken by selecting the longest job). Iteratively, the next scheduling time is either the completion time of the latest job assigned so far to the machine or the minimum release time of an unassigned job, whichever is more (no job can be started before the machine becomes idle, nor can it be started before it gets released). Among all jobs released by every newly- determined (as just specified) scheduling time, a job with the minimum due date is assigned to the machine. Note that since there are $n$ scheduling times and at each scheduling time, a minimal element in an ordered list is searched for, the time complexity of the heuristics is $O(n \log n)$.

A gap in an ED-schedule is its maximal consecutive time interval in which the machine is idle (with our convention, there occurs a zero-length gap $(c_j, t_i)$ whenever job $i$ starts at its release time immediately after the completion of job $j$).

A block in an ED-schedule is its consecutive part consisting of the successively scheduled jobs without any gap in between, which is preceded and succeeded by a (possibly a zero-length) gap.

Suppose job $i$ preceding job $j$ in ED-schedule $S$ is said to push job $j$ in $S$ if $j$ will be rescheduled earlier whenever $i$ is forced to be scheduled behind $j$ (it follows that jobs $i$ and $j$ belong to the same block).

**Example 1.** *We have seven jobs in our first problem instance. The processing times, the release times and the due dates of these jobs are defined as follows:*
*$p_1 = p_3 = 5, p_2 = 30, p_4 = p_5 = p_6 = p_7 = 10$.*
*$r_3 = 11, r_5 = r_6 = r_7 = 42$, whereas the rest of the jobs are released at Time 0, except Job 4, released at Time 36.*
*$d_1 = 75, d_2 = 80, d_3 = 11, d_4 = 53, d_5 = d_6 = d_7 = 52$.*
*It can be readily verified that the ED heuristics will assign the jobs in increasing order of their indexes creating an ED-schedule $S = (1, 0)(2, 5)(3, 35)(4, 40)(5, 50)(6, 60)(7, 70)$, as depicted in Figure 1 (in every pair in brackets, the first number is the job index, and the second number is its starting time). In schedule $S$, consisting of a single block (there is no gap in it), Job 2 pushes Job 3 and the succeeding jobs.*

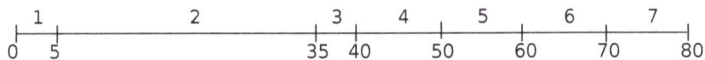

**Figure 1.** The initial Earliest Due date (ED)-schedule $S$ for the problem instance of Example 1.

Given an ED-schedule $S$, let $i$ be a job that realizes the maximum job lateness in that schedule, i.e., $L_i(S) = \max_j\{L_j(S)\}$, and let $B$ be the block in $S$ containing job $i$. Among all jobs $i \in B$ with $L_i(S) = \max_j\{L_j(S)\}$, the latest scheduled one is said to be an overflow job in schedule $S$. Clearly, every schedule contains at least one overflow job, whereas every block in schedule $S$ may contain one or more overflow jobs (two or more overflow jobs from the same block will then be "separated" by non-kernel jobs).

A kernel in schedule $S$ is the longest sequence of jobs ending with an overflow job $o$, such that no job from this sequence has a due date more than $d_o$.

We shall normally use $K(S)$ for the earliest kernel in schedule $S$, and abusing the terminology, we shall also use $K(S)$ for the set of jobs in the sequence. For kernel $K$, let $r(K) = \min_{i \in K}\{r_i\}$.

Note that every kernel is contained within some block in schedule $S$, and hence, it may contain no gap. The following observation states a sufficient optimality condition for the single-criterion problem $1|r_j|L_{\max}$ (the reader may have a look at [25] for a proof):

**Observation 1.** *The maximum job lateness in a kernel K cannot be reduced if the earliest scheduled job in K starts at time $r(K)$. Hence, if an ED-schedule S contains a kernel with this property, it is optimal for problem $1|r_j|L_{max}$.*

Thus, we may ensure that there exists no feasible schedule with the value of our objective function less than that in solution $S$ whenever the condition in Observation 1 is met. Then, we immediately proceed to the reconstruction of solution $S$ aiming at the minimization of our second objective function, as we describe in the next section.

Suppose the condition in Observation 1 is not satisfied. Then, there must exist job $e$ with $d_e > d_o$ (i.e., less urgent than the overflow job $o$), scheduled before all jobs of kernel $K$ imposing a forced delay for the jobs of that kernel. We call job $e$ an emerging job for kernel $K$. Note that an emerging job $e$ and kernel $K$ belong to the same block.

If the earliest scheduled job of kernel $K$ does not start at its release time, it is immediately preceded and pushed by an emerging job, which we call the delaying (emerging) job for kernel $K$.

We can easily verify that in ED-schedule $S$ of Example 1 (see Figure 1), there is a single kernel $K_1 = K(S)$ consisting of a single (overflow) Job 3, with $L_{max}(S) = L_{3,S} = 40 - 11 = 29$ (note that $L_{7,S} = 80 - 52 = 28$). There are two emerging Jobs 1 and 2 for kernel $K$, and Job 2 is the delaying emerging job for that kernel.

Clearly, the maximum job lateness in schedule $S$ may only be decreased by restarting the jobs of kernel $K = K(S)$ earlier. Note that if we reschedule an emerging job $e$ behind the jobs of kernel $K$, we may restart these jobs earlier. As a result, the maximum job lateness in kernel $K$ may be reduced. We shall refer to the operation of the rescheduling of emerging job $e$ after kernel $K$ as the activation of that emerging job for kernel $K$. To provide the early restarting for jobs of kernel $K$, we maintain any job scheduled behind kernel $K$ in $S$ scheduled behind kernel $K$ (all these jobs are said to be in the state of activation for that kernel). We call the resultant ED-schedule a complementary to the $S$ schedule and denote it by $S_e$. In general, a complementary schedule $S_E$ is defined for a set of emerging jobs $E$ in schedule $S$. In schedule $S_E$, all jobs from set $E$ and all the jobs scheduled behind kernel $K$ in $S$ are in the state of activation for kernel $K$. It is easy to see that by activating a single emerging job $e$, kernel $K$ will already be restarted earlier; hence, the maximum job lateness realized by a job in kernel $K$ will be reduced.

**Observation 2.** *Suppose the release time of all the emerging jobs in a set E is increased to the magnitude $r(K)$, and the ED heuristics is newly applied to the modified instance. Then, if some job j scheduled behind kernel K in schedule S gets rescheduled before kernel K, $d_j > d_o$, where o is the corresponding overflow job in kernel K.*

**Proof.** By the ED heuristics, any job $i$ released before the jobs of kernel $K$ with $d_i \le d_o$ would have been included within kernel $K$ in schedule $S$. Hence, if a job scheduled behind kernel $K$ in schedule $S$ becomes included before kernel $K$, then $d_j > d_o$ (i.e., it is less urgent than all kernel jobs). □

By the above observation, if we merely increase the release time of any above available job $j$ with $d_j > d_o$ and that of any emerging job from set $E$ to magnitude $r(K)$, then the ED heuristics will create complementary schedule $S_E$ for the modified instance.

**Observation 3.** *Let l be the delaying emerging job in schedule S. Then, in complementary schedule $S_l$, the earliest scheduled job of kernel $K = K(S)$ starts at its release time $r(K)$ and is preceded by a gap.*

**Proof.** By the definition of complementary schedule $S_l$, no job $j$ with $d_j > d_o$ may be included before kernel $K(S)$ in schedule $S_l$. As we have already observed, no available job $i$ with $d_i \le d_o$ may be included before kernel $K$ in complementary schedule $S_l$. Then, clearly, the time interval before the earliest released job of kernel $K$ will remain idle in complementary schedule $S_l$, and the observation follows. □

For the problem instance of Example 1, we form set $E$ with a single emerging Job 1 and obtain a complementary schedule $S_1 = (2,0)(3,30)(1,35)(4,40)(5,50)(6,60)(7,70)$, as depicted in Figure 2. There is a single kernel $K_2 = K(S_1)$ consisting of Jobs 5, 6 and 7 with $L_{\max}(S_1) = L_{7,S_1} = 80 - 52 = 28$ in schedule $S_1$ (note that $L_{3,S_1} = 35 - 11 = 24$).

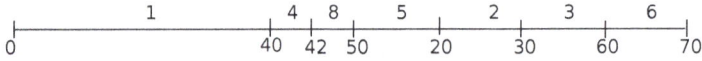

**Figure 2.** Complementary ED-schedule $S_1$ for Example 1.

The proof of our next observation can be seen in [26].

**Property 1.** *Suppose in schedule $S_E$ there arises a (possibly zero-length) gap before kernel $K = K(S)$. Then, that schedule is optimal for problem $1|r_j|L_{\max}$ if $K(S_E) = K(S)$.*

Note that the above property yields another halting condition for problem $1|r_j|L_{\max}$. The following example illustrates that activation of an emerging job does not always yield an improvement in terms of the minimization of the maximum job lateness.

**Example 2.** *Now, we form an instance with six jobs as follows. Jobs 4 and 6 are kernel jobs, each defining a distinct kernel with the delaying emerging Jobs 3 and 5, respectively. Jobs 1 and 2 are other emerging jobs. All emerging jobs are released at Time 0, whereas kernel Jobs 4 and 6 are released at Times 20 and 50 respectively; $p_1 = p_2 = p_4 = p_6 = 5$, $p_3 = 25$, $p_5 = 20$. The due dates of emerging jobs are large enough numbers, Jobs 1 and 2 being slightly more urgent than the delaying emerging Jobs 3 and 5; $d_4 = 25$ and $d_6 = 55$. Then ED heuristics will deliver solution $S = (1,0)(2,5)(3,10)(4,35)(5,40)(6,60)$ with the delaying emerging Job 3. The lateness of Job 4 is 15, and that of Job 6 is 10; Job 4 forms the first kernel with the only overflow Job 4. In the complementary schedule $S_4 = (1,0)(2,5)(4,20)(3,25)(5,50)(6,70)$, there arises a gap $(10,20)$. In that schedule, a new kernel is formed by Job 6, which completes at Time 75 and has the lateness of 20. This example suggests that, instead of creating schedule $S_1$, it could have been better to consider schedule $S_E$, for a properly selected set $E$.*

## 4. Partitioning the Scheduling Horizon

In this section, we start the construction of a bridge between the two scheduling problems $1|r_j|L_{\max}$ and $1|r_j|C_{\max}$ establishing a helpful relationship between these problems and a version of a well-studied bin packing problem. To begin with, we introduce our bins. The notion of a bin is closely tied to that of a kernel. Given an ED-schedule with a set of kernels, we define the bins in that ED-schedule as the segments or the intervals left outside the kernel intervals. The bins will be packed with non-kernel jobs, and since, the interval within which every kernel can be scheduled has some degree of flexibility, the bins are defined with the same degree of flexibility. Abusing the terminology, we will use the term "bin" for both the time interval and the corresponding schedule portion, as well. Note that there is a bin between two adjacent kernel intervals and a bin before the first and after the last kernel interval; any kernel defines two neighboring bins adjacent to that kernel, one preceding and one succeeding that kernel.

We first describe how we obtain a partition of the scheduling horizon into the kernel and bin intervals. In the next section, we explain how this partition is used for the solution of our bi-criteria scheduling problem that is reduced to the distribution of non-kernel jobs within the bins in some (near) optimal fashion (that takes into account both objective criteria).

The initial partition is constructed based on the earliest kernel identified in the initial ED-schedule (one constructed for the originally given problem instance). This kernel already defines the corresponding two bins in that initial partition. For instance, in schedule $S$ of Example 1 (Figure 1), kernel $K_1 = K(S)$ is detected, which already defines the two bins in that schedule. In the next generated complementary schedule $S_1$ (Figure 2), there arises a new kernel within the second bin. Note that the execution interval of kernel $K_1$ in schedules $S$ and $S_1$ is different (it is $[35, 40)$ in schedule $S$ and $[30, 35)$ in schedule $S_1$). Correspondingly, the first bin is extended up to time moment 35 in schedule $S$, and it is extended up to Time 30 in schedule $S_1$. The newly arisen kernel $K_2 = K(S_1)$ defines two new bins in schedule $S_1$; hence, that schedule contains three bins, in total.

In general, the starting and completion times of every bin are not a priori fixed; they are defined in accordance with the allowable flexibility for the corresponding kernel intervals. Our scheme proceeds in a number of iterations. To every iteration, a complete schedule with a particular distribution of jobs into the bins corresponds, whereas the set of jobs scheduled in every kernel interval remains the same. In this way, two or more complete schedules for the same partition might be created (for instance, schedules $S_1$ and $(S_1)_1$ of Example 1; see Figures 2 and 3). At an iteration $h$, during the scheduling of a bin, a new kernel may arise, which is added to the current set of kernels $\mathcal{K}$ (kernel $K_2$ arises within the second bin in complementary schedule $S_1$ of Example 1): the updated set contains all the former kernels together with the newly arisen one (since schedule $S$ of Figure 1 contains only one kernel, $\mathcal{K} = \{K_1\}$, there are only two bins in it; the next generated schedule $S_1$ of Figure 2 contains already two kernels $K_1$ and $K_2$, $\mathcal{K} = \{K_1, K_2\}$ and three bins). Note that the partition of the scheduling horizon is changed every time a new kernel arises.

**Adjusting time intervals for kernel and bin segments:** Although the time interval within which each kernel can be scheduled is restricted, it has some degree of flexibility, as we have illustrated in Example 1. In particular, the earliest scheduled job of a kernel $K$ might be delayed by some amount without affecting the current maximum job lateness. Denote this magnitude by $\delta(K)$. Without loss of generality, the magnitude $\delta(K)$ can take the values from the interval $\delta(K) \in [0, p_l]$, where $l$ is the delaying emerging job for kernel $K$ in the initially constructed ED-schedule $\sigma$. Indeed, if $\delta(K) = 0$, the kernel will not be delayed, whereas $\delta(K) = p_l$ corresponds to the delay of the kernel in the initial ED-schedule $S$ (which is 19 for kernel $K_1$ in schedule $S$ of Figure 1). In particular, we let $p_{max}$ be the maximum job processing time and shall assume, without loss of generality, that $\delta(K) \leq p_{max}$.

To define $\delta(K)$, let us consider a partial schedule constructed by the ED heuristics for only jobs of kernel $K \in \mathcal{K}$. The first job in that schedule starts at time $r(K)$ (we assume that there is no emerging job within that sequence, as otherwise we continue with our partitioning process). Note that the lateness of the overflow job of that partial schedule is a lower bound on the maximum job lateness; denote this magnitude by $L^*(K)$. Then, $L^* = \max_{K \in \mathcal{K}}\{L^*(K)\}$ is also a lower bound on the same objective value. Now, we let $\delta(K) = L^* - L^*(K)$, for every kernel $K \in \mathcal{K}$. The following observation is apparent:

**Observation 4.** *For $\delta(K) = 0$, the lateness of the latest scheduled job of kernel $K$ is a lower bound on the optimal job lateness, and for $\delta(K) = p_l$, it is a valid upper bound on the same objective value.*

**Observation 5.** *In a schedule $S_{opt}$ minimizing the maximum job lateness, every kernel $K$ starts either no later than at time $r(K) + \delta(K)$ or it is delayed by some $\delta \geq 0$, where $\delta \in [0, p_{max}]$.*

**Proof.** First note that in any feasible schedule, every kernel $K$ can be delayed by the amount $\delta(K)$ without increasing the maximum job lateness. Hence, kernel $K$ does not need to be started before time $r(K) + \delta(K)$. Furthermore, in any created ED-schedule, the delay of any kernel cannot be more than $p_{max}$, as for any delaying emerging job $l$, $p_l \leq p_{max}$. Hence, $\delta \in [0, p_{max}]$. □

Thus for a given partition, the starting and completion time of every bin in a corresponding complete schedule may vary according to Observation 5. Our scheme incorporates a binary search procedure in which the trial values for $\delta$ are drown from interval $[0, p_{max}]$. To every iteration in the binary search procedure, some trial value of $\delta$ corresponds, which determines the maximum allowable job lateness, as we describe below.

We observe that not all kernels are tight in the sense that, for a given kernel $K \in \mathcal{K}$, $\delta(K)$ might be strictly greater than zero. By Observation 5, in any generated schedule with trial value $\delta$, every kernel

$K$ is to be started within the interval $[r(K) + \delta(K), r(K) + \delta(K) + \delta]$, the corresponding bin intervals being determined with respect to these starting times. Notice that if the latest job of kernel $K$ completes at or before time moment $r(K) + P(K) + \delta$, then the lateness of no job of that kernel will surpass the magnitude $L(\delta) = r(K) + P(K) + \delta - d$, where $d$ is the due date of the corresponding overflow job and $P(K)$ is the total processing time of the jobs in kernel $K$.

We shall refer to the magnitude $L(\delta) = r(K) + P(K) + \delta - d$ as the threshold on the maximum allowable job lateness corresponding to the current trial value $\delta$. Ideally, not only jobs of kernel $K$, but no other job is to have the lateness more than $L(\delta)$. Our scheme verifies if there exists a feasible schedule with the maximum job lateness no more than $L(\delta)$. If it succeeds (fails, respectively) to create a feasible schedule with the maximal job lateness no more than threshold $L(\delta)$, the next smaller (larger, respectively) value of $\delta$ is taken in the binary partition mode. According to our partition, the lateness of no kernel job may surpass the threshold $L(\delta)$. Hence, it basically remains to schedule the bin intervals (see the next section). As already noted, if while scheduling some bin, the next incoming job results in a lateness greater than $L(\delta)$, then a new kernel $K$ including this job is determined (see the next section for the details). Then, the current set of the kernels is updated as $\mathcal{K} := \mathcal{K} \cup \{K\}$, and the scheme proceeds with the updated partition.

A positive delay for a kernel might be unavoidable in an optimal job arrangement. For instance, the time interval of kernel $K_1$ in schedules $S$ and $S_1$ (Figures 1 and 2) is $[35, 40)$ and $[30, 35)$, respectively. The time interval of kernel $K_1$ remains to be $[30, 35)$ in the optimal solution. Indeed, let us form our next set $E$ again with the same emerging Job 1 constructing complementary schedule $(S_1)_1 = (2, 0)(3, 30)(4, 36)(5, 46)(6, 56)(7, 66)(1, 76)$; see Figure 3 (this complementary schedule has a gap $[35, 36)$ marked as a dark region in the figure).

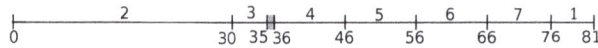

**Figure 3.** A complementary ED-schedule $(S_1)_1$.

It is easily verified that schedule $(S_1)_1$ with kernels $K_1$ and $K_2$ and the corresponding three arranged bins is optimal with $L_{\max}((S_1)_1) = L_{3,(S_1)_1} = 35 - 11 = 24 = L_{7,(S_1)_1} = 76 - 52 = 24$. In that schedule, we have achieved an optimal balance between the delays of the two kernels by redistributing the available jobs within the bins. As a result, the makespan is reduced without affecting the maximum job lateness. Note that the jobs of kernel $K_1$ and also those of kernel $K_2$ are delayed by an appropriate amount of time units without affecting the maximum job lateness. Our framework determines such an appropriate delay for every detected kernel, resulting in an optimal balance between kernel and bin intervals. Observe that, although the partition of the scheduling horizon in schedules $S_1$ and $(S_1)_1$ is the same, the bins are packed in different ways in these two complementary schedules.

Instead of forming the complementary schedules, we shall construct a direct algorithm for scheduling the bins in the following section.

## 5. Scheduling the Bins

At the iteration in the binary search procedure with the trial value $\delta$ and a given partition, we have reduced the scheduling problem to that of filling in the bin intervals by non-kernel jobs. We schedule the bins according to the order in which they appear in the current partition of bin and kernel segments. If all the bins from the current partition are scheduled without surpassing current threshold value $L(\delta)$ (see the previous section), these bins together with the corresponding kernels are merged respectively, forming the overall complete schedule (we have the successful outcome for trial $\delta$). The created feasible solution respecting threshold $L(\delta)$ will have an optimal or sub-optimal makespan among all feasible solutions respecting $L(\delta)$ (see Theorem 2 and Lemma 3).

In this section, we describe how the bins can efficiently be filled in with non-kernel jobs and how during this process, new kernels and the corresponding bins may be declared, yielding an updated partition of the scheduling horizon. While scheduling the bins, our (secondary) goal is to reduce the length of the total idle time interval, which yields a reduced makespan. Although most bin scheduling problems are NP-hard, there are heuristic algorithms with good approximation for these problems. Adopting such a heuristics for our scheduling model, we may pack our bins with close to the minimal total gap length (makespan), as we describe below. Whenever strict optimality is important, an implicit enumerative algorithm can be applied. First, we need a few additional notions.

Let $B$ be the bin immediately preceding kernel $K$ (our construction is similar for the last bin of the current partition). While scheduling bin $B$, we distinguish two types of the available jobs for that bin. We call jobs that may only be feasibly scheduled within that bin y-jobs and ones that may also be feasibly scheduled within a succeeding bin the x-jobs. We have two sub-types of the y-jobs for bin $B$: a Type (a) y-job may only be feasibly scheduled within that bin, unlike the Type (b) y-jobs, which could have also been included in some preceding bin (a Type (b) y-job is released before the interval of bin $B$ and can potentially be scheduled within a preceding bin).

The procedure for scheduling bin $B$ has two phases: Phase 1 consists of two passes. In the first pass, y-jobs are scheduled. In the second pass, x-jobs are included within the remaining available space of the bin by a specially-modified decreasing next fit heuristics, commonly used for bin packing problems. Note that all Type (a) y-jobs can be feasibly scheduled only within bin $B$. Besides Type (a) y-jobs, we are also forced to include all the Type (b) y-jobs in bin $B$, unless we carry out a global job rearrangement and reschedule a Type (b) y-job within some preceding bins. We consider such a possibility in Phase 2.

Phase 1, first pass (scheduling y-jobs): The first pass of Phase 1 consists of two steps. In Step 1, the y-jobs are scheduled in bin $B$ by the ED heuristics resulting in a partial ED-schedule, which we denote by $S(B, y)$. It consists of all the y-jobs that the ED heuristics could include without surpassing the current threshold $L(\delta)$ (recall that $L(\delta)$ is the currently maximum allowable lateness).

If schedule $S(B, y)$ is y-complete, i.e., it contains all the y-jobs for bin $B$, then it is the output of the first pass. Otherwise, the first pass continues with Step 2. We need the following discussion before we describe that step.

If schedule $S(B, y)$ is not y-complete, then the lateness of the next considered y-job at Step 1 surpasses the current $L(\delta)$. If that y-job is a Type (b) y-job, then an instance of Alternative (b2) with that Type (b) y-job (abbreviated IA(b2)) is said to occur (the behavior alternatives were introduced in a wider context earlier in [26]). IA(b2) is dealt with in Phase 2, in which at least one Type (b) y-job is rescheduled within the interval of some earlier constructed bin.

**Lemma 1.** *If no IA(b2) at the first pass of Phase 1 occurs, i.e., the earliest arisen y-job $y$ that could not have been included in schedule $S(B, y)$ is a Type (a) y-job, then a new kernel consisting of Type (a) y-jobs including job $y$ occurs.*

**Proof.** First, note that all Type (a) y-jobs can be feasibly scheduled without surpassing the current allowable lateness and be completed within bin $B$. Hence, if the earliest arisen y-job $y$ that could not have been included turns out to be a Type (a) y-job, a forced right-shift by some other y-job and/or the order change in the sequence of the corresponding Type (a) y-jobs must have occurred. Therefore, there must exist a corresponding delaying emerging job, and hence, a new kernel can be declared. □

If schedule $S(B, y)$ is not y-complete and no IA(b2) at the first pass of Phase 1 occurs, then Step 2 at Phase 1 is invoked. At that step, the new kernel from Lemma 3 is declared, and the current set of kernels and bins is respectively updated. Then, the first pass is called recursively for the newly formed partition; in particular, Step 1 is invoked for the earliest of the newly-arisen bins. This completes the description of the first pass.

**Lemma 2.** *If during the scheduling of bin $B$ at the first pass, no IA(b2) arises, then this pass outputs a y-complete schedule $S(B, y)$ in which the lateness of no y-job is greater than $L(\delta)$.*

**Proof.** By Lemma 1 and the construction of the first pass, if at that pass, no IA(b2) arises, the constructed partial schedule $S(B, y)$ contains all the y-jobs with the lateness no more than $L(\delta)$, which proves the lemma. □

Phase 1, second pass (scheduling x-jobs): The second pass is invoked if no IA(b2) during the execution of the first pass occurs. The first pass will then output a y-complete schedule $S(B, y)$ (Lemma 2), which is the input for the second pass. At the second pass, the available x-jobs are included within the idle time intervals remaining in schedule $S(B, y)$. Now, we describe a modified version of a common Decreasing Next Fit heuristics, adapted to our problem for scheduling bin $B$ (DNF heuristics for short). The heuristics deals with a list of the x-jobs for kernel $K$ sorted in non-increasing order of their processing times, whereas the jobs with the same processing time are sorted in the non-decreasing order of their due dates (i.e., more urgent jobs appear earlier in that list). Iteratively, the DNF heuristics selects the next job $x$ from the list and tries to schedule it at the earliest idle time moment $t'$ at or behind its release time. If time moment $t' + p_i$ is greater than the allowable upper endpoint of the interval of bin $B$, job $x$ is ignored, and the next x-job from the list is similarly considered (if no unconsidered x-job is left, then the second pass for bin $B$ completes). Otherwise ($t' + p_i$ falls within the interval of bin $B$), the following steps are carried out.

(A) If job $x$ can be scheduled at time moment $t'$ without pushing an y-job included in the first pass (and be completed within the interval of the current bin), then it is scheduled at time $t'$ (being removed from the list), and the next x-job from the list is similarly considered.

(B) Job $x$ is included at time $t'$ if it pushes no y-job from schedule $S(B, y)$. Otherwise, suppose job $x$, if included at time $t'$, pushes a y-job, and let $S(B, +x)$ be the corresponding (auxiliary) partial ED-schedule obtained by rescheduling (right-shifting) respectively all the pushed y-jobs by the ED heuristics.

    (B.1) If none of the right-shifted y-jobs in schedule $S(B, +x)$ result in a lateness greater than the current threshold $L(\delta)$ (all these jobs being completed within the interval of bin $B$), then job $i$ is included, and the next x-job from the list is similarly considered.

We need the following lemma to complete the description of the second pass.

**Lemma 3.** *If at Step (B), the right-shifted y-job y results in a lateness greater than the threshold value $L(\delta)$, then there arises a new kernel K consisting of solely Type (a) y-jobs including job y.*

**Proof.** Since the lateness of job $y$ surpasses $L(\delta)$, it forms part of a newly arisen kernel, the delaying emerging job of which is $x$. Furthermore (similarly as in Lemma 1), kernel $K$ cannot contain any Type (b) y-job, as otherwise, such a y-job would have been included within the idle time interval in which job $x$ is included. Hence, that Type (b) y-job could not succeed job $x$. □

    (B.2) If the condition in Lemma 3 is satisfied, then the corresponding new kernel is declared, and the current set of kernels and bins is updated. Similarly, as in Step 2, the first pass is called for the first newly declared bin (initiated at time moment $t'$) from the newly updated partition.

    (B.3) If the lateness of some y-job surpasses the threshold $L(\delta)$ in schedule $S(B, +x)$, then the next x-job from the list is similarly processed from Step (A).

    (B.4) If all the x-jobs from the list have been processed, then the second pass for scheduling bin $B$ completes.

The following theorem tackles a frontier between the polynomially solvable instances and intractable ones for finding a Pareto-optimal frontier (the two related NP-hard problems are the scheduling problem with our primary objective and the derived bin packing problem):

**Theorem 2.** *Suppose, in the iteration of the binary search procedure with trial $\delta$, all the bins are scheduled at Phase 1, i.e., no IA(b2) at Phase 1 arises. Then, a complete schedule in which the lateness of no job is greater than $L(\delta)$ is constructed in time $O(n^2 \log n)$. This schedule is Pareto sub-optimal, and it is Pareto optimal if no idle time interval in any bin is created.*

**Proof.** For every bin $B$, since no IA(b2) during its construction at Phase 1 arises, this phase will output a partial schedule for that bin, in which the lateness of no job is greater than $L(\delta)$ (Lemmas 2 and 3). At the same time, by our construction, no kernel job may result in a lateness greater than $L(\delta)$. Then, the maximum job lateness in the resultant complete schedule is no greater than $L(\delta)$. This schedule is Pareto sub-optimal due to the DNF heuristics (with a known sub-optimal approximation ratio). Furthermore, if no idle time interval in any bin is created, there may exist no feasible solution with the maximum job lateness $L(\delta)$ such that the latest scheduled job in it completes earlier than in the above generated solution; hence, it is Pareto optimal.

Let us now estimate the cost of Phase 1. For the purpose of this estimation, suppose $n_1$ ($n_2$, respectively) is the number of y-jobs (x-jobs, respectively). In the first pass, y-jobs are scheduled by the ED heuristics in time $O(n_1 \log n_1)$ if all the scheduled bins are y-complete. Otherwise, since no IA(b2) arises, each time the scheduling of a bin cannot be completed by all the y-jobs, a new kernel is declared that includes the corresponding y-jobs (Lemma 1). Since there may arise less than $n_1$ different kernels, the number of times a new kernel may arise is less than $n_1$. Thus, the same job will be rescheduled less than $n_1$ times (as a bin or a kernel job), which yields the time complexity of $O(n_1^2 \log n_1)$ for the first pass. Since the time complexity of the DNF heuristics is the same as that of ED heuristics used at the first pass and new kernels are similarly declared (Lemma 3), the cost of the second pass is similarly $O(n_2^2 \log n_2)$. Thus, the overall cost for the construction of all the bins is $O(n^2 \log n)$, which is also the cost of the overall scheme since the kernel jobs are scheduled by the ED heuristics. $\square$

Note that the second condition in the above theorem is sufficient, but not necessary for minimizing the makespan; i.e., some bins in a feasible schedule respecting threshold $L(\delta)$ may have idle time intervals, but that schedule may minimize the makespan. In general, we rely on the efficiency of the heuristics used for packing our bins for the minimization of the makespan criterion. We also straightforwardly observe that by the DNF heuristicss, no unscheduled job may fit within any available gap in any bin without forcing some other job to surpass the current allowable lateness $L(\delta)$ at Phase 1, whereas the DNF-heuristicss tends to keep reserved "short" unscheduled jobs.

Continuing with the issue of the practical behavior, as we have mentioned in the Introduction, the ED heuristics turns out to be an efficient tool for a proper distribution of the jobs within the bins. In [9], computational experiments were conducted for 200 randomly generated instances of our problem. For the majority of these instances, the forced delay of kernel $K(S)$ in the initial ED-schedule $S$ (i.e., the difference between the completion time of the corresponding delaying emerging job and $r(K)$) was neglectable. Therefore, the lateness of the corresponding overflow job was either optimal or very close to the optimal maximum job lateness (as no job of kernel $K$ can be started earlier than at time $r(K)$ in any feasible solution). At the same time, since the ED heuristics creates no avoidable machine idle time interval, the minimum makespan is also achieved. Thus, for the majority of the tested problem instances, both objective criteria were simultaneously minimized or the maximum jobs lateness was very close to the optimum, whereas the makespan was optimal.

The computational experiments from [9] also yield that, for most of the tested instances, already for small trial $\delta$'s, the bins would have been scheduled in Phase 1 (i.e., no IA(b2) would arise). Moreover, there will be no avoidable gap left within the bins for these $\delta$'s. In other words, the conditions in Theorem 2 would be satisfied, and Phase 1 would successfully completed. In theory, however, we cannot guarantee such a behavior, and we proceed with our construction in Phase 2.

Phase 2: If the first condition in Theorem 2 is not satisfied, i.e., an IA(b2) at the first pass of Phase 1 occurs, then Phase 2 is invoked. Let $y$ be the earliest arisen Type (b) y-job that could not have been included in schedule $S(B, y)$. In general, job $j$ may be pushed by one or more other Type (b) y-jobs. Since the lateness of job $y$ surpasses $L(\delta)$, a complete schedule respecting this threshold value cannot be created unless the lateness of job $y$ is decreased. The next observation follows:

**Observation 6.** *Suppose an IA(b2) with (a Type (b) y-job) y at the first pass of Phase 1 in bin B arises. Then, the resultant maximum job lateness in bin B yielded by job y cannot be decreased unless job y and/or other Type (b) y-jobs currently pushing job y in bin B are rescheduled to some earlier bin(s).*

According to the above observation, in Phase 2, either job $y$ is to be rescheduled to an earlier bin or a subset of the above Type (b) y-jobs (with an appropriate total processing time) is to be formed and all jobs from that subset are to be rescheduled to some earlier bin(s). In particular, if the lateness of job $y$ surpasses $L(\delta)$ by amount $\lambda$, then the total processing time of jobs in such a subset is to be at least $\lambda$. By the construction, no new job can feasibly be introduced into any already scheduled bin $B$ unless some jobs (with an appropriate total length) from that bin are rescheduled behind all the above Type (b) y-jobs. Such a job, which we call a substitution job for job $y$, should clearly be an emerging job for job $y$, i.e., $d_s > d_y$ (as otherwise, once rescheduled, its lateness will surpass $L(\delta)$).

**Theorem 3.** *If for an IA(b2) with a (Type (b) y-job) y, there exists no substitution job for job y, then there exists no feasible schedule respecting threshold value $L(\delta)$. Hence, there exists no Pareto-optimal solution with maximum job lateness $L(\delta)$.*

**Proof.** Due to Observation 6, job $y$ or some other Type (b) y-job(s) pushing job $y$ must be rescheduled to a bin preceding bin $B$. By the way we construct our bins, no new job can feasibly be introduced into any bin preceding bin $B$ unless some (sufficiently long) job $s$ from that bin is rescheduled behind all the above Type (b) y-jobs. $d_s < d_y$ must hold as otherwise job $s$ or some job originally included before job $y$ will surpass the current threshold; i.e., $s$ is a substitution job. It follows that if there exists no substitution job, the lateness of at least one job from bin $B$ or from its preceding bin will surpass threshold value $L(\delta)$, and hence, there may exist no feasible schedule respecting $L(\delta)$. The second claim immediately follows. □

Due to the above theorem, if these exists no substitution job, the call from the binary search procedure for the current $\delta$ halts with the failure outcome, and the procedure proceeds with the next (larger) trial value for $\delta$; if there remains no such $\delta$, the whole scheme halts.

The case when there exists a substitution job for job $y$ is discussed now. We observe that there may exist a Pareto-optimal solution. To validate whether it exists, we first need to verify if there exists a feasible solution respecting threshold $L(\delta)$ (given that the call to the bin packing procedure was performed from the iteration in the binary search procedure with that threshold value). This task is easily verified if, for instance, there is a single substitution job $s$. Indeed, we may construct an auxiliary ED-schedule in which we activate that substitution job for job $y$ (similarly as before, we increase artificially the release time of job $s$ to that of job $y$, so that once the ED heuristics is newly applied to that modified problem instance, the substitution job $s$ will be forced to be rescheduled behind job $y$). In the resultant ED-schedule with the activated job $s$, if the lateness of none of the rescheduled y-jobs is no greater than $L(\delta)$, Phase 1 can be resumed for bin $B$. Otherwise, there may exit no feasible solution respecting threshold $L(\delta)$ (hence, no Pareto-optimal solution with the maximum job lateness not exceeding $L(\delta)$ may exist), and the scheme can proceed with the next trial $\delta$.

In practice, with a known upper limit on the total number of substitution jobs, the above activation procedure can be generalized to another procedure that enumerates all the possible subsets of substitution jobs in polynomial time. In general, if the number of substitution jobs is bounded only by $n$, the number of possible subsets of substitution jobs with total processing time $\lambda$ or more might be an exponent of $n$. In theory and practice, we may take one of the two alternative options: either we try all such possible non-dominated subsets building an implicit enumerative algorithm, or we may rather choose a heuristic approach generating a fixed number of "the most promising" subsets of substitution jobs in polynomial time. Based on our general framework, both implicit enumeration and heuristic algorithms might be implemented in different ways. The study of such algorithms, which would rely on the results of numerical experiments, is beyond the scope of this paper and could rather be a subject of independent research.

## 6. Conclusions

We have proposed a general framework for the solution of a strongly NP-hard bi-criteria scheduling problem on a single machine with the objectives to minimize the maximum job lateness and the makespan. Theorem 2 specifies explicitly the nature of the problem instances for which the Pareto optimal frontier can be found in polynomial time. Finding a feasible solution with the maximum job lateness not exceeding some constant threshold becomes intractable if the first condition in the theorem is not satisfied. Relying on the computational experiments from [9], we have observed that already for small trial $\delta$'s, the bins would have been scheduled in Phase 1, i.e., no IA(b2) would arise in Phase 1. In addition, since there will be no avoidable gap left within the bins for these $\delta$'s, the second condition in Theorem 2 would be satisfied, and Phase 1 would successfully complete in polynomial time. We have also observed that for the majority of the tested problem instances from [9], both objective criteria were simultaneously minimized or the maximum job lateness was very close to the optimum, whereas the makespan was optimal. Theoretically, we are unable to guarantee such a behavior, and hence, we described how the solution process can proceed in Phase 2. We have shown that it suffices to enumerate some subsets of the substitution jobs and generate the resultant feasible schedules. With a complete enumeration, we guarantee that one of these solutions will respect a given threshold, unless there exists no such a feasible schedule. In practice, if a complete enumeration of all dominant subsets turns out to take an admissible time, we may "switch" to heuristic methods that enumerate a constant number of the "most promising" subsets and the corresponding feasible solutions (In the latter case, theoretically, it is possible that we fail to create a feasible schedule respecting the threshold where such a feasible schedule exists. Then, the Pareto frontier that the algorithm will generate will miss the corresponding Pareto-optimal solution, i.e., it will not be "complete".).

By partitioning the scheduling horizon into kernel and bin intervals, we have reduced the problem of minimizing the makespan in a feasible schedule respecting a given threshold to that of minimizing the gaps in the bins in that schedule. Bin packing problems with this objective have been extensively studied in the literature, and a number of efficient enumeration algorithms and also heuristicss with good theoretical and also practical behavior exist. We showed how such a heuristic method can be adopted for packing our bins with jobs with release times and due dates (the latter job parameters impose additional restrictions, not present in bin packing problems). In this way, our framework has reduced the bi-criteria scheduling problem to the above two single-criterion problems, inducing four basic types of algorithms for the bi-criteria problem (depending on whether an exact or heuristic method is incorporated for either of the two problems). For the future work, it would be interesting to test particular implementations of these algorithms and to verify their performance for problem instances that significantly differ from those tested in [9].

**Conflicts of Interest:** The authors declare no conflict of interest.

## References

1.  Garey, M.R.; Johnson, D.S. *Computers and Intractability: A Guide to the Theory of NP—Completeness;* W. H. Freeman and Company: San Francisco, CA, USA, 1979.
2.  Jackson, J.R. Schedulig a Production Line to Minimize the Maximum Tardiness. In *Manegement Scince Research Project, University of California;* Office of Technical Services: Los Angeles, CA, USA, 1955.

3. Schrage, L. Obtaining Optimal Solutions to Resource Constrained Network Scheduling Problems. Unpublished work, 1971.
4. Potts, C.N. Analysis of a heuristic for one machine sequencing with release dates and delivery times. *Oper. Res.* **1980**, *28*, 1436–1441. [CrossRef]
5. Hall, L.A.; Shmoys, D.B. Jackson's rule for single-machine scheduling: Making a good heuristic better. *Mathem. Oper. Res.* **1992**, *17*, 22–35. [CrossRef]
6. Nowicki, E.; Smutnicki, C. An approximation algorithm for single-machine scheduling with release times and delivery times. *Discret Appl. Math.* **1994**, *48*, 69–79. [CrossRef]
7. Larson, R.E.; Dessouky, M.I. heuristic procedures for the single machine problem to minimize maximum lateness. *AIIE Trans.* **1978**, *10*, 176–183. [CrossRef]
8. Kise, H.; Ibaraki, T.; Mine, H. Performance analysis of six approximation algorithms for the one-machine maximum lateness scheduling problem with ready times. *J. Oper. Res. Soc. Japan* **1979**, *22*, 205–223. [CrossRef]
9. Vakhania, N.; Perez, D.; Carballo, L. Theoretical Expectation versus Practical Performance of Jackson's heuristic. *Math. Probl. Eng.* **2015**, *2015*, 484671. [CrossRef]
10. McMahon, G.; Florian, M. On scheduling with ready times and due dates to minimize maximum lateness. *Oper. Res.* **1975**, *23*, 475–482. [CrossRef]
11. Carlier, J. The one–machine sequencing problem. *Eur. J. Oper. Res.* **1982**, *11*, 42–47. [CrossRef]
12. Sadykov, R.; Lazarev, A. Experimental comparison of branch-and-bound algorithms for the $1|r_j|L_{max}$ problem. In Proceedings of the Seventh International Workshop MAPSP'05, Siena, Italy, June 6–10 2005; pp. 239–241.
13. Grabowski, J.; Nowicki, E.; Zdrzalka, S. A block approach for single-machine scheduling with release dates and due dates. *Eur. J. Oper. Res.* **1986**, *26*, 278–285. [CrossRef]
14. Larson, M.I. Dessouky and Richard E. DeVor. A Forward-Backward Procedure for the Single Machine Problem to Minimize Maximum Lateness. *IIE Trans.* **1985**, *17*, 252–260, doi:10.1080/07408178508975300. [CrossRef]
15. Pan, Y.; Shi, L. Branch-and-bound algorithms for solving hard instances of the one-machine sequencing problem. *Eur. J. Oper. Res.* **2006**, *168*, 1030–1039. [CrossRef]
16. Liu, Z. Single machine scheduling to minimize maximum lateness subject to release dates and precedence constraints. *Comput. Oper. Res.* **2010**, *37*, 1537–1543. [CrossRef]
17. Gharbi, A.; Labidi, M. Jackson's Semi-Preemptive Scheduling on a Single Machine. *Comput. Oper. Res.* **2010**, *37*, 2082–2088, doi:10.1016/j.cor.2010.02.008. [CrossRef]
18. Lenstra, J.K.; Rinnooy Kan, A.H.G.; Brucker, P. Complexity of machine scheduling problems. *Ann. Discret. Math.* **1977**, *1*, 343–362.
19. Kacem, I.; Kellerer, H. Approximation algorithms for no idle time scheduling on a single machine with release times and delivery times. *Discret. Appl. Math.* **2011**, *164*, 154–160, doi:10.1016/j.dam.2011.07.005. [CrossRef]
20. Lazarev, A. The Pareto-optimal set of the NP-hard problem of minimization of the maximum lateness for a single machine. *J. Comput. Syst. Sci. Int.* **2006**, *45*, 943–949. [CrossRef]
21. Lazarev, A.; Arkhipov, D.; Werner, F. Scheduling Jobs with Equal Processing Times on a Single Machine: Minimizing Maximum Lateness and Makespan. *Optim. Lett.* **2016**, *11*, 165–177. [CrossRef]
22. Carlier, J.; Hermes, F.; Moukrim, A.; Ghedira, K. Exact resolution of the one-machine sequencing problem with no machine idle time. *Comp. Ind. Eng.* **2010**, *59*, 193–199, doi:10.1016/j.cie.2010.03.007. [CrossRef]
23. Chrétienne, P. On single-machine scheduling without intermediate delays. *Discrete Appl. Math.* **2008**, *156*, 2543–2550. [CrossRef]
24. Ehrgott, M. *Multicriteria Optimization*; Springer Science & Business Media: Berlin, Germany, 2005; Volume 491.
25. Chinos, E.; Vakhania, N. Adjusting scheduling model with release and due dates in production planning. *Cogent Eng.* **2017**, *4*, 1–23. [CrossRef]
26. Vakhania, N. A better algorithm for sequencing with release and delivery times on identical processors. *J. Algorithms* **2003**, *48*, 273–293. [CrossRef]

*algorithms*

MDPI

*Article*

# Near-Optimal Heuristics for Just-In-Time Jobs Maximization in Flow Shop Scheduling

**Helio Yochihiro Fuchigami** [1,*] (ID)**, Ruhul Sarker** [2] **and Socorro Rangel** [3] (ID)

[1]    Faculty of Sciences and Technology (FCT), Federal University of Goias (UFG),
      74968-755 Aparecida de Goiânia, Brazil
[2]    School of Engineering and Information Technology (SEIT), University of New South Wales (UNSW),
      Canberra ACT 2610, Australia; r.sarker@adfa.edu.au
[3]    Instituto de Biociências, Letras e Ciências Exatas (IBILCE), Universidade Estadual Paulista (UNESP),
      19014-020 São Paulo, Brazil; socorro@ibilce.unesp.br
*    Correspondence: heliofuchigami@ufg.br; Tel.: +55-62-3209-6550

Received: 28 February 2018; Accepted: 4 April 2018; Published: 6 April 2018

**Abstract:** The number of just-in-time jobs maximization in a permutation flow shop scheduling problem is considered. A mixed integer linear programming model to represent the problem as well as solution approaches based on enumeration and constructive heuristics were proposed and computationally implemented. Instances with up to 10 jobs and five machines are solved by the mathematical model in an acceptable running time (3.3 min on average) while the enumeration method consumes, on average, 1.5 s. The 10 constructive heuristics proposed show they are practical especially for large-scale instances (up to 100 jobs and 20 machines), with very good-quality results and efficient running times. The best two heuristics obtain near-optimal solutions, with only 0.6% and 0.8% average relative deviations. They prove to be better than adaptations of the NEH heuristic (well-known for providing very good solutions for makespan minimization in flow shop) for the considered problem.

**Keywords:** just-in-time scheduling; flow shop; heuristics

## 1. Introduction

Permutation flow shop scheduling, a production system in which jobs follow the same flow for all machines in the same order, is one of the most important production planning problems [1]. This type of manufacturing environment is very often encountered in intermittent industrial production systems. In this context, the just-in-time scheduling aims to achieve a solution that minimizes the cost functions associated with the earliness and tardiness of jobs. Therefore, it is more common to find research addressing the sum of the earliness and tardiness of jobs or the penalties caused by a deviation from a previously established due date for the delivery of a product. In practice, these performance measures are very important for companies as both the earliness and tardiness of completing jobs entail relatively higher costs of production, such as increases in inventory levels, fines, cancellations of orders or even loss of customers.

A different approach is to consider an objective related to the number of early/tardy jobs rather than earliness/tardiness duration [2]. Lann and Mosheiov [3] introduced in 1996 a class of problems in the just-in-time area that aims to maximize the number of jobs completed exactly on their due dates, which are called just-in-time jobs. However, up to date, there are still few works in the literature that consider as an optimization criterion the maximization of the number of just-in-time jobs, despite its applicability. Examples of applications in which such structure may arise include: chemical or hi-tech industries, where parts need to be ready at specific times in order to meet certain required conditions (arrival of other parts, specific temperature, pressure, etc.); production of perishable items (e.g., food and drugs) under deterministic demand; maintenance services agencies which handle the preferable due date for customers; and rental agencies (hotels amd car rental), where reservations schedule must meet exactly the time requested by all clients.

In classical notation of three fields, the maximization of the number of just-in-time jobs in a permutation flow shop problem can be denoted by $F_m \mid prmu, d_j \mid n_{JIT}$, where $F_m$ indicates a generic flow shop environment with $m$ machines, $prmu$ the permutation constraint, $d_j$ the existence of a due date for each job and $n_{JIT}$ the objective function that maximizes the number of jobs completed just-in-time. In some environments, such as a single-machine problem and permutation flow shop, the schedule is provided by only job sequencing. In others, it is also necessary to allocate jobs to machines, such as in the ones with parallel machines and in hybrid flow shop. Although this research addresses the problem of the permutation flow shop, in addition to sequencing, there is also the possibility of inserting idle time into some jobs to adjust their completions to the due dates (if there is slack). That is, the solution (schedule) comprises sequencing and timing phases; besides an order of jobs, it is also necessary to define the starting and ending stages of each operation. In the literature, it is known as a schedule with inserted idle time [4].

The purpose of this work is to provide insights for a theme in the area of scheduling that, despite its importance for several industrial contexts, is relatively unexplored. This study proposes a mathematical model to represent the permutation flow shop scheduling with the total number of just-in-time jobs being maximized and a set of heuristic solution approaches to solve it in a relatively short execution time. A comprehensive computational study is presented. As shown in the literature review, to the best of our knowledge, all but one of the papers dealing with the maximization of the total number of just-in-time jobs only present theoretical results. Thus, this research helps to fill the gap of computational results in the literature of scheduling problems. We are concerned with applications for which there is no interest in jobs that are finished before or after their due dates.

The remainder of this paper is organized as follows. A review of the just-in-time scheduling problems literature is presented in Section 2. A formal description of the problem considered and the mixed integer linear programming model developed is given in Section 3. The proposed constructive heuristics methods are discussed in Section 4. Analyses of the results from all the solution approaches computationally implemented in this study are presented in Section 5. Final remarks are given in Section 6.

## 2. Literature Review

Although the just-in-time philosophy involves broader concepts, until recent decades, scheduling problems in this area were considered variations of earliness and tardiness minimization, as can be seen in the review by [5]. Many problems are presented in [6,7]. The most common just-in-time objective functions found in the literature are related the total weighted earliness and tardiness, with equal, asymmetric or individual weights.

Recently, Shabtay and Steiner [8] published a review of the literature addressing the problem of maximizing the number of just-in-time jobs which demonstrates that this theme has been very little explored. The work presented in [7], which deals with various types of just-in-time scheduling problems, mentions only one paper that considers the number of just-in-time jobs, that is, the survey presented in [8] of single-, parallel- and two-machine flow shop environments.

There are some publications in the relevant literature that discuss the problem considering the criteria of maximizing the number of just-in-time jobs in different production systems, specifically with flow shops. Choi and Yoon [9] showed that a two-machine weighted problem is classified as NP-complete and a three-machine identical weights one as NP-hard. These authors proposed and demonstrated several dominance conditions for an identical weights problem. Based on these conditions, they presented an algorithm for a two-machine problem. In addition, they proved that an optimal solution to a two-machine problem is given by the earliest due date (EDD) priority rule.

Several two-machine weighted problems, a flow shop, job shop and open shop, are considered by [10]. They propose pseudo-polynomial time algorithms and a way of converting them to polynomial time approach schemes. Shabtay [11] examined four different scenarios for a weighted problem: firstly, two machines and identical weights for jobs; secondly, a proportional flow shop, which the processing times on all machines are the same; thirdly, a set of identical jobs to be produced for different clients (the processing times are equal but due dates different for each client); and, lastly, the no-wait flow shop with no waiting times between the operations of a job. A dynamic programming algorithm and, for a two-machine problem, an algorithm of less complexity than that were proposed by [9]. A proportional flow shop problem was addressed by [12] who considered options of with and without a no-wait restriction, and proposed dynamic programming algorithms.

Given the difficulty of directly solving this class of problems, some authors preferred to convert them into other types of optimization problems, such as the partition problem [9] and the modeling of acyclic directed graphs [11]. A bi-criteria problem of maximizing the weighted number of just-in-time jobs and minimizing the total resource consumption cost in a two-machine flow shop was considered by [2].

Focusing on maximizing the number of just-in-time jobs, only one work [2] presented computational experiments. Gerstl et al. [12] only mentioned that a model is implemented. Unlike most studies in the shop scheduling area employing experimental research methodologies, all other papers examined used theoretical demonstrations and properties to attest to the validity of their proposed methods, as well as analyses of algorithmic complexity, but did not present computational results.

Yin et al. [13] considered the two-machine flow shop problem with two agents, each of them having its own job set to process. The agents need to maximize individually the weighted number of just-in-time jobs of their respective set. The authors provided two pseudo-polynomial-time algorithms and their conversion into two-dimensional fully polynomial-time approximation schemes (FPTAS) for the problem.

In general, the research studies about the maximization of the number of just-in-time jobs encountered in the literature are limited to a two-machine flow shop problem. No constructive heuristic was found for a $m$-machine flow shop or computational experiments, which are important to evaluate the effectiveness of the method in terms of solution quality and computational efficiency. One of the main contributions of this research is the presentation of efficient and effective solution methods for maximizing the number of just-in-time jobs which are demonstrated to be applicable to practical multiple-machine flow shop problems. Preliminary results of this research are presented in [14,15].

## 3. A Mixed Integer Linear Programming Model

The flow shop scheduling problem can be generically formulated as follows. Consider a set of $n$ independent jobs ($J = \{J_1, J_2, ..., J_n\}$), all of which have the same weight or priority, cannot be interrupted, are released for processing at zero time, have to be executed in $m$ machines ($M = \{M_1, M_2, ..., M_m\}$) and are physically arranged to follow identical unidirectional linear flows (i.e., all the jobs are processed through the machines at the same sequence). Each job ($J_j$) requires a processing time ($p_{jk}$) in each machine ($M_k$) and have its due date represented by $d_j$, both considered known and fixed. The solution consists of finding a schedule that maximizes the number of jobs finished at exactly their respective due dates.

To formulate a mathematical model to represent the problem, consider the following parameters, indices and variables.

*Parameters and indices:*

| | | |
|---|---|---|
| $n$ | number of jobs | |
| $m$ | number of machines | |
| $i, j$ | job index | $i = 0, \ldots, n, j = 0, \ldots, n$ |
| $k$ | machine index | $k = 0, \ldots, m$ |
| $p_{jk}$ | processing time of job $J_j$ on machine $M_k$ | $j = 1, \ldots, n, k = 1, \ldots, m$ |
| $d_j$ | due date of job $J_j$ | $j = 1, \ldots, n$ |
| $B$ | very large positive integer, the value considered in Section 5 is: $B = 100\sum_{j=1}^{n}\sum_{k=1}^{m} p_{ij}$ | |
| $J_0$ | dummy job—the first in the scheduling | |
| $M_0$ | dummy machine—theoretically considered before the first (physical) machine | |

*Decision variables:*

| | | |
|---|---|---|
| $C_{jk}$ | completion time of job $J_j$ on machine $M_k$ | $j = 0, \ldots, n, k = 0, \ldots, m$ |
| $U_j$ | equals 1 if job $J_j$ is just-in-time or 0 otherwise | $j = 1, \ldots, n$ |
| $x_{ij}$ | equals 1 if job $J_i$ is assigned immediately before job $J_j$ or 0, otherwise | |
| | | $j = 0, \ldots, n, j = 1, \ldots, n, i \neq j.$ |

The mixed integer programming model given by Expressions (1)–(12) is based on the approach proposed by Dhouib et al. [16].

$$\text{Max } Z = \sum_{j=1}^{n} U_j \tag{1}$$

Subject to:

$$C_{jk} - C_{ik} + B(1 - x_{ij}) \geq p_{jk}, \; i = 0, \ldots, n, \; j = 1, \ldots, n, \; i \neq j, \; k = 1, \ldots, m \tag{2}$$

$$C_{jk} - C_{j(k-1)} \geq p_{jk}, \; j = 1, \ldots, n, \; k = 1, \ldots, m \tag{3}$$

$$C_{jm} - B(1 - U_j) \leq d_j, \; j = 1, \ldots, n \tag{4}$$

$$C_{jm} + B(1 - U_j) \geq d_j, \; j = 1, \ldots, n \tag{5}$$

$$\sum_{j=1}^{n} x_{0j} = 1, \tag{6}$$

$$x_{0j} + \sum_{i=1, i \neq j}^{n} x_{ij} = 1, \; j = 1, \ldots, n \tag{7}$$

$$\sum_{j=1, j \neq i}^{n} x_{ij} \leq 1, \; i = 0, \ldots, n \tag{8}$$

$$C_{j0} = 0, \; j = 0, \ldots, n \tag{9}$$

$$C_{jk} \in \Re_{+}, \; j = 0, \ldots, n, \; k = 0, \ldots, m \tag{10}$$

$$U_j \in \{0, 1\}, \; j = 1, \ldots, n \tag{11}$$

$$x_{ij} \in \{0, 1\}, \; i = 0, \ldots, n, \; j = 1, \ldots, n \tag{12}$$

The optimization criterion expressed in Equation (1) is to maximize the number of just-in-time jobs. Expressions in Equation (2) ensure the consistency of the completion times of jobs, that is, if job $J_i$ immediately precedes job $J_j$ (i.e., $x_{ij} = 1$), the difference between their completion times on each machine must be at least equal to the processing time of job $J_j$ on the machine considered; otherwise (if $x_{ij} = 0$), if there is no relationship between the completion times of this pair of jobs, then the constraints in Equation (2) are redundant. Constraints in Equation (3) require that the $k$th operation of job $J_j$ be completed after the $(k - 1)$th operation plus the processing time ($p_{jk}$). Generally, the value of $B$ is an upper bound to the makespan.

The expressions in Equations (4) and (5) jointly establish that the variable $U_j$ equals 1 if job $J_j$ finishes on time or 0 otherwise. When job $J_j$ does not finish on time, i.e., $U_j = 0$, these two sets of constraints become redundant. The constraints in Equation (6) ensure that only one job is assigned to the first position in the sequence (not considering the dummy job $J_0$). The expressions in Equation (7) establish that job $J_j$ ($j \neq 0$) either occupies the first position (after $J_0$) or is necessarily preceded by another job. The constraints in Equation (8) ensure that one job, at most, immediately precedes another. The expressions in Equation (9) define that all the jobs are ready to be initiated on machine $M_1$, that is the completion time for all jobs in the dummy machine ($k = 0$) is zero, and so ensure the consistency of the restrictions in Equation (3). The expressions in Equations (10)–(12) define the domain of the variables.

The model's size in relation to its numbers of variables and constraints are presented in Table 1. For example, an instance with $n = 5$ jobs and $m = 3$ machines has 69 variables and 91 constraints and another one with $n = 10$ jobs and $m = 5$ machines has 236 variables and 268 constraints.

**Table 1.** Model's size in relation to numbers of variables and constraints.

| Variables | Binary | $n^2 + 2n$ |
|---|---|---|
| | Integer | $n^2 + m + n + 1$ |
| | Total | $n(2n + 3) + m + 1$ |
| Constraints | (2) | $n^2 m + nm$ |
| | (3) | $nm$ |
| | (4) | $n$ |
| | (5) | $n$ |
| | (6) | 1 |
| | (7) | $n$ |
| | (8) | $n + 1$ |
| | (9) | $n + 1$ |
| | Total | $n(n + 2m + 6) + m + 3$ |

## 4. Constructive Heuristics

To efficiently solve the problem described in Section 3, ten constructive heuristic methods are proposed based on an investigation of the problem structure and inspired by relevant classical algorithms, such as the NEH heuristic [17], Hodgson's algorithm [18] and well-known priority rules, such as the EDD which is the ascending order of due dates, and the minimum slack time (MST) which sequences jobs according to the smallest amount of slack ($d_j - \sum_{k=1}^{m} p_{jk}$). It is known that, in a single-machine problem, the EDD and MST rules minimize the maximum tardiness and earliness, respectively [19].

All the heuristics require a timing adjustment procedure that consists of checking the best instant to start the operations for a given sequence to complete a greater number of jobs on time. The simple shifting of an early job with a slack time to match its conclusion to its due date could result in an overall improvement in the solution. A pseudo-code of the timing adjustment procedure is given in Algorithm 1. Note that a shift is applied only in the last operation of a job (Step 3 in Algorithm 1) because keeping each previous operation starting at its earliest possible instant can be an advantage in this problem as it enables jobs that could be late because of their first operations to be anticipated. In addition, this shift is maintained when there is an improvement in the solution (or at least a tie); otherwise, the replacement is reversed to anticipate operations which contribute to eliminating possible tardiness in subsequent jobs.

*Timing adjustment procedure*

Step 1. For the given sequence, considering starting each operation as early as possible, compute the number of just-in-time jobs ($n_{JIT}$) and consider $J_{initial} = J_{[1]}$, where $J_{[j]}$ is the job in position $j$ of the sequence.

Step 2. From $J_{initial}$, identify the first early job in the sequence ($J_E$) and go to Step 3. If there are no early jobs (from $J_{initial}$), STOP.

Step 3. Move the last operation of job $J_E$ to eliminate earliness and make its conclusion coincide with its due date. Properly reschedule the last operations of the jobs after $J_E$.

Step 4.  Compute the new number of just-in-time jobs ($n_{JIT}'$).

If $n_{JIT}' < n_{JIT}$ (the new solution is worse than the previous one), return both the last operation of $J_E$ and the operations of the following jobs to their previous positions, set $J_{initial} = J_{[initial]} + 1$ and go to Step 2.

Else (the new solution is better than or equal to the previous one), keep the new schedule, set $J_{initial} = J_{[E]} + 1$ and go to Step 2.

**Algorithm 1.** Pseudo-code of timing adjustment procedure.

The first four heuristics proposed are adaptations of the NEH algorithm's insertion method and are given in Algorithms 2–5. H1 and H2 employ the EDD rule as the initial order while H3 and H4 consider the MST rule. Another feature is that H2 and H4 improve on H1 and H3, respectively, by using neighborhood search in the partial sequence.

Two types of neighborhood search are employed, insertion and permutation, in the same way as [20]. Given a sequence (or a partial sequence) of n jobs, its insertion neighborhood consists of all $(n-1)^2$ sequences obtained by removing a job from its place and relocating it to another position. The permutation neighborhood is composed by all $n(n-1)/2$ sequences obtained by permuting the positions of two jobs (see Example 1).

Table 2 summarizes the main procedures used in each one of the constructive heuristics.
**Example 1.** Consider the initial sequence with four jobs: $\{J_3, J_2, J_1, J_4\}$. The insertion neighborhood results in $(n-1)^2 = 9$ sequences:

-  Initially, inserting the first job $J_3$: $\{J_2, \boldsymbol{J_3}, J_1, J_4\}$, $\{J_2, J_1, \boldsymbol{J_3}, J_4\}$, $\{J_2, J_1, J_4, \boldsymbol{J_3}\}$;
-  Then, inserting the second job $J_2$ only in the positions not considered yet: $\{J_3, J_1, \boldsymbol{J_2}, J_4\}$, $\{J_3, J_1, J_4, \boldsymbol{J_2}\}$; for example, it is not necessary to insert $J_2$ in the first position because the sequence resulting was already listed; and
-  Next, the third job $J_1$ is inserted and, lastly, the fourth one $J_4$: $\{\boldsymbol{J_1}, J_3, J_2, J_4\}$, $\{J_3, J_2, J_4, \boldsymbol{J_1}\}$, $\{\boldsymbol{J_4}, J_3, J_2, J_1\}$, $\{J_3, \boldsymbol{J_4}, J_2, J_1\}$.

Starting again from the initial sequence $\{J_3, J_2, J_1, J_4\}$ of the same example, the permutation neighborhood results in $n(n-1)/2 = 6$ sequences:

-  First, the first two jobs are permuted: $\{\boldsymbol{J_2}, \boldsymbol{J_3}, J_1, J_4\}$; then the first and third jobs: $\{\boldsymbol{J_1}, J_2, \boldsymbol{J_3}, J_4\}$; and the first and fourth jobs: $\{\boldsymbol{J_4}, J_2, J_1, \boldsymbol{J_3}\}$;
-  Next, from the initial sequence, the second and third are permutated: $\{J_3, \boldsymbol{J_1}, \boldsymbol{J_2}, J_4\}$; and the second and fourth: $\{J_3, \boldsymbol{J_4}, J_1, \boldsymbol{J_2}\}$; and
-  Lastly, from the initial sequence, the third and fourth jobs are permutated: $\{J_3, J_2, \boldsymbol{J_4}, \boldsymbol{J_1}\}$.

*Heuristic H1*

Step 1.  Order jobs according to the EDD rule (in the case of a tie, use the lower $\sum p_{jk}$).
Step 2.  For the first two jobs, apply the timing adjustment procedure to find the best partial sequence (between these two possibilities) with the lower $n_{JIT}$.
Step 3.  For h = 3 to $n$, do:

Keeping the relative positions of the jobs of the partial sequence, insert the hth job of the order defined in Step 1 in all possible positions and apply the timing adjustment procedure in each insertion; consider the new partial sequence with the best $n_{JIT}$ (in the case of a tie, use the upper position).

**Algorithm 2.** Pseudo-code of heuristic H1.

*Heuristic H2*

Step 1.  Order jobs by the EDD rule (in the case of a tie, use the lower $\sum p_{jk}$).

Step 2. For the first two jobs, apply the timing adjustment procedure to find the best partial sequence (between these two possibilities) with the lower $n_{JIT}$.

Step 3. For h = 3 to $n$, do:

Add the $h$th job of order defined in Step 1 in the last position of the partial sequence;

Considering the insertion neighborhood with $(h − 1)2$ partial sequences and applying the timing adjustment procedure, determine that with the best $n_{JIT}$;

From the best solution obtained so far, considering the permutation neighborhood with $h(h − 1)/2$ partial sequences and applying the timing adjustment procedure, determine that with the best $n_{JIT}$.

**Algorithm 3.** Pseudo-code of heuristic H2.

*Heuristic H3*

Step 1. Order jobs by the MST rule (in the case of a tie, use the lower $\sum p_{jk}$).

Steps 2 and 3. The same as in heuristic H1.

**Algorithm 4.** Pseudo-code of heuristic H3.

*Heuristic H4*

Step 1. Order jobs by the MST rule (in the case of a tie, use the lower $\sum p_{jk}$).

Steps 2 and 3. The same as in heuristic H2.

**Algorithm 5.** Pseudo-code of heuristic H4.

The next four heuristics, H5, H6, H7 and H8, employ ideas from the classic Hodgson's algorithm which provides the optimal solution for minimizing the number of tardy jobs in a single-machine problem. They are presented in Algorithms 6–9. Again, the first two consider the EDD rule and the last two the MST. In addition, H6 and H8 are improved versions of H5 and H7, respectively, which use neighborhood search methods at the end of execution.

*Heuristic H5*

Step 1. Order jobs by the EDD rule (in the case of a tie, use the lower $\sum p_{jk}$).
Step 2. Apply the timing adjustment procedure.
Step 3. Identify the first tardy job in the sequence ($J_T$). If there are no tardy jobs, STOP.
Step 4. Replace/Place job $J_T$ as the final in the sequence and go to Step 2.

**Algorithm 6.** Pseudo-code of heuristic H5.

*Heuristic H6*

Step 1. Order jobs by the EDD rule (in the case of a tie, use the lower $\sum p_{jk}$).
Step 2. Apply the timing adjustment procedure.
Step 3. Identify the first tardy job in the sequence ($J_T$). If there are no tardy jobs, STOP.
Step 4. Replace job $J_T$ as the final in the sequence.
Step 5. Considering the insertion neighborhood with $(h − 1)2$ partial sequences and applying the timing adjustment procedure, determine that with the best $n_{JIT}$; and

From the best solution obtained so far, considering the permutation neighborhood with $h(h − 1)/2$ partial sequences and applying the timing adjustment procedure, determine that with the best $n_{JIT}$.

**Algorithm 7.** Pseudo-code of heuristic H6.

*Heuristic H7*

Step 1. Order jobs by the MST rule (in the case of a tie, use the lower $\sum p_{jk}$).

Steps 2–4. The same as in heuristic H5.

> **Algorithm 8.** Pseudo-code of heuristic H7.

*Heuristic H8*

Step 1. Order jobs by the MST rule (tie-break by the lower $\sum p_{jk}$).

Steps 2–5. The same as in heuristic H6.

> **Algorithm 9.** Pseudo-code of heuristic H8.

For the last two heuristics, H9 (Algorithm 10) uses the EDD rule as the initial order and H10 (Algorithm 11) the MST rule, both adopting a different form of neighborhood search, the forward/backward procedure. The new procedure re-insert one job at a time in the last position in a sequence by considering the best solution found so far, and then iteratively re-insert the last job in all other positions retaining the one with the best $n_{JIT}$. This corresponds to a diverse way to test neighbor solutions (see Example 2).

**Example 2.** Consider again the sequence: $\{J_3, J_2, J_1, J_4\}$. For position $h = 1$, the first job $J_3$ is replaced in the last position; if the new solution is improved, the new sequence is kept (and also the value of $h = 1$), otherwise, the previous one is recovered (and the h is incremented). Thus, if the current sequence is $\{J_2, J_1, J_4, J_3\}$, then the new first job $J_2$ is replaced in the last position and the test is repeated for $\{J_1, J_4, J_3, J_2\}$. Considering now that the new solution is not better than the previous one, the sequence is kept $\{J_2, J_1, J_4, J_3\}$ and $h = h + 1 = 2$. In the next step, the second job (i.e., the one in the $h$th position) is replaced in the last position and the test is redone. This forward procedure is repeated until the $(n - 1)$th position. Then, a backward procedure is applied, reinserting the last job in all previous positions, and so on.

*Heuristic H9*

Step 1. Order jobs by the EDD rule (in the case of a tie, use the lower $\sum p_{jk}$).
Step 2. Apply the timing adjustment procedure.
Step 3. Set $h = 1$. While $h < n$, do (from position 1 to $n - 1$):

> Forward procedure: replace the $h$th job defined in Step 1 in the last position of the partial sequence and apply the timing adjustment procedure. If the new $n_{JIT}$ is better than the previous one, keep the new schedule (and the $h$ value); otherwise, go back to the previous one and set $h = h + 1$.

Step 4. Set $h = n$. While $h > 1$, do (from the last position to the second):

> Backward procedure: Replace the $h$th job in all the previous positions and apply the timing adjustment procedure considering the best solution found. If a new best solution is found, keep the new schedule (and the $h$ value); otherwise, go back to the previous solution and set $h = h - 1$.

Step 5. STOP.

> **Algorithm 10.** Pseudo-code of heuristic H9.

*Heuristic H10*

Step 1. Order jobs by the MST rule (in the case of a tie, use the lower $\sum p_{jk}$).

Steps 2–5. The same as in heuristic H9.

> **Algorithm 11.** Pseudo-code of heuristic H10.

**Table 2.** Composition of heuristic's structure.

| Heuristic | | Initial Ordering | | Neighborhood Search | | |
|---|---|---|---|---|---|---|
| | | EDD | MST | Insertion | Permutation | Fwd/Bwd |
| NEH-based | H1 | x | | | | |
| | H2 | x | | x | x | |
| | H3 | | x | | | |
| | H4 | | x | x | x | |
| Hodgson-based | H5 | x | | | | |
| | H6 | x | | x | x | |
| | H7 | | x | | | |
| | H8 | | x | x | x | |
| Other | H9 | x | | | | x |
| | H10 | | x | | | x |

## 5. Computational Experiments and Results

In this section, we describe the computational experiments conducted to study the behavior of the mathematical model presented in Section 3 and the heuristics described in Section 4. The codes for the heuristics were implemented in the Delphi programming environment, the mathematical model was written in the syntax of the AMPL modeling language and the instances solved with the branch-and-cut algorithm included in the IBM-CPLEX 12.6.1.0. All tests were conducted on a Pentium Dual-Core with a 2.0 GHz processor, 3.0 GB RAM and Windows Operating System.

A total of 15,600 instances were generated. They are separated into Group 1 of small instances and Group 2 of medium and large ones as described in Section 5.1. The computational study was divided in two parts: Experiment 1 and Experiment 2. Section 5.2 presents the comparative results of the procedures described in Section 4 (Experiment 1). The quality of the heuristics solutions is reinforced when they are compared to the best solution obtained by solving the instances of the mathematical model with CPLEX, as described in Section 5.3 (Experiment 2). The computational efficiency of the solution's strategies, computed in terms of CPU time, is discussed in Section 5.4.

### 5.1. Problem Instances

The instances were separated into two groups, with Group 1 consisting of small instances and Group 2 of medium and large ones. In each group, the instances are divided into classes defined by the number of jobs ($n$), number of machines ($m$) and scenarios of due dates, with 100 instances randomly generated for each class to reduce the sampling error.

The processing times were generated in the interval $U[1,99]$, as in the most production scheduling scenarios found in the literature (e g , [21,22]). In Group 1, the parameters were $n \in \{5, 6, 7, 8, 10\}$ and $m \in \{2, 3, 5\}$ and, in Group 2, $n \in \{15, 20, 30, 50, 80, 100\}$ and $m \in \{5, 10, 15, 20\}$. These values were chosen to cover a significant range of instances of various sizes.

The generation of due dates followed the method used by [23], with a uniform distribution in the interval $[P(1 - T - R/2), P(1 - T + R/2)]$, where $T$ and $R$ are the tardiness factor of jobs and dispersion range of due dates, respectively, and $P$ the lower bound for the makespan which is defined as in Taillard [24]:

$$P = \max\left\{ \max_{1 \leq k \leq m} \left\{ \sum_{j=1}^{n} p_{jk} + \min_{j} \sum_{q=1}^{k-1} p_{jq} + \min_{j} \sum_{q=k+1}^{m} p_{jq} \right\}, \max_{j} \sum_{k=1}^{m} p_{jk} \right\} \qquad (13)$$

The following scenarios represent the configurations obtained by varying the values of $T$ and $R$:

- Scenario 1: low tardiness factor ($T = 0.2$) and small due date range ($R = 0.6$);
- Scenario 2: low tardiness factor ($T = 0.2$) and large due date range ($R = 1.2$);
- Scenario 3: high tardiness factor ($T = 0.4$) and small due date range ($R = 0.6$); and

- Scenario 4: high tardiness factor ($T = 0.4$) and large due date range ($R = 1.2$).

Using these parameters, 6000 instances were generated in Group 1, divided in 60 classes. That is, five levels of number of jobs, three levels of number of machines, four scenarios and 100 instances per class ($5 \times 3 \times 4 \times 100 = 6000$). For Group 2, 9600 instances were generated, divided in 96 classes, six levels of number of jobs, four levels of number of machines, four scenarios and 100 instances per class ($6 \times 4 \times 4 \times 100 = 9600$). A total of 15,600 instances were generated and their parameters are summarized in Table 3.

**Table 3.** Parameters for generation of the instances.

|  | Group 1 | Group 2 |
|---|---|---|
| Number of jobs | 5, 6, 7, 8, 10 | 15, 20, 30, 50, 80, 100 |
| Number of machines | 2, 3, 5 | 5, 10, 15, 20 |
| Scenario configurations |  | Scenario 1: $T = 0.2$ and $R = 0.6$; Scenario 2: $T = 0.2$ and $R = 1.2$; Scenario 3: $T = 0.4$ and $R = 0.6$; Scenario 4: $T = 0.4$ and $R = 1.2$ |
| Number of instances per class | 100 | 100 |
| Number of instances per group | 6000 | 9600 |
| Total number of instances solved | 15,600 | |

### 5.2. Experiment 1: Relative Comparison of the Proposed Heuristics

In this first part of the results evaluation, a relative comparison of the proposed heuristics was conducted. The most common measure used in the literature to compare the performances of solution methods is the relative percentage deviation (RPD) (e.g., [22,25]), which, adapted to the maximization problem addressed, is calculated as:

$$RPD = (\frac{n_{JIT}^{best} - n_{JIT}^{h}}{n_{JIT}^{best}})100, \tag{14}$$

where $n_{JIT}^{best}$ is the best solution found and $n_{JIT}^{h}$ the heuristic solution evaluated. The lower is a method's RPD, the better is its performance, with a RPD of zero indicating that the method provided the best solution found (or achieved a tie).

To compare the heuristics performances for Group 1's instances, the solution of an enumeration method (EM), based on [25], was used as a reference (see the pseudo-code in Algorithm 12). It is important to note that the EM provides a reference solution to the problem and, although it listed $n!$ possible permutations (sequences) of jobs, it does not warrant an optimal solution since the schedule is composed by sequencing and timing. That is, for the same sequence, it is possible to find several schedules with different starting times for each operation, thereby resulting in many possibilities for the same number of just-in-time jobs, as can be seen in Figure 1.

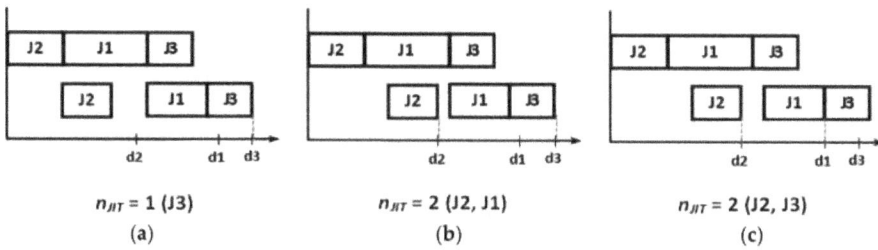

**Figure 1.** Example of different schedules with the same sequence of jobs: (**a**) only $J_3$ is just-in-time; (**b**) $J_2$ and $J_1$ are just-in-time; and (**c**) $J_2$ and $J_3$ are just-in-time.

Therefore, the EM enumerates all possible sequences, but it is impractical to probe all possible schedules, especially for medium and large scale instances. Given that, the EM was not applied for Group 2's instances, and the best solution found by the heuristics H1–H10 was considered as reference for the instances in this group.

*Enumeration Method*

Step 1Enumerate the $n!$ possible sequences and consider the one with the best $n_{JIT}$.
Step 2Apply the timing adjustment procedure.

    **Algorithm 12.** Pseudo-code of the enumeration method.

Figure 2 graphically presents the results, with 95% confidence intervals of the average RPD obtained by each heuristic method for both Groups 1 and 2. The 95% confidence interval means that the results (RPD) of 95 cases among 100 are within the displayed range.

The global analysis of results revealed that H6 clearly outperformed the other heuristics, with a RPD of 0.2% for both Groups 1 and 2 (as can be seen in Table 4) which indicates that it provided the best or very close to the best solutions for many cases.

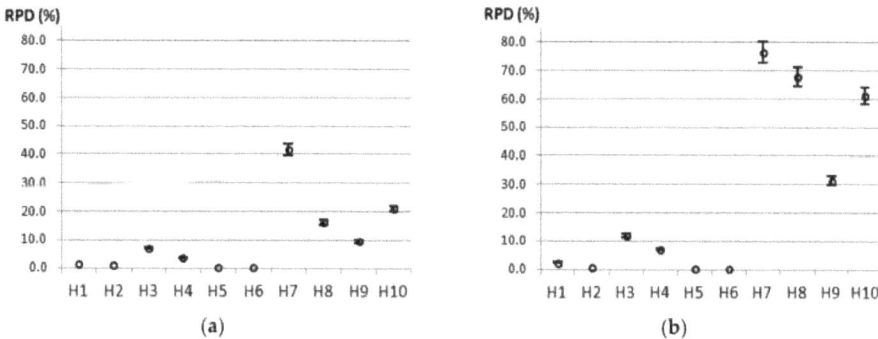

**Figure 2.** Comparison of performances of heuristics by group with 95% confidence interval of average RPD: (**a**) related to EM for Group 1; and (**b**) related to the best H1–H10 solution for Group 2.

**Table 4.** Overall performance rankings (RPD) of proposed heuristics in relation to EM for Group 1 and to the best found solution for Group 2.

| Group 1 | H6 | H5 | H2 | H1 | H4 | H3 | H9 | H8 | H10 | H7 |
|---|---|---|---|---|---|---|---|---|---|---|
| | 0.2 | 0.4 | 1.1 | 1.2 | 3.7 | 7.0 | 9.6 | 16.2 | 21.1 | 41.7 |
| Group 2 | H6 | H5 | H2 | H1 | H4 | H3 | H9 | H10 | H8 | H7 |
| | 0.2 | 0.3 | 0.6 | 2.4 | 7.0 | 11.9 | 31.4 | 61.4 | 68.1 | 76.6 |

The results from H6 were similar to those from H5 which ranked second for overall performance with RPDs of 0.4% in Group 1 and 0.3% in Group 2, as shown in Table 4. Both these heuristics considered the EDD rule as the initial order and iteratively placed the first tardy job at the end of the sequence. However, H6 also employed insertion and permutation neighborhood searches. Therefore, it can be seen that, although this improvement phase provided better performance, H6 had already produced very good results before the neighborhood searches (explained by the behavior of H5).

Following the rankings, H2 and H1 were third and fourth, respectively, and both exhibited similar behavior by considering the EDD rule as the initial order and then employing the insertion method to construct the sequence. The difference between them is that H2 employs neighborhood searches before attempting to insert the new job in the partial sequence.

In addition, it may be noted in Table 4 that the rankings for Groups 1 and 2 were almost the same, with the sole exception of the inversion between heuristics H8 and H10 in eighth and ninth places. Another observation is that, in the five worst methods, the deviations in Group 1 were much lower than those in Group 2 which indicated that there is a greater dispersion of solution quality for medium and large instances, as can be seen in Figure 2.

The worst method was H7 which considered the MST rule as the initial order and iteratively placed the first tardy job at the end of the sequence. Although it is very similar to H5, as the only difference between them was in their initial orders (the EDD rule was used in H5), the discrepancies in their results were quite significant. This demonstrated that, of the two options used for the initial rule, the EDD was always more advantageous than the MST, as was obvious for every pair of heuristics which were only differentiated by these rules, that is, H1 and H3, H2 and H4, H5 and H7, H6 and H8, and H9 and H10.

Tables 5–8 provide detailed average results for the instances according to the group and size for each heuristic considering different numbers of jobs and machines.

As the number of jobs grows, the four best methods, H6, H5, H2 and H1, had relatively stable performances for Groups 1 (Table 5) and Group 2 (Table 6), with very low RPD values (below 1.5). The exception was H1 for Group 2 (Table 6), which had decreasing relative deviations ranging from 3.9% with 15 jobs to 0.6% with 100. The other heuristics had increasing RPD values with increasing numbers of jobs, except H7 and H8 for Group 2 which had decreasing values for instances with up to 100 jobs. Furthermore, H3 showed a slight decrease in instances with 15 to 50 jobs, and an increase in those with 80 and 100.

**Table 5.** Comparison of performances (RPD) of heuristics by number of jobs for Group 1.

| $n$ | H1 | H2 | H3 | H4 | H5 | H6 | H7 | H8 | H9 | H10 |
|---|---|---|---|---|---|---|---|---|---|---|
| 5 | 1.0 | 1.1 | 5.4 | 1.8 | 0.4 | 0.2 | 29.1 | 5.5 | 5.7 | 12.7 |
| 6 | 1.2 | 0.9 | 7.4 | 3.4 | 0.4 | 0.1 | 35.8 | 9.4 | 7.3 | 17.7 |
| 7 | 1.2 | 1.2 | 6.4 | 3.4 | 0.4 | 0.2 | 43.9 | 16.2 | 8.9 | 20.0 |
| 8 | 1.5 | 1.0 | 7.8 | 4.8 | 0.3 | 0.3 | 49.3 | 21.8 | 11.6 | 25.3 |
| 10 | 1.2 | 1.1 | 8.3 | 4.9 | 0.4 | 0.3 | 50.6 | 28.3 | 14.5 | 29.8 |

**Table 6.** Comparison of performances (RPD) of heuristics by number of jobs for Group 2.

| $n$ | H1 | H2 | H3 | H4 | H5 | H6 | H7 | H8 | H9 | H10 |
|-----|-----|-----|------|-----|-----|-----|------|------|------|------|
| 15 | 3.9 | 1.5 | 10.9 | 6.4 | 0.7 | 0.4 | 59.6 | 41.3 | 26.2 | 40.1 |
| 20 | 3.6 | 1.1 | 9.4 | 6.2 | 0.5 | 0.4 | 69.4 | 55.7 | 28.3 | 47.5 |
| 30 | 3.3 | 0.7 | 8.9 | 6.7 | 0.3 | 0.2 | 77.7 | 69.0 | 30.5 | 57.5 |
| 50 | 2.1 | 0.3 | 8.5 | 6.7 | 0.3 | 0.1 | 84.6 | 79.6 | 33.4 | 69.3 |
| 80 | 1.0 | 0.1 | 8.8 | 7.3 | 0.1 | 0.1 | 89.4 | 86.3 | 34.8 | 76.3 |
| 100 | 0.6 | 0.0 | 25.1 | 8.7 | 0.2 | 0.1 | 78.9 | 76.6 | 35.1 | 77.6 |

In terms of the numbers of machines, as can be seen in Tables 7 and 8, the results indicate, in each group, high stability with relatively small variations in their RPD amplitudes. This may suggest that the number of machines was not a relevant factor in performances for the problem addressed.

**Table 7.** Comparison of performances (RPD) of heuristics by number of machines for Group 1.

| $m$ | H1 | H2 | H3 | H4 | H5 | H6 | H7 | H8 | H9 | H10 |
|-----|-----|-----|-----|-----|-----|-----|------|------|------|------|
| 2 | 0.3 | 0.5 | 7.1 | 3.7 | 0.2 | 0.1 | 42.2 | 16.5 | 7.5 | 20.8 |
| 3 | 0.9 | 0.8 | 7.4 | 3.7 | 0.3 | 0.1 | 42.3 | 16.7 | 9.6 | 21.3 |
| 5 | 2.4 | 1.9 | 6.7 | 3.5 | 0.7 | 0.5 | 40.7 | 15.5 | 11.8 | 21.1 |

**Table 8.** Comparison of performances (RPD) of heuristics by number of machines for Group 2.

| $m$ | H1 | H2 | H3 | H4 | H5 | H6 | H7 | H8 | H9 | H10 |
|-----|-----|-----|------|-----|-----|-----|------|------|------|------|
| 5 | 0.5 | 0.3 | 12.2 | 7.2 | 0.2 | 0.1 | 78.0 | 69.8 | 28.6 | 61.8 |
| 10 | 2.1 | 0.7 | 13.0 | 7.1 | 0.4 | 0.2 | 75.9 | 68.0 | 30.7 | 61.3 |
| 15 | 3.1 | 0.6 | 11.0 | 7.1 | 0.3 | 0.2 | 76.9 | 68.2 | 32.5 | 61.3 |
| 20 | 3.9 | 0.8 | 11.6 | 6.6 | 0.5 | 0.3 | 75.5 | 66.4 | 33.7 | 61.1 |

Table 9 presents each group's deviations for the four defined scenarios, varying the values of $T$ and $R$ due date factors. These results were consistent with those obtained from the previous analyses of ranking methods, with H6 presenting the best result followed by H5. The results of different scenarios suggest that variations in the tardiness factor and due date range did not exert any relevant influence. It is interesting to note the two top heuristics, H6 and H5, obtained identical results for both groups in Scenarios 2 and 4 characterized by their wide due date ranges. That is, when the interval between due dates was large, neighborhood searches did not provide improvements.

**Table 9.** Performances (RPD) of heuristics by group and scenario in relation to EM for Group 1 and to the best found solution for Group 2.

| | Scenario | H1 | H2 | H3 | H4 | H5 | H6 | H7 | H8 | H9 | H10 |
|----------|-----|-----|-----|------|-----|-----|-----|------|------|------|------|
| | 1 | 0.9 | 0.8 | 8.7 | 4.6 | 0.4 | 0.3 | 42.2 | 16.5 | 10.0 | 20.8 |
| Group 1 | 2 | 1.0 | 0.7 | 4.7 | 2.2 | 0.2 | 0.2 | 45.7 | 19.2 | 4.3 | 20.0 |
| | 3 | 1.9 | 1.7 | 9.3 | 5.1 | 0.8 | 0.5 | 39.3 | 14.7 | 15.0 | 22.1 |
| | 4 | 1.0 | 1.0 | 5.5 | 2.7 | 0.1 | 0.1 | 39.8 | 14.5 | 9.2 | 21.5 |
| | 1 | 1.7 | 0.7 | 15.0 | 8.9 | 0.4 | 0.2 | 76.3 | 68.1 | 39.4 | 61.6 |
| Group 2 | 2 | 1.6 | 0.5 | 7.8 | 4.5 | 0.1 | 0.1 | 78.6 | 71.3 | 13.8 | 62.0 |
| | 3 | 3.7 | 1.0 | 15.2 | 9.4 | 0.7 | 0.4 | 75.1 | 65.4 | 43.0 | 60.9 |
| | 4 | 2.5 | 0.3 | 9.7 | 5.3 | 0.1 | 0.1 | 76.3 | 67.6 | 29.3 | 61.0 |

*5.3. Experiment 2: Quality of the Heuristic Solutions in Relation to the Optimal Solution*

In the second part of the computational experimentation, the quality of each heuristic solution was measured by the RPD in relation to the optimal solution provided by solving the instances of mathematical model by CPLEX. The RPD is calculated by Expression (14), where $n_{JIT}^{best}$ is the optimal solution given by CPLEX.

The optimality of the CPLEX solution was proven for the 6000 instances of Group 1. The analysis of quality the heuristics' results in relation to those optimal solutions, using 95% confidence intervals of the average RPDs, is depicted in Figure 3.

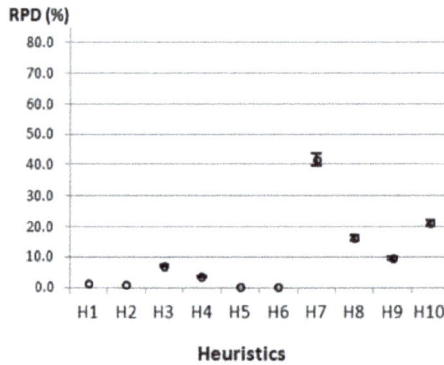

**Figure 3.** Comparisons of performances of heuristics in relation to those of model (95% confidence intervals of average RPD).

It is remarkable that the values of graphs in Figure 2a (Group 1) and Figure 3 are very similar. Table 10 presents rankings of the solution quality of the heuristics, i.e., the values of the RPD of the optimal solutions.

**Table 10.** Overall performance rankings (average RPD) of heuristics in relation to optimal solution for instances in Group 1.

|     | H6 | H5 | H2 | H1 | H4 | H3 | H9 | H8 | H10 | H7 |
| --- | --- | --- | --- | --- | --- | --- | --- | --- | --- | --- |
| RPD | 0.6 | 0.8 | 1.4 | 1.6 | 4.0 | 7.4 | 10.0 | 16.6 | 21.4 | 41.9 |

Table 10 shows the same ranking and values very close to the ones presented in Table 4 for relative comparisons of Group 1, which validated and reinforced the results from the previous analyses. It is important to highlight the excellent performance of the best heuristic (H6) that had a deviation from the optimal solution of just 0.6% which meant that it provided a near-optimal solution.

Of the 6000 instances optimally solved, H6 reached the optimal solution in 5830 cases (97.2% of instances) and, in the other 170 instances, the difference between its result and the optimal solution was only one just-in-time job in 168 cases and two just-in-time jobs in two other cases. It is also relevant to emphasize that the average running times of H6 were 0.1 ms and 0.39 s for the instances in Groups 1 and 2, respectively.

According to the previous comparative analysis, the results for H5 and H6 were very close and H2 and H1 also achieved significant performances. This confirmed that the EDD was better than the MST rule for the instances considered and the procedure based on Hodgson's approach (used in H6 and H5) outperformed the insertion method of NEH (applied in H2 and H1). Similarly, in all cases of each pair of heuristics, those which applied a neighborhood search produced improved results.

Another interesting observation is that the enumeration method, as explained in Section 5.3, did not guarantee the optimal solution in the case of a flow shop with just-in-time jobs. Of the instances optimally solved, the deviations from the enumeration method were on average 0.4%, as expected, with an average running time of 1.48 s. The average running time of CPLEX was 199.88 s (3.33 min).

Given the difficulties to prove optimality, for Group 2, only one instance per class was solved by CPLEX with the CPU time was limited to 3600 s. Thus, 96 medium and large instances were executed.

For 22 instances, no integer solution was found or the model provided a solution with zero value. In most cases, the heuristic solutions were much better than the lower bound given by CPLEX, leading to very negative RPD, as can be observed in Table 11.

**Table 11.** Overall performance (average RPD) of heuristics in relation to lower bound given by CPLEX by number of jobs for instances of Group 2.

| n | H1 | H2 | H3 | H4 | H5 | H6 | H7 | H8 | H9 | H10 |
|---|---|---|---|---|---|---|---|---|---|---|
| 15 | 6.5 | 1.0 | 10.3 | 6.9 | 1.0 | 1.0 | 56.7 | 41.1 | 28.2 | 36.8 |
| 20 | −5.0 | −6.5 | 4.0 | −2.8 | −8.8 | −8.8 | 66.5 | 49.6 | 23.1 | 41.1 |
| 30 | −49.0 | −52.5 | −38.0 | −40.1 | −52.0 | −52.0 | 68.4 | 57.2 | −7.4 | 38.5 |
| 80 | −1080.2 | −1082.0 | −1001.7 | −1032.6 | −1082.0 | −1082.0 | −13.5 | −93.2 | −764.4 | −251.9 |
| 100 | −1276.7 | −1276.7 | −990.8 | −1159.3 | −1276.7 | −1276.7 | −460.6 | −518.8 | −865.0 | −276.1 |

The more negative is the RPD of a heuristic, the better is its result in relation to the lower bound. It could be noted that even the worst heuristic (H7) provided results that were better than the lower bound given by CPLEX. These results confirm all previous inferences. The coincidence of the results for H5 and H6 is remarkable, suggesting that neighborhood searches do not improve the solution of medium and large instances.

### 5.4. Computational Efficiency

The comparison of computational efficiency, i.e., the average consumption of CPU time measured in milliseconds (ms), of each heuristic for Groups 1 and 2, the enumeration method and CPLEX for Group 1 are shown in Table 12.

**Table 12.** Computational efficiency (average CPU times in milliseconds).

| Solution Method | Group 1 | Group 2 |
|---|---|---|
| H1 | 0.08 | 67.99 |
| H2 | 0.33 | 7895.49 |
| H3 | 0.03 | 72.55 |
| H4 | 0.32 | 7820.17 |
| H5 | 0.01 | 1.21 |
| H6 | 0.13 | 391.51 |
| H7 | 0.01 | 1.28 |
| H8 | 0.12 | 190.49 |
| H9 | 0.06 | 79.02 |
| H10 | 0.05 | 56.29 |
| EM | 1483.65 | – |
| Model | 199,882.99 | – |

As expected, the EM and CPLEX consumed much more CPU time than the heuristics. For the small instances (Group 1), the computational times of all the heuristics were almost zero and, for the medium and large ones (Group 2), H2 and H4 took the longest times, nearly 8 s on average, which were relatively high compared with those of other methods but does not preclude their use. All other heuristics required far less time than one second which demonstrated the viability of using them in practice. It is worth noting that H5, which ranked second with a solution quality very close to that of H6, consumed less than a half second on average for large instances, thereby indicating its high computational efficiency.

Finally, in an overall analysis of the results and the solution approaches proposed in this paper (exact and heuristic), the applicability of the developed heuristics were justified and demonstrated in terms of both solution quality (an average of 0.6% deviation from the optimum with the best heuristic H6) and computational efficiency (an average of 0.4 s for large instances also with H6). The mathematical model and the enumeration method were useful as quality certificate. Moreover, the mathematical model can be useful if other constraints or requirements are added to the problem.

## 6. Final Remarks

This research achieves its proposed goals of developing and implementing effective and efficient methods for solving a flow shop scheduling problem by maximizing the number of just-in-time jobs, as demonstrated in the computational experiments. A MIP model is proposed to represent the problem and, together with an enumeration algorithm, is useful as quality certificate for the solution values given by the constructive heuristics proposed.

The CPLEX system solves instances of the MIP model with up to 10 jobs and five machines in at most 47 min. It provides a lower bound for the optimal solution for instances with up to 100 jobs and 20 machines. The enumeration method does not guarantee optimality because the solution is formed by job sequencing and timing (the starting times of jobs). However, it shows relative applicability and considerable quality (0.2% deviations in small instances) with an average running time of 1.5 s.

The practicability and applicability of all proposed heuristic methods are demonstrated, in particular for large-scale instances, with very good quality results and non-prohibitive runtimes. The best heuristic, H6, demonstrates a near-optimal solution, with just a 0.5% average relative deviation from the exact solution and optimal solutions for more than 98% of instances, while the performance of the second best, H5, is very close. In total, 15,600 instances are solved, with the average relative deviation of H6 only 0.2% and that of H5 approximately 0.3%. The H6 and H5 heuristics consider the EDD rule as the initial solution and then iteratively place the first tardy job at the end of the sequence. Although their results are very close, H6 improves on H5 by using neighborhood searches.

In this study, the focus is on solving a flow shop scheduling problem by reducing the interval between the completion time of the last operation of a job and its due date. This enables an adjustment in the timing of jobs which results in the possibility of inserting idle time between operations. Therefore, there is no concern about the first operations of each job, i.e., their executions could be approximated. Rescheduling these operations could reduce the idle time between them and possibly also minimize the total time required to complete the schedule (makespan). Therefore, it is suggested that future work consider multiple-criteria functions, including flow measures (as a makespan and/or flow time), in scenarios with earliness and tardiness.

**Acknowledgments:** This work was supported by CNPq (502547/2014-6, 443464/2014-6, and 233654/2014-3), CAPES (BEX 2791/15-3), FAPESP (2013/07375-0 and 2016/01860-1) and FAPEG (201510267000983).

**Author Contributions:** Helio conceived, designed and performed the experiments; Helio, Ruhul and Socorro analyzed the data and wrote the paper.

**Conflicts of Interest:** The authors declare no conflict of interest.

## References

1.  Pinedo, M.L. *Scheduling: Theory, Algorithms and Systems*, 5th ed.; Prentice-Hall: Upper Saddle River, NJ, USA, 2016; ISBN 978-3319265780.
2.  Shabtay, D.; Bensoussan, Y.; Kaspi, M. A bicriteria approach to maximize the weighted number of just-in-time jobs and to minimize the total resource consumption cost in a two-machine flow-shop scheduling system. *Int. J. Prod. Econ.* **2012**, *136*, 67–74. [CrossRef]
3.  Lann, A.; Mosheiov, G. Single machine scheduling to minimize the number of early and tardy jobs. *Comput. Oper. Res.* **1996**, *23*, 769–781. [CrossRef]
4.  Kanet, J.J.; Sridharan, V. Scheduling with inserted idle time: Problem taxonomy and literature review. *Oper. Res.* **2000**, *48*, 99–110. [CrossRef]
5.  Baker, K.R.; Scudder, G.D. Sequencing with earliness and tardiness penalties: A review. *Oper. Res.* **1990**, *38*, 22–36. [CrossRef]
6.  Józefowska, J. *Just-in-Time Scheduling: Models and Algorithms for Computer and Manufacturing Systems*; Springer Science: New York, NY, USA, 2007; ISBN 978-387-71717-3.

7. Ríos-Solís, Y.A.; Ríos-Mercado, R.Z. *Just-In-Time Systems*; Springer Sciences: New York, NY, USA, 2012; ISBN 978-1-4614-1122-2.
8. Shabtay, D.; Steiner, G. Scheduling to maximize the number of just-in-time jobs: A survey. In *Just-in-Time Systems*; Ríos-Solís, Y.A., Ríos-Mercado, R.Z., Eds.; Springer Sciences: New York, NY, USA, 2012; ISBN 978-1-4614-1122-2.
9. Choi, B.-C.; Yoon, S.-H. Maximizing the weighted number of just-in-time jobs in flow shop scheduling. *J. Sched.* **2007**, *10*, 237–243. [CrossRef]
10. Shabtay, D.; Bensoussan, Y. Maximizing the weighted number of just-in-time jobs in several two-machine scheduling systems. *J. Sched.* **2012**, *15*, 39–47. [CrossRef]
11. Shabtay, D. The just-in-time scheduling problem in a flow-shop scheduling system. *Eur. J. Oper. Res.* **2012**, *216*, 521–532. [CrossRef]
12. Gerstl, E.; Mor, B.; Mosheiov, G. A note: Maximizing the weighted number of just-in-time jobs on a proportionate flowshop. *Inf. Process. Lett.* **2015**, *115*, 159–162. [CrossRef]
13. Yin, Y.; Cheng, T.C.E.; Wang, D.-J.; Wu, C.-C. Two-agent flowshop scheduling to maximize the weighted number of just-in-time jobs. *J. Sched.* **2017**, *20*, 313–335. [CrossRef]
14. Fuchigami, H.Y.; Rangel, S. Métodos heurísticos para maximização do número de tarefas just-in-time em flow shop permutacional. In Proceedings of the Simpósio Brasileiro de Pesquisa Operacional, Porto de Galinhas, Brazil, 25–28 August 2015.
15. Fuchigami, H.Y.; Rangel, S. Um estudo computacional de um modelo matemático para *flow shop* permutacional com tarefas *just-in-time*. In Proceedings of the Simpósio Brasileiro de Pesquisa Operacional, Vitória, Brazil, 27–30 September 2016.
16. Dhouib, E.; Teghem, J.; Loukil, T. Minimizing the number of tardy jobs in a permutation flowshop scheduling problem with setup times and time lags constraints. *J. Math. Model. Algorithm* **2013**, *12*, 85–99. [CrossRef]
17. Nawaz, M.; Enscore, E.E., Jr.; Ham, I. A heuristic algorithm for the *m*-machine *n*-job flow-shop sequencing problem. *OMEGA–Int. J. Manag. Sci.* **1983**, *11*, 91–95. [CrossRef]
18. Hodgson, T.J. A note on single machine sequencing with random processing times. *Manag. Sci.* **1977**, *23*, 1144–1146. [CrossRef]
19. Baker, K.R.; Trietsch, D. *Principles of Sequencing and Scheduling*; John Wiley & Sons: New York, NY, USA, 2009; ISBN 978-0-470-39165-5.
20. Nagano, M.S.; Branco, F.J.C.B.; Moccellin, J.V. Soluções de alto desempenho para programação da produção flow shop. *GEPROS* **2009**, *4*, 11–23.
21. Li, X.; Chen, L.; Xu, H.; Gupta, J.N.D. Trajectory scheduling methods for minimizing total tardiness in a flowshop. *Oper. Res. Perspect.* **2015**, *2*, 13–23. [CrossRef]
22. Vallada, E.; Ruiz, R.; Minella, G. Minimizing total tardiness in the *m*-machine flowshop problem: A review and evaluation of heuristics and metaheuristics. *Comput. Oper. Res.* **2008**, *35*, 1350–1373. [CrossRef]
23. Ronconi, D.P.; Birgin, E.G. Mixed-integer programming models for flow shop scheduling problems minimizing the total earliness and tardiness. In *Just-in-Time Systems*; Ríos-Solís, Y.A., Ríos-Mercado, R.Z., Eds.; Springer Sciences: New York, NY, USA, 2012; ISBN 978-1-4614-1122-2.
24. Taillard, E. Benchmarks for basic scheduling problems. *Eur. J. Oper. Res.* **1993**, *64*, 278–285. [CrossRef]
25. Laha, D.; Sarin, S.C. A heuristic to minimize total flow time in permutation flow shop. *OMEGA–Int. J. Manag. Sci.* **2009**, *37*, 734–739. [CrossRef]

*algorithms*

MDPI

Article

# Hybrid Flow Shop with Unrelated Machines, Setup Time, and Work in Progress Buffers for Bi-Objective Optimization of Tortilla Manufacturing

Victor Hugo Yaurima-Basaldua [1], Andrei Tchernykh [2,3,*], Francisco Villalobos-Rodríguez [1] and Ricardo Salomon-Torres [1]

[1]   Software Engineering, Sonora State University, San Luis Rio Colorado, Sonora 83455, Mexico;
      victor.yaurima@ues.mx (V.H.Y.-B.); fco.vr1@gmail.com (F.V.-R.); ricardo.salomon@uabc.edu.mx (R.S.-T.)
[2]   Computer Science Department, CICESE Research Center, Ensenada 22860, Mexico
[3]   School of Electrical Engineering and Computer Science, South Ural State University, Chelyabinsk 454080,
      Russia
*    Correspondence: chernykh@cicese.mx; Tel.: +521-646-178-6994

Received: 27 February 2018; Accepted: 1 May 2018; Published: 9 May 2018

**Abstract:** We address a scheduling problem in an actual environment of the tortilla industry. Since the problem is NP hard, we focus on suboptimal scheduling solutions. We concentrate on a complex multistage, multiproduct, multimachine, and batch production environment considering completion time and energy consumption optimization criteria. The production of wheat-based and corn-based tortillas of different styles is considered. The proposed bi-objective algorithm is based on the known Nondominated Sorting Genetic Algorithm II (NSGA-II). To tune it up, we apply statistical analysis of multifactorial variance. A branch and bound algorithm is used to assert obtained performance. We show that the proposed algorithms can be efficiently used in a real production environment. The mono-objective and bi-objective analyses provide a good compromise between saving energy and efficiency. To demonstrate the practical relevance of the results, we examine our solution on real data. We find that it can save 48% of production time and 47% of electricity consumption over the actual production.

**Keywords:** multiobjective genetic algorithm; hybrid flow shop; setup time; energy optimization; production environment

---

## 1. Introduction

Tortillas are a very popular food as a favorite snack and meal option in various cultures and countries. There are two types of tortillas: wheat-based and corn-based. Their overall consumption is growing all over the world. Originally, tortillas were made by hand: grinding corn into flour, mixing the dough, and pressing to flatten it. In fact, many small firms today still use processes that are more labor-intensive, requiring people to perform a majority of the tasks such as packaging, loading, baking, and distributing. This trend is expected to change over the next years. Due to its considerable practical significance, optimization of tortilla production is important. To improve the production timing parameters (setup, changeover, waiting), operational cost (energy consumption, repairing, service), throughput, etc., careful analysis of the process and of advance scheduling approaches is needed.

This industry is a typical case of a hybrid flow shop, which is a complex combinatorial optimization problem that arises in many manufacturing systems. In the classical flow shop, a set of jobs has to pass through various stages of production. Each stage can have several machines. There is also the flexibility of incorporating different-capacity machines, turning this design into hybrid flow shop scheduling (HFS).

The production decisions schedule work to machines in each stage to determine the processing order according to one or more criteria. Hybrid flow shop scheduling problems are commonly encountered in manufacturing environments. A large number of heuristics for different HFS configurations considering realistic problems have been proposed [1–6].

In spite of significant research efforts, some aspects of the problem have not yet been sufficiently explored. For example, previous works deal mostly with improving the efficiency of production, where objective functions are usually used to minimize the total production time (makespan), the total tardiness, etc. [7]. Although the production efficiency is essential, it is not the only factor to consider in manufacturing operations. One of the main motivations of this work is to reduce energy consumption. In recent years, it has been recognized that high energy consumption can increase operational cost and cause negative environmental impacts [8].

In this paper, we propose a genetic algorithm that contributes to the solution of a bi-objective HFS problem of a real tortilla production environment, with unrelated machines, setup time, and work in progress buffers taking into account the total completion time and energy consumption of the machines.

The rest of the paper is organized as follows. After presenting the description and characteristics of the production line and jobs in Section 2, we discuss related work in Section 3. We describe the model of the problem in Section 4. We present the details of the proposed solution in Section 5 and the algorithm calibration in Section 6. We discuss computational experiments and results in Section 7. Finally, we conclude with a summary and an outlook in Section 8.

## 2. Production Process

Figure 1 shows the eight main stages of the production process: (1) Preparation, (2) Mixing, (3) Division and Rounding, (4) Repose, (5) Press, (6) Bake, (7) Cooling, (8) Stacking and Packing.

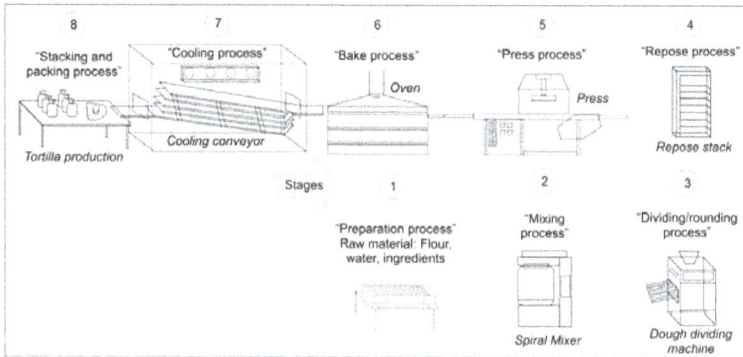

**Figure 1.** Tortilla production line.

In this paper, we consider a hybrid flow shop with six stages (from 2 to 7), with different numbers of machines in the stages 2, 4, and 6. In the production of wheat flour tortillas, the raw ingredients prepared in Stage 1 are transferred to a large mixer (Stage 2), where they are combined to create the dough.

Corn tortilla production starts with a process called nixtamalization, in which the corn is soaked in an alkali solution of lime (calcium hydroxide) and hot water. The dough is divided into fragments with round shapes (Stage 3). These fragments repose in a controlled environment to obtain a certain consistency as to the temperature and humidity (Stage 4). Next, they are pressed to take the shape of a tortilla (Stage 5). Rows of tortillas are heated in stoves (Stage 6). The next steps are to cool (Stage 7), stack, and pack them (Stage 8).

The process is distributed among flow lines with several machines per stage (Figure 2). It is operated in a semi-automated way controlled by operators.

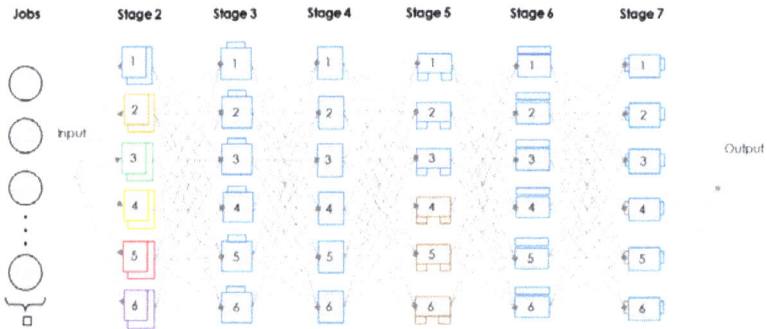

**Figure 2.** Six-stage hybrid flow shop.

Tables 1 and 2 present the capacities of the machines in each stage. These are represented by the amount of work (tortillas in kilograms) that can be processed on each machine.

**Table 1.** The processing capacity of machines in Stage 2 (kg per minute).

| Stage | | | 2 | | | |
|---|---|---|---|---|---|---|
| Machines | 1 | 2 | 3 | 4 | 5 | 6 |
| Jobs 1–5 | 80 | 80 | 100 | 130 | 160 | 200 |

**Table 2.** The processing capacity of machines in Stages 3–7 (kg per minute).

| Stage | | 3 | 4 | 5 | 6 | 7 |
|---|---|---|---|---|---|---|
| Machines | | 1–6 | 1–6 | 1–3 | 4–6 | 1–6 | 1–6 |
| | 1 | 1.3 | 200 | 7.2 | 9.6 | 9.6 | 9.6 |
| | 2 | 1.3 | 200 | 6.4 | 8.8 | 8.8 | 8.8 |
| Jobs | 3 | 0.44 | 200 | 2.4 | 4 | 4 | 4 |
| | 4 | 0.31 | 200 | 3.2 | 4.8 | 4.8 | 4.8 |
| | 5 | 1.3 | 200 | 4 | 8.8 | 8.8 | 8.8 |

Table 1 demonstrates that Stage 2 has six machines with processing capabilities of 80, 80, 100, 130, 160, and 200 kg per minute that can process jobs from 1 to 5. Machines 1 and 2 have the same capacities.

Table 2 shows that Stages 3, 4, 6, and 7 have identical machines, and Stages 2 and 5 have unrelated machines. Table 3 shows the power of the machines in each stage in kW per minute. The consumption is independent of the type of job to be processed. Machines in Stage 4 do not consume energy. The equipment is cabinets, where dough balls remain for a given period.

**Table 3.** Energy consumption of machines (kW per minute).

| Stage | | 2 | 3 | 4 | 5 | 6 | 7 |
|---|---|---|---|---|---|---|---|
| | 1 | 0.09 | 0.03 | 0 | 0.62 | 0.04 | 0.12 |
| | 2 | 0.09 | 0.03 | 0 | 0.62 | 0.04 | 0.12 |
| Machine | 3 | 0.11 | 0.03 | 0 | 0.62 | 0.04 | 0.12 |
| | 4 | 0.14 | 0.03 | 0 | 0.9 | 0.04 | 0.12 |
| | 5 | 0.23 | 0.03 | 0 | 0.9 | 0.04 | 0.12 |
| | 6 | 0.23 | 0.03 | 0 | 0.9 | 0.04 | 0.12 |

Table 4 shows characteristics of the jobs. We consider a workload with five jobs for production of the five types of tortillas most demanded by industry: Regular, Integral, Class 18 pcs, Taco 50, and MegaBurro. Each job is characterized by its type, group, and batches required for production. It has several assigned batches. This quantity must be within a certain range that is obtained from the standard deviations of the average monthly production in kilograms of tortillas for a year.

**Table 4.** Job characteristics.

| Job $j$ | Name | Type | Group | Batch Limits |
|---------|------|------|-------|--------------|
| 1 | Regular | 1 | 1 | 356–446 |
| 2 | Integral | 2 | 2 | 34–61 |
| 3 | Class 18 pcs | 3 | 1 | 7–10 |
| 4 | Taco 50 | 4 | 1 | 13–22 |
| 5 | MegaBurro | 5 | 1 | 12–20 |

The batch indicates the number of 10 kg tasks that have to be processed in the job. A job with one batch indicates processing 10 kg, and with five batches indicates processing 50 kg.

The group indicates the compatibility of the job tasks. Job tasks belonging to the same group may be processed simultaneously. For instance, they can be processed on all unrelated machines of Stage 2, if their sizes are less than or equal to the capacity in kilograms of the machines (Table 1). The job tasks of the same type of dough can be combined, forming compatibility groups. In this case, the jobs of Types 1, 3, 4, and 5 belong to Group 1, and jobs of Type 2 belong to Group 2.

Table 5 shows processing times in minutes for every 10 kg of tortilla for a job type. The 10 kg represents a task. We see that in Stages 2 and 4, all machines process jobs for the same durations of 1 and 45 min, respectively, due to the fact that the time does not depend on the job type.

**Table 5.** Processing time $P_{i_k,j}$ of job $j$ at machine $k$ in stage $i$ for each 10 kg.

| Stage | | 2 | 3 | 4 | 5 | | 6 | 7 |
|-------|---|---|---|---|---|---|---|---|
| Machines | | 1–6 | 1–6 | 1–6 | 1–3 | 4–6 | 1–6 | 1–6 |
| | 1 | 1 | 1.25 | 45 | 1.43 | 1.07 | 12.15 | 17.25 |
| | 2 | 1 | 1.25 | 45 | 1.48 | 1.11 | 12.15 | 17.25 |
| Jobs | 3 | 1 | 4.38 | 45 | 3.69 | 2.77 | 21.38 | 26.48 |
| | 4 | 1 | 3.13 | 45 | 3.00 | 2.25 | 24.50 | 29.60 |
| | 5 | 1 | 0.42 | 45 | 2.28 | 1.14 | 18.24 | 23.34 |

Table 6 shows the setup time of machines at each stage. At stages from 3 to 7, the setup time is independent of the task sequence, job type, and weight.

**Table 6.** Sequence setup time for all machines in stage $i$.

| Stage | 2 | 3 | 4 | 5 | 6 | 7 |
|-------|---|---|---|---|---|---|
| Setup time | $st_2$ | 0.7 | 0.7 | 0.5 | 0 | 0 |

At Stage 2, the setup time is calculated as follows:

$$st_2 = \frac{\left(35 + \left(\frac{t_w}{10} \times 5\right)\right)}{60} \tag{1}$$

where $t_w$ is the total weight in kilograms of the tasks to be processed by machines in the same time period. These tasks have to belong to jobs of the same group.

## 3. Related Works

### 3.1. Energy-Aware Flow Shop Scheduling

Investment in new equipment and hardware can certainly contribute to energy savings [9]. The use of "soft" techniques to achieve the same objective is also an effective option [10].

The advance schedule could play an important role in reducing the energy consumption of the manufacturing processes. Meta-heuristics, e.g., genetic algorithms (GA) [11], particle swarm optimization [12], and simulated annealing [13] are greatly popularity for use in the design of production systems.

However, studies of flow shop scheduling problems for energy saving appear to be limited [14,15]. In fact, most production systems that allow changing of the speed of the machines belong to the mechanical engineering industry and usually take the configuration of a work shop instead of a flow shop [16].

One of the first attempts to reduce energy consumption through production scheduling can be found in the work of Mouzon et al. [17]. The authors collected operational statistics for four machines in a flow shop and found that nonbottleneck machines consume a considerable amount of energy when left idle. As a result, they propose a framework for scheduling power on and off events to control machines for achieving a reduction of total energy consumption. Dai et al. [18] applied this on/off strategy in a flexible flow shop, obtaining satisfactory solutions to minimize the total energy consumption and makespan.

Zhang and Chiong [16] proposed a multiobjective genetic algorithm incorporated into two strategies for local improvements to specific problems, to minimize power consumption, based on machine speed scaling.

Mansouri et al. [19] analyzed the balance between minimizing the makespan, a measure of the level of service and energy consumption, in a permutation flow shop with a sequence of two machines. The authors developed a linear model of mixed-integer multiobjective optimization to find the Pareto frontier composed of energy consumption and total makespan.

Hecker et al. [20] used evolutionary algorithms for scheduling in a non-wait hybrid flow shop to optimize the allocation of tasks in production lines of bread, using particle swarm optimization and ant colony optimization. Hecker et al. [21] used a modified genetic algorithm, ant colony optimization, and random search procedure to study the makespan and total time of idle machines in a hybrid permutation flow model.

Liu and Huang [14] studied a scheduling problem with batch processing machines in a hybrid flow shop with energy-related criteria and total weighted tardiness. The authors applied the Nondominated Sorting Genetic Algorithm II (NSGA-II).

Additionally, in time completion problems, Yaurima et al. [3] proposed a heuristic and meta-heuristic method to solve hybrid flow shop (HFS) problems with unrelated machines, sequence-dependent setup time (SDST), availability constraints, and limited buffers.

### 3.2. Multiobjective Optimization

Different optimization criteria are used to improve models of a hybrid flow shop. An overview of these criteria can be found in [4]. Genetic algorithms have received considerable attention as an approach to multiobjective optimization [6,22,23].

Deb et al. [24] proposed a computationally efficient multiobjective algorithm called NSGA-II (Nondominated Sorting Genetic Algorithm II), which can find Pareto-optimal solutions. The computational complexity of the algorithm is $O(MN^2)$, where $M$ is the number of targets and $N$ is the size of the data set.

In this article, we consider it to solve our problem with two criteria. An experimental analysis of two crossover operators, three mutation operators, three crossover and mutation probabilities, and three populations is presented. For each case of machines per stage, the goal is to find the most desirable solutions that provide the best results considering both objectives.

*3.3. Desirability Functions*

Getting solutions on the Pareto front does not resolve the multiobjective problem. There are several methodologies to incorporate preferences in the search process [25].

These methodologies are responsible for giving guidance or recommendations concerning selecting a final justifiable solution from the Pareto front [26]. One of these methodologies is Desirability Functions (DFs) that are responsible for mapping the value of each objective to desirability, that is, to values in a unitless scale in the domain [0, 1].

Let us consider that the objective $f_i$ is $\mathcal{Z}_i \subseteq \mathbb{R}$; then a DF is defined as any function $d_i : \mathcal{Z}_i \to [0, 1]$ that specifies the desirability of different regions of the domain $\mathcal{Z}_i$ for objective $f_i$ [25].

The Desirability Index (DI) is the combination of the individual values of the DFs in one preferential value in the range [0, 1]. The higher the values of the DFs, the greater the DI. The individual in the final front with the highest DI will be the best candidate to be selected by the decision maker [27]. The algorithm calibration of desirability functions is applied to the final solutions of the Pareto front to get the best combination of parameters.

## 4. Model

Many published works address heuristics for flow shops that are considered production environments [3,23,28–30]. We consider the following problem: A set $J = \{1, 2, 3, 4, 5\}$ of jobs available at time 0 must be processed on a set of 6 consecutive stages $S = \{2, 3, 4, 5, 6, 7\}$. of the production line, subject to minimization of total processing time $C_{max}$ and energy consumption of the machines $E_{op}$.

Each stage $i \in S$ consists of a set $M_i = \{1, \ldots, m_i\}$ of parallel machines, where $|M_i| \geq 1$. The sets $M_2$ and $M_5$ have unrelated machines and the sets $M_3$, $M_4$, $M_6$, and $M_7$ have identical machines. We consider three different cases of $m_i = \{2, 4, 6\}$ machines in each stage.

Each job $j \in J$ belongs to a particular group $G_k$. Jobs 1, 3, 4, 5 $\in J$ belong to $G_1$, job 2 $\in J$ belongs to $G_2$. Each job is divided into a set $T_j$ of $t_j$ tasks, $T_j = \{1, 2, \ldots, t_j\}$. Jobs and tasks in the same group can be processed simultaneously only in stage 2 $\in S$ according to the capacity of the machine. In stages 3, 4, 5, 6, 7 $\in S$, the tasks must be processed by exactly one machine in each stage (Figure 2).

Each task $t \in T_j$ consists of a set of batches of 10 kg each. The tasks can have different numbers of batches. The set of all the tasks together of all the jobs is given by $T = \{1, 2, \ldots, Q\}$.

Each machine must process an amount of "$a$" batches at the time of assigning a task.

Let $p_{i,l,q}$ be the processing time of the task $q \in T_j$ on the machine $l \in M_i$ at stage $i$. Let $S_{i,l,qp}$ be the setup (adjustment) time of the machine $l$ from stage $i$ to process the task $p \in T_j$ after processing task $q$. Each machine has a process buffer for temporarily storing the waiting tasks called "work in progress". The setup times in stage 2 $\in S$ are considered to be sequence dependent; in stages 3, 4, 5, 6, 7 $\in S$, they are sequence independent.

Using the well-known three fields notation $\alpha \mid \beta \mid \gamma$ for scheduling problems introduced in [31], the problem is denoted as follows:

$$FH6, \left(RM^{(2)}, (PM^{(i)})_{i=3^4}, RM^{(5)}, (PM^{(i)})_{i=6^7}\right) | S_{sd2}, S_{si3^7} \mid \{C_{max}, E_{op}\}. \tag{2}$$

*Encoding*

Each solution (chromosome) is encoded as a permutation of integers. Each integer represents a batch (10 kg) of a given job. The enumeration of each batch is given in a range specified by the order of the jobs and the number of batches.

Batches are listed according to the job to which they belong. Table 7 shows an example the five jobs. If Job 1 has nine batches, they are listed from 1 to 9. If Job 2 has 12 batches, they are listed from 10 to 21. Reaching Job 5, we have accumulated 62 batches.

**Table 7.** Example of enumeration of batches for jobs.

| Job | 1 | 2 | 3 | 4 | 5 |
|---|---|---|---|---|---|
| Batches | 9 | 12 | 16 | 15 | 10 |
| Batch index | 1–9 | 10–21 | 22–37 | 38–52 | 53–62 |

Figure 3 shows an example of the production representation. Groups are composed according to the number of machines in the stage $(m_i)$. The batches of each job are randomly distributed between the groups. To obtain the number of batches per group (a), we divide the total amount of batches by the number of machines per stage.

| Machine 1 | | | | | Machine 2 | | | | | Machine $m_i$ | | | | |
|---|---|---|---|---|---|---|---|---|---|---|---|---|---|---|
| 56 | 23 | ... | 19 | 16 | 1 | 55 | ... | 62 | 27 | ... | 11 | 24 | ... | 37 | 17 |
| 1 | 2 | ... | $a-1$ | $a$ | 1 | 2 | ... | $a-1$ | $a$ | | 1 | 2 | ... | $a-1$ | $a$ |

**Figure 3.** Example of the representation of the chromosome for $m_i$ machines per stage.

Let us consider six stages with $m_1 = 4$ machines per stage. The chromosome is composed of 4 groups. Let us assume that we have 62 batches for 5 jobs. This is equal to 62 (batches)/4 (machines per stage) = 15.5 (batches per group). The result is rounded to the nearest higher integer, ending with a = 16 batches per group. An example of processing tasks on Machine 1 is shown in Figure 4.

| | Machine 1 | | | | | | | | | | | | | | | | |
|---|---|---|---|---|---|---|---|---|---|---|---|---|---|---|---|---|---|
| Original Sequence | 56 | 23 | 38 | 35 | 1 | 42 | 61 | 14 | 51 | 11 | 5 | 59 | 48 | 15 | 19 | 16 | Batches |
| Ordered Sequence | 1 | | | | | 2 | | | | 3 | | | | | | | Proc. Order |
| | 1 | 5 | 23 | 35 | 38 | 42 | 51 | 48 | 56 | 61 | 59 | 14 | 11 | 15 | 19 | 16 | Batches |
| | 1 | 3 | | 4 | | | 4 | | 5 | | | 2 | | | | | Jobs |

**Figure 4.** Example of tasks on Machine 1.

In the ordered sequence, "Jobs" indicate jobs to which the batches belong according to their enumeration. These groups indicate the number of total tasks for each job from left to right. For example, in the ordered sequence, Job 4 has two tasks; the first one "4-1" has three batches (38, 42, 51) and the second one "4-2" has one batch (48).

"Proc. Order" indicates the order of processing of the task groups. Assuming that a machine in Stage 1 has to process up to 70 kg (7 batches), Group 1 is composed of sets of tasks 1-1 (1, 5), 3-1 (23, 35), 4-1 (38, 42, 51) with seven batches. When the capacity of the machine is filled, Group 2 is formed for the tasks 4-2 (48) and 5-1 (56, 61, 59). Assuming that the batches of Job 2 are not compatible with any of the remaining four jobs, the batches are processed separately from the other two groups, forming the task 2-1 with five batches (14, 11, 15, 19, 16). When the first group's tasks have completed their processing in the machine, the second group is processed, and so on.

The tasks processed in the group are assigned to the machines of the following stages individually according to their order of completion in the first stage. When the tasks of a group—for example, Group 1 (1-1, 3-1, 4-1)—are processed together, the completion time is the same for all. Therefore, the three tasks will be ready at the same time to be assigned individually to machines available in the next stage, until the last (6th) stage.

In the post-second stages, the tasks that are ready to be assigned are in the process buffer. The task that has the longest processing time is assigned first to an available machine. The calculation of the total completion time $C_{max}$ is as follows. The time $C_{i,p}$ for completing task $p$ in stage $i$ is calculated according to the formula

$$C_{i,p} = \min_{1 \le l \le m} \{\max\{C_{i,p} + S_{i_l,qp}; C_{i-1,p}\} + p_{i_l,q}\}. \tag{3}$$

The maximum completion time of all tasks and jobs is calculated as

$$C_{max} = max_{p=1}^{Q} \{C_{7,p}\}. \tag{4}$$

$Q$ indicates the total number of tasks for all jobs. $C_{m,p}$ is the completion time of task $p \in T_j$ in the last stage $7 \in S$. The total energy consumption $E_{op}$ of the execution of all tasks is calculated as

$$E_{op} = \sum_{q=1}^{Q} p_{i_l,q} \cdot E_{i_l}. \tag{5}$$

where $E_{i_l}$ indicates the electrical power consumption of machine $l$ in stage $i$, and $p_{i_l,q}$ refers to the processing time of the task $q \in T$.

## 5. Bi-Objective Genetic Algorithm

### 5.1. Nondominated Sorting Genetic Algorithm II (NSGA-II)

The NSGA-II algorithm is used to assign tasks to machines so that $C_{max}$ and $E_{op}$ are minimized. NSGA-II is a multiobjective genetic algorithm characterized by elitism and stacking distance to maintain the diversity of the population to find as many Pareto-optimal solutions as possible [24]. It generates a population of $N$ individuals, where each represents a possible solution. Individuals evolve through genetic operators to find optimal or near-optimal solutions. Three operators are usually applied: tournament selection (using a stacking tournament operator), crossover, and mutation to generate another $N$ individuals or children.

From the mixture of these two populations, a new population of size $2N$ is created. Then, the best individuals are taken according to their fitness value by ordering the population on nondominated fronts. Individuals from the best nondominated fronts are first taken, one by one, until $N$ individuals are selected

The crowding distance is then compared to preserve diversity in the population. This operator compares two solutions and chooses a tournament winner by selecting the setting that is located on the best Pareto front. If the participating tournament configurations are on the same front, the best crowding distance (highest) to determine the winning setting is used. Later, the algorithm applies the basic genetic operators and promotes the next generation cycle with the configurations that occupy the best fronts, preserving the diversity through the crowding distance. A job is assigned to a machine at a given stage by taking into consideration processing speeds, setup times, machine availability, and energy consumption. Each solution is encoded in a permutation of integers.

### 5.2. Crossover Operators

Crossover operators allow the obtaining of new solutions (children) by combining individuals (parents) of the population. The crossover operator is applied under a certain probability. In this paper, we consider two operators: the partially mapped crossover and the order crossover.

Partially mapped crossover (PMX) [32,33]: This operator uses two cut points (Figure 5). The part of the first parent between the two cut points is copied to the children. Then, the following parts of the children are filled by a mapping between the two parents, so that their absolute positions are inherited where possible from the second parent.

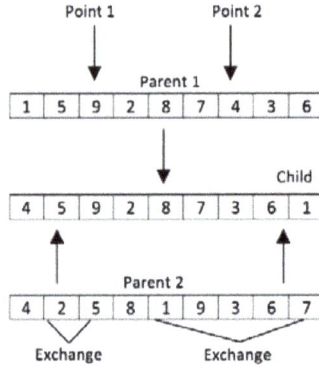

**Figure 5.** Partially mapped crossover (PMX) operator.

Figure 6 shows an example of the PMX crossover. The cross points in both parents serve to form the child chromosome. The 16 listed lots of the 4 jobs are randomly distributed on each parent chromosome. When crossed, they form the child chromosome with 16 lots.

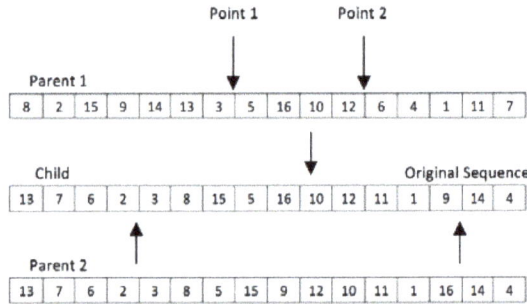

**Figure 6.** PMX crossover operator example with batch index.

The cross parents are composed of the batches of the jobs listed in Table 8.

**Table 8.** Enumeration of the batches for jobs.

| Job | Batches | Batch Index |
|-----|---------|-------------|
| 1 | 4 | 1–4 |
| 2 | 4 | 5–8 |
| 3 | 4 | 9–12 |
| 4 | 4 | 13–16 |

Figures 7 and 8 show representations of tasks for two and four machines per stage. The original sequence is ordered considering the enumeration of the jobs and their batches. Each machine is assigned batches to process, which form the tasks of each job. Each machine must process a certain set of tasks; each set is assigned a processing order (Proc. Order). These sets are formed according to the processing compatibility.

| | m1 | | m2 | | | |
|---|---|---|---|---|---|---|
| | 1 | 2 | 1 | | 2 | Proc. Order |
| Ordered | 2 3 13 15 | 5 6 7 8 | 1 | 4 9 10 11 12 14 | 16 | Batch |
| | 1 4 | 2 | 1 | 3 | 4 | Jobs |

**Figure 7.** Chromosome representation example with two machines per stage.

| | m1 | m2 | m3 | m4 | |
|---|---|---|---|---|---|
| | 1 | 2 | 1 | 1 | Proc. Order |
| Ordered | 2 3 6 7 | 3 15 5 8 | 10 11 12 16 | 1 4 9 14 | Batch |
| | 1 4 2 | 1 4 2 | 3 | 4 1 3 4 | Jobs |

**Figure 8.** Chromosome representation example with four machines per stage.

The tasks of Jobs 1, 3, and 4 can be processed in the same set at the same time. The tasks of Job 2 are put into a separate set because they are not compatible with each other. For example, in Figure 7, the first two tasks (1-1, 4-1) are processed by Machine 1, then the second set with one task (2-1). In Machine 2, two tasks of the first set (1-2 and 3-1) are processed, assuming that the machine has only the capacity to process up to 6 batches (60 kg). The second set will then be followed by one task (4-1).

Figure 8 shows four machines per stage. Four batches are assigned to each machine. Machines 1 and 2 process two sets of tasks. Machines 3 and 4 process only one set. Upon completion of the tasks processed in sets in Step 1, in the later stages, they will be processed individually.

Order crossover (OX) [32,34–37]: Two cut points are selected randomly; the part of the first parent located between these two cut points is copied to the child. The child's other spaces are left blank. The values copied to the child are deleted in the second parent. The following positions in the child are filled starting in the blank spaces and considering the order found in the second parent (Figure 9).

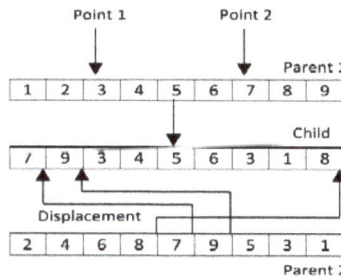

**Figure 9.** Order crossover (OX) operator.

*5.3. Mutation Operators*

Mutation operators produce small changes in individuals according to a probability. This operator helps to prevent falling into local optima and to extend the search space of the algorithm. Three mutation operators are considered: Displacement, Exchange, and Insertion.

Displacement is a generalization of the insertion mutation, in which instead of moving a single value, several values are changed (Figure 10). The exchange operator selects two random points and these position values are exchanged (Figure 11). In insertion, a value is selected randomly and will be inserted at an arbitrary position (Figure 12).

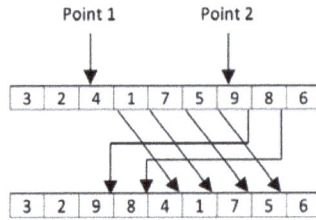

**Figure 10.** Displacement mutation operator.

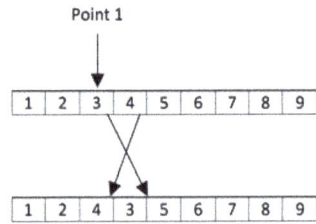

**Figure 11.** Exchange mutation operator.

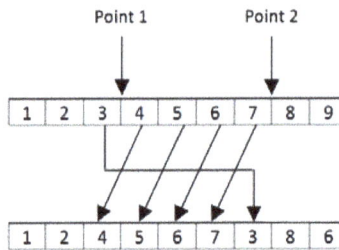

**Figure 12.** Insertion mutation operator.

## 6. Tuning Up the Parameters

### 6.1. Parameter Calibration

Calibration is a procedure for choosing the algorithm parameters that provide the best result in the response variables. A computational experiment for calibration includes the following steps: (a) each instance or workload is run with all possible combinations of parameters; (b) the best solution is obtained; (c) the relative difference of each algorithm on the best solution is calculated; (d) multifactorial Analysis of Variance (ANOVA) with a 95% confidence level is applied to find the parameter that most influences the solution; (e) the set of the best values is selected for each parameter.

Table 9 shows the parameters used for calibration. A total of $3 \times 2 \times 3 \times 3 \times 3 = 162$ different algorithms or combinations are considered. Thirty runs for each combination are performed, with $162 \times 30 = 4860$ experiments in total. For each of the 30 runs, a different workload will be taken. For each of the five jobs, 30 batches are generated randomly from a uniform distribution according to their limits (Table 4).

Each batch is assigned to a workload, obtaining 30 loads from 5 jobs. The variation of processing capabilities (in minutes) of the machines in Stages 2 and 5 is generated according to Tables 1 and 2. Stage 2 has 6 machines with different processing capabilities in kilograms. Stage 5 has 6 machines.

According to the selection of a number of machines per stage, the set of machines is chosen. The machines are numbered from 1 to 6. By considering different numbers of machines per stage (2, 4, or 6), we conduct 4860 × 3 = 14,580 experiments. In each of the 30 runs of each combination, the best individual applying desirability function is obtained. Subsequent values of each objective ($C_{max}$ and $E_{op}$) are used to calculate an average of 30, obtaining a single value for each objective.

**Table 9.** Parameters used for calibration.

| Factors | Levels |
|---|---|
| Population | 20, 30, 50 |
| Crossover operators | OX, PMX |
| Crossover probability | 0.5, 0.7, 0.9 |
| Mutation operators | Displacement, Exchange, Insertion |
| Mutation probability | 0.05, 0.1, 0.2 |
| Selection | Binary tournament |
| Stop criterion | 50 iterations |

The performance of each algorithm is calculated as the relative increment of the best solution (RIBS). The RIBS is calculated with the following formula:

$$RIBS = \frac{Heu_{sol} - Best_{sol}}{Best_{sol}} \times 100 \tag{6}$$

where $Heu_{sol}$ is the value of the objective function obtained by the algorithm, and $Best_{sol}$ is the best value obtained during the execution of all possible combinations of parameters.

All experiments were performed on a PC with Intel Core i3 CPU and 4 GB RAM. The programming language used to encode the algorithm is R. It provides advantages, including the scalability and libraries [38]. The calibration of the algorithm lasted approximately 15 days and 9 h.

*6.2. Residual Analysis*

To verify that the data are normalized, the supposition of the suitability of the model using the normality, homoscedasticity, and independence of residues is verified. The residuals are calculated according to the following formula [39]:

$$e_i = y_i - \overline{y_l}, \, i = 1, \, 2, \, 3, \, \ldots, \, n \tag{7}$$

where $y_i$ is the RIBS for the run $i$ and $\overline{y_l}$ is the average of the experiment. Figure 3 shows the graph of the normal probability of residuals for 6 machines per stage. As can be seen (Figure 13), the graph complies with the assumption of normality. The same results are obtained for 2 and 4 machines per stage.

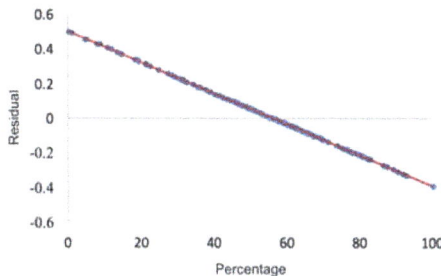

**Figure 13.** Graph of the normal probability of waste for six machines per stage.

*6.3. Variance Analysis*

Variance analysis is applied to evaluate the statistical difference between the experimental results and to observe the effect of the parameters on the quality of the results. It is used to determine factors that have a significant effect and to discover the most important factors. The parameters of the problem are considered as factors and their values as levels. We assume that there is no interaction between factors.

The *F*-Ratio is the ratio between Factor Mean Square and Mean Square residue. A high *F*-Ratio means that this factor significantly affects the response. The *p*-value shows the statistical significance of the factors: *p*-values that are less than 0.05 have a statistically significant effect on the response variable (RIBS) with a 95% confidence level. According to the *F*-Ratio and *p*-value, the most important factors in the case of two machines per stage are the mutation operator and the crossover operator variable. DF shows the number of degrees of freedom (Table 10).

**Table 10.** Relative increment of the best solution (RIBS) analysis of variance for two machines per stage.

| Source | Sum of Squares | DF | Mean Square | F-Ratio | p-Value |
|---|---|---|---|---|---|
| A: Crossover | 1747 | 1 | 1747 | 9.755 | 0.00213 |
| B: Mutation | 11,967 | 1 | 11,967 | 66.833 | $9.58 \times 10^{-14}$ |
| C: Crossover probability | 1519 | 1 | 1519 | 8.481 | 0.00411 |
| D: Mutation probability | 654 | 1 | 654 | 3.652 | 0.05782 |
| E: Population | 1732 | 1 | 1732 | 9.673 | 0.00222 |
| Residuals | 27,934 | 156 | 179 | | |

In the case of four machines per stage, the most important factors are the mutation operator and crossover probability (Table 11). For six machines per stage, the most important factors are the mutation operator and mutation probability (Table 12).

**Table 11.** RIBS analysis of variance for four machines per stage.

| Source | Sum of Squares | DF | Mean Square | F-Ratio | p-Value |
|---|---|---|---|---|---|
| A: Crossover | 760 | 1 | 760 | 4.602 | 0.03349 |
| B: Mutation | 13,471 | 1 | 13,471 | 81.565 | $6.09 \times 10^{-16}$ |
| C: Crossover probability | 1540 | 1 | 1540 | 31.223 | $1.00 \times 10^{-7}$ |
| D: Mutation probability | 5157 | 1 | 5157 | 9.326 | 0.00266 |
| E: Population | 3395 | 1 | 3395 | 20.557 | $1.15 \times 10^{-5}$ |
| Residuals | 25,765 | 156 | 165 | | |

**Table 12.** RIBS analysis of variance for six machines per stage.

| Source | Sum of Squares | DF | Mean Square | F-Ratio | p-Value |
|---|---|---|---|---|---|
| A: Crossover | 2587 | 1 | 2587 | 18.39 | $3.14 \times 10^{-5}$ |
| B: Mutation | 17,785 | 1 | 17,785 | 126.47 | $2 \times 10^{-16}$ |
| C: Crossover probability | 2886 | 1 | 2886 | 20.52 | $1.16 \times 10^{-5}$ |
| D: Mutation probability | 4840 | 1 | 4840 | 34.42 | $2.58 \times 10^{-8}$ |
| E: Population | 2881 | 1 | 2881 | 20.49 | $1.18 \times 10^{-5}$ |
| Residuals | 21,937 | 156 | 165 | | |

Figures 14–17 show means and 95% Least Significant Difference (LSD) confidence intervals. Figure 14 shows the results obtained for mutation operators for two, four, and six machines per stage. We can see that operator Displacement is the best mutation among the three tested operators. Figure 15 shows the results for the crossover operators, where we observe that the PMX operator is the best. The results for four and six stages are similar.

Figure 16 shows plots for the crossover probability. It can be seen that the best crossover probability occurring is 0.9. Figure 17 presents plots for the mutation probability, where the probability of 0.2 is statistically more significant.

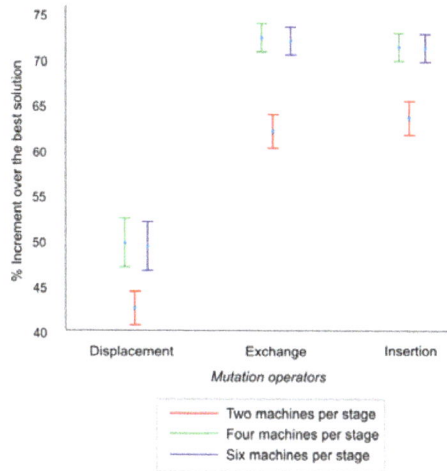

**Figure 14.** Means and 95% LSD confidence intervals of mutation operators.

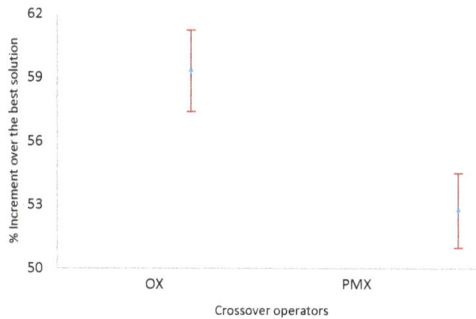

**Figure 15.** Means and 95% LSD confidence intervals of crossover operators—two machines per stage.

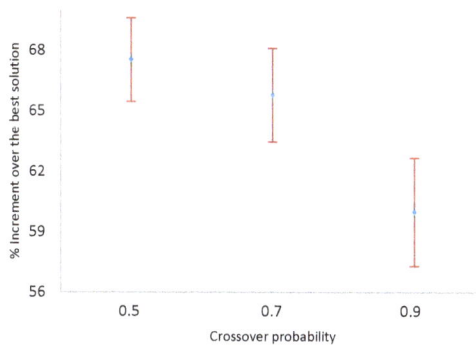

**Figure 16.** Means and 95% LSD confidence intervals of crossover probability—four machines per stage.

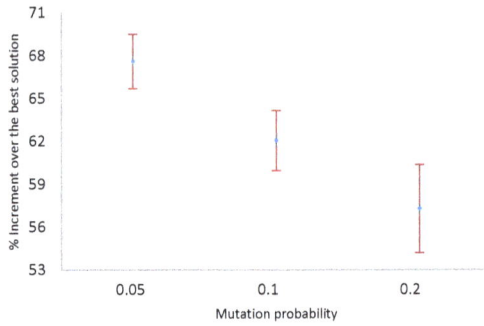

**Figure 17.** Means and 95% LSD confidence intervals of mutation probability—six machines per stage.

*6.4. Pareto Front Calibration Analysis*

Figures 18–20 show the solution space obtained by calibration experiments. The horizontal axis represents the energy $E_{op}$ consumed by the machines. The vertical axis represents the time completion $C_{max}$ of jobs. To determine the best individuals of all experiments, the Pareto front is calculated for each case. Each point represents a combination of parameters from 162 experiments explained in Section 6.1. Each front consists of numbered points from lowest to highest according to their DI. The closer to 1, the better the place in the enumeration is (see Section 3.3).

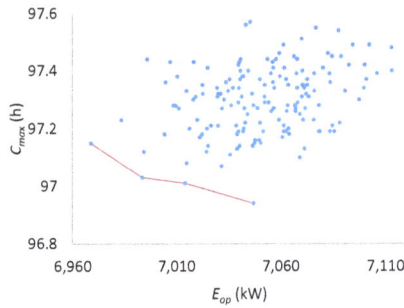

**Figure 18.** Pareto front for two machines per stage.

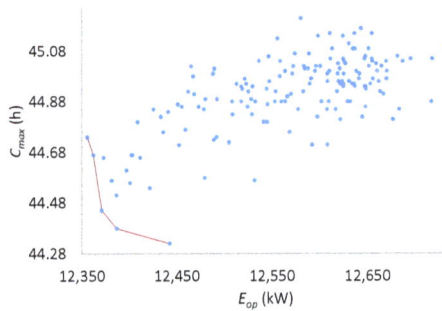

**Figure 19.** Pareto front for four machines per stage.

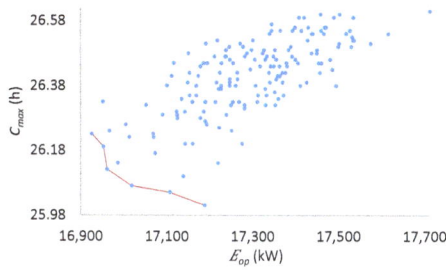

**Figure 20.** Pareto front for six machines per stage.

Tables 13–15 show the ordering of the front points regarding a combination of $E_{op}$, $C_{max}$, and Desirability Index (DI) for each Pareto front. We see, for example, that the parameter combination 156 corresponds to a higher DI, followed by 129, etc.

**Table 13.** Optimum parameter combination selection for bi-objective genetic algorithm (GA) for two machines per stage.

| No. | Comb. | $E_{op}$ | $C_{max}$ | DI |
|-----|-------|----------|-----------|-----|
| 1 | 156 | 6993.41 | 97.03 | 0.843 |
| 2 | 129 | 6968.70 | 97.15 | 0.816 |
| 3 | 128 | 7014.08 | 97.01 | 0.782 |
| 4 | 120 | 7046.87 | 96.94 | 0.677 |

**Table 14.** Optimum parameter combination selection for bi-objective GA for four machines per stage.

| N | Comb. | $E_{op}$ | $C_{max}$ | DI |
|---|-------|----------|-----------|-----|
| 1 | 156 | 12,381.55 | 44.38 | 0.926 |
| 2 | 129 | 12,370.82 | 44.45 | 0.906 |
| 3 | 147 | 12,442.75 | 44.32 | 0.878 |
| 4 | 128 | 12,362.17 | 44.67 | 0.775 |
| 5 | 120 | 12,355.79 | 44.74 | 0.730 |

**Table 15.** Optimum parameter combination selection for bi-objective GA for six machines per stage.

| N | Comb. | $E_{op}$ | $C_{max}$ | DI |
|---|-------|----------|-----------|-----|
| 1 | 156 | 17,018.73 | 26.07 | 0.891 |
| 2 | 129 | 16,961.46 | 26.12 | 0.883 |
| 3 | 120 | 17,108.33 | 26.05 | 0.847 |
| 4 | 147 | 16,953.30 | 26.19 | 0.823 |
| 5 | 138 | 17,189.52 | 26.01 | 0.816 |
| 6 | 75 | 16,925.85 | 26.23 | 0.796 |

Table 16 shows the parameters of these combinations. The crossover PMX has higher values of DI. The mutation operator Displacement is used in each of the top three combinations. The crossover probability is maintained at 0.9 in the two most significant combinations. The mutation probability remains at the highest with 0.2 in all three, as does the population with 50 individuals.

The most influential parameters correspond to the combination 156, which has greater DI in three cases (see Tables 13–15).

Our analysis shows that crossover PMX and mutation operator Displacement are the best among those tested. The best probability of crossover is 0.9 and that of mutation is 0.2. The population of 50 individuals is statistically more significant.

**Table 16.** Optimum parameter combination for bi-objective GA.

| No. | Comb. | Cr | Mu | Pcr | Pmu | Pob | |
|-----|-------|-----|--------------|-----|-----|-----|----------|
| 1 | 156 | PMX | Displacement | 0.9 | 0.2 | 50 | Selected |
| 2 | 129 | OX | Displacement | 0.9 | 0.2 | 50 | |
| 3 | 120 | OX | Displacement | 0.7 | 0.2 | 50 | |

## 7. Experimental Analysis

### 7.1. Experimental Setup

In multiobjective optimization, there is usually no single best solution, but a set of solutions that are equally good. Our objective is to obtain a good approximation of the Pareto front regarding two objective functions [40].

Our bi-objective genetic algorithm is tuned up using the following parameters obtained during the calibration step: crossover, PMX; mutation, Displacement; crossover probability, 0.9; mutation probability, 0.2; population size, 50.

We consider five jobs in each of the 30 workloads, obtaining the number of batches for each job based on a uniform random distribution (Table 4).

### 7.2. Bi-Objective Analysis

Figures 21–23 show the solution sets and the Pareto fronts. In each front, 1500 solutions are included, being 50 (individuals) × 30 (workloads).

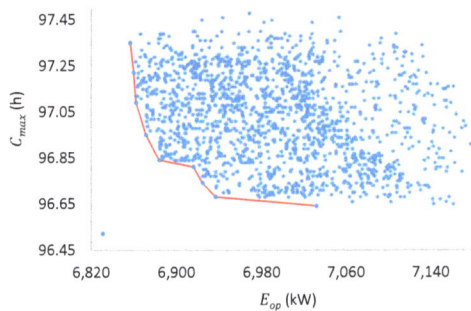

**Figure 21.** Pareto front for two machines per stage.

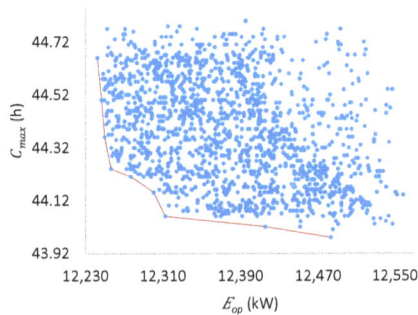

**Figure 22.** Pareto front for four machines per stage.

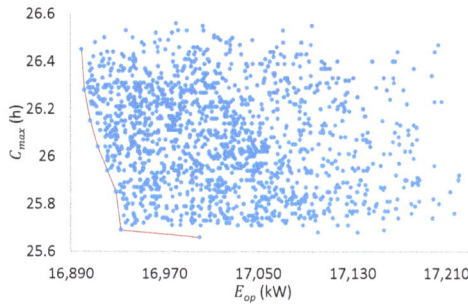

**Figure 23.** Pareto front for six machines per stage.

Figure 21 shows the solution space (two objectives) for two machines per stage. This two-dimensional solution space represents a feasible set of solutions that satisfy the problem constraints. The Pareto front covers a wide range of $E_{op}$ from 6856 to 7178 kW, while $C_{max}$ is in the range of 96.64 to 97.48 h.

Figure 22 shows the solution space for four machines per stage. The Pareto front covers $E_{op}$ from 12,242 to 12,559 kW, while $C_{max}$ is in the range of 43.98 to 44.8 h. Finally, Figure 23 shows the solution space for 6 machines per stage. $E_{op}$ ranges from 16,899 to 17,221 kW, while $C_{max}$ is in the range of 25.66 to 26.66 h.

Although the Pareto fronts are of good quality, it is observed that many of the generated solutions are quite far from it. Therefore, a single execution of the algorithm can produce significantly different results. The selection of the solutions included in the Pareto front depends on the preference of the decision maker.

### 7.3. Branch and Bound Analysis

A Branch and Bound algorithm (B&B) is an exact method for finding an optimal solution to an NP-hard problem. It is an enumerative technique that can be applied to a wide class of combinatorial optimization problems. The solution searching process is represented by a branching tree. Each node in the tree corresponds to a subproblem, which is defined by a subsequence of tasks that are placed at the beginning of the complete sequence.

This subsequence is called partial sequence (PS). The set of tasks not included in PS is called NPS. When a node is branched, one or more nodes are generated by adding a task ahead of the partial sequence associated with the node being branched. To avoid full enumeration of all task permutations, the lower bound of the value of the objective functions is calculated in each step for each partial schedule. In the case of ties, the algorithm selects a node with the lowest lower bound.

The B&B algorithm was run on a PC with an Intel Core i3 CPU and 4 GB of RAM. The programming language used to encode the algorithm was R. The algorithm lasted approximately 12 days, 3 h.

A comparison of the B&B algorithm concerning the Pareto front is shown in Figure 21 as a lowest left point. The B&B result is considered as the best result that could be obtained, known as the global optimum.

Table 17 shows the relative degradation of our bi-objective algorithm over the B&B best solutions. Results are obtained by averaging 30 experiments. We see that the results for each objective are not more than 3.44% worst for $C_{max}$ and 2.94% for $E_{op}$. Table 18 shows the computation times for obtaining the Pareto front solutions.

According to the results comparison, we conclude that the proposed bi-objective algorithm can produce satisfactory results close to a global optimum in short computational time.

**Table 17.** RIBS of the solution of the Branch and Bound (B&B) algorithm concerning the Pareto front solutions.

| $30 \times 2$ | | $30 \times 4$ | | $30 \times 6$ | |
|---|---|---|---|---|---|
| $E_{op}$ | $C_{max}$ | $E_{op}$ | $C_{max}$ | $E_{op}$ | $C_{max}$ |
| 2.94 | 0.12 | 1.72 | 0.46 | 0.50 | 1.8 |
| 1.55 | 0.17 | 0.87 | 0.55 | 1.02 | 0.35 |
| 0.77 | 0.33 | 0.41 | 0.96 | 0.42 | 3.44 |
| 1.37 | 0.23 | 0.33 | 1.55 | 0.46 | 2.27 |
| 1.24 | 0.30 | 0.58 | 0.89 | 0.62 | 0.47 |
| 0.58 | 0.45 | 0.77 | 0.75 | 0.60 | 1.1 |
| 0.45 | 0.59 | 0.35 | 1.23 | 0.55 | 1.45 |
| 0.42 | 0.73 | 0.29 | 1.92 | 0.43 | 2.78 |
| 0.37 | 0.86 | 2.27 | 0.37 | | |

**Table 18.** Computational time for obtaining Pareto front solutions.

| | Time (min) | | Time (min) | | Time (min) |
|---|---|---|---|---|---|
| | 2.85 | | 3.77 | | 7.55 |
| | 3.79 | | 5.81 | | 8.6 |
| | 1.74 | | 6.78 | | 6.7 |
| 2 machines per | 2.76 | 4 machines per | 5.67 | 6 machines per | 6.6 |
| stage | 3.73 | stage | 3.85 | stage | 7.62 |
| | 3.86 | | 4.82 | | 7.66 |
| | 2.85 | | 4.85 | | 8.61 |
| | 1.77 | | 4.94 | | 6.76 |
| | 2.85 | | 5.8 | | |

*7.4. Comparison with Industry*

Figures 24 and 25 show $E_{op}$ and $C_{max}$ of the industry and the B&B algorithm. The term "Algorithm" in these graphs represents the results of the B&B algorithm. The term "Industry" represents the actual data of the production company. The industry results shown in Table 19 are obtained considering the average monthly production load in kilograms of tortillas per year. The results of the B&B algorithm are obtained for a different number of machines per stage. Table 20 shows the same results as in Table 19 in percentages.

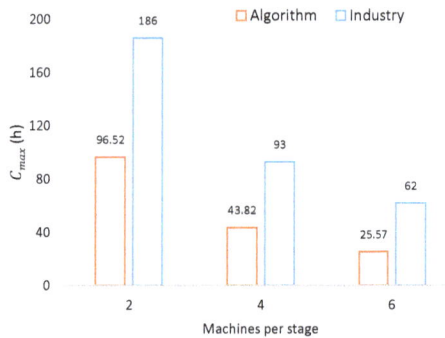

**Figure 24.** $C_{max}$ of B&B algorithm and industry.

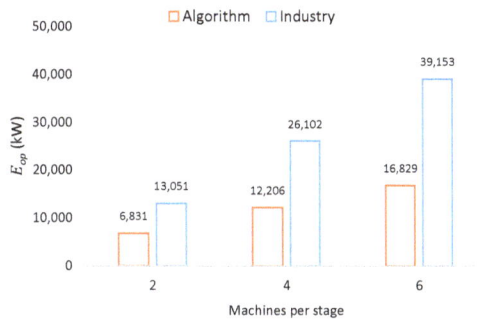

**Figure 25.** $E_{op}$ of B&B algorithm and industry.

**Table 19.** Results of the industry and B&B algorithm.

| Machines per Stage | Industry | | B&B Algorithm | |
|---|---|---|---|---|
| | $E_{op}$ (kW) | $C_{max}$ (h) | $E_{op}$ (kW) | $C_{max}$ (h) |
| 2 | 13,050.90 | 186 | 6,831.40 | 96.52 |
| 4 | 26,101.80 | 93 | 12,206.42 | 43.82 |
| 6 | 39,152.70 | 62 | 16,828.72 | 25.57 |

**Table 20.** The industry and bi-objective GA degradation over B&B algorithm (%).

| Machines per Stage | Industry | | Bi-Objective GA | |
|---|---|---|---|---|
| | % $E_{op}$ | % $C_{max}$ | % $E_{op}$ | % $C_{max}$ |
| 2 | 47.65 | 48.11 | 1.53 | 0.17 |
| 4 | 53.24 | 52.88 | 0.87 | 0.54 |
| 6 | 57.02 | 58.75 | 0.62 | 0.47 |

Figure 24 shows $C_{max}$ considering working hours. We see that the results of the algorithm are significantly improved compared with industry. For two machines per stage, the difference is almost a halving of $C_{max}$ from 186 to 96.52 h. In the case of four machines, it remains below half (93 and 43.82 h). In the case of six machines, the results are better by almost three times with regard to $C_{max}$ (62 h versus 25.57 h).

Figure 25 shows $E_{op}$ according to the processing time of machines (Table 3). We see that the B&B algorithm significantly improves upon the results obtained in industry. For two machines per stage, the $E_{op}$ objective is reduced to almost half, and for four and six machines, the $E_{op}$ objective of our algorithm is reduced to less than half.

Table 20 shows the percentage of degradation of the results of industry and those selected from the Pareto front obtained from the bi-objective GA compared with results of the B&B algorithm.

We observe that the degradation of $E_{op}$ and $C_{max}$ observed in industry are closer to or higher than 50%. Comparing B&B with our bi-objective GA, we observe that our results are less than 1.53% worse for $E_{op}$ and 0.54% for $C_{max}$. Due to the fact that the B&B algorithm finds the global optimum, we demonstrate the quality of our algorithm.

## 8. Conclusions

Our main contributions are multifold:

(1)   We formulated the complex problem of the real-life industry environment of tortilla production considering two optimization criteria: total completion time and energy consumption;

(2)  We proposed a bi-objective solution to solve a hybrid flow shop with unrelated machines, setup time, and work in progress buffers. It is based on the known NSGA-II genetic algorithm;

(3)  We calibrated our algorithm using a statistical analysis of multifactorial variance. The ANOVA technique is used to understand the impact of different factors on the solution quality. A branch and bound algorithm was used to assert the obtained performance.

(4)  We provided a comprehensive experimental analysis of the proposed solution based on data from a tortilla production company in Mexico.

(5)  We demonstrated that our solutions are not more than 3.5 percent off the global optimum for both criteria. They save more than 48 percent of production time and 47 percent of energy consumption compared with the actual production plan followed by the company.

**Author Contributions:** All authors contributed to the analysis of the problem, designing algorithms, performing the experiments, analysis of data, and writing the paper.

**Funding:** This work is partially supported by Russian Foundation for Basic Research (RFBR), project No. 18-07-01224-a.

**Conflicts of Interest:** The authors declare no conflict of interest.

## References

1.  Linn, R.; Zhang, W. Hybrid flow shop scheduling: A survey. *Comput. Ind. Eng.* **1999**, *37*, 57–61. [CrossRef]
2.  Quadt, D.; Kuhn, H. A taxonomy of flexible flow line scheduling procedures. *Eur. J. Oper. Res.* **2007**, *178*, 686–698. [CrossRef]
3.  Yaurima, V.; Burtseva, L.; Tchernykh, A. Hybrid flowshop with unrelated machines, sequence-dependent setup time, availability constraints and limited buffers. *Comput. Ind. Eng.* **2009**, *56*, 1452–1463. [CrossRef]
4.  Ruiz, R.; Vázquez-Rodríguez, J.A. The hybrid flow shop scheduling problem. *Eur. J. Oper. Res.* **2010**, *205*, 1–18. [CrossRef]
5.  Pei-Wei, T.; Jeng-Shyang, P.; Shyi-Ming, C.; Bin-Yih, L. Enhanced parallel cat swarm optimization based on the Taguchi method. *Expert Syst. Appl.* **2012**, *39*, 6309–6319.
6.  Javanmardi, S.; Shojafar, M.; Amendola, D.; Cordeschi, N.; Liu, H.; Abraham, A. Hybrid Job Scheduling Algorithm for Cloud Computing Environment. In *Proceedings of the Fifth International Conference on Innovations in Bio-Inspired Computing and Applications IBICA 2014*; Komer, P., Abraham, A., Snasel, V., Eds.; Advances in Intelligent Systems and Computing Book Series; Springer: Berlin/Heidelberg, Germany, 2014; Volume 303, pp. 43–52.
7.  Ribas, I.; Leisten, R.; Framiñan, J.M. Review and classification of hybrid flow shop scheduling problems from a production system and a solutions procedure perspective. *Comput. Oper. Res.* **2010**, *37*, 1439–1454. [CrossRef]
8.  Luo, H.; Du, B.; Huang, G.; Chen, H.; Li, X. Hybrid flow shop scheduling considering machine electricity consumption cost. *Int. J. Prod. Econ.* **2013**, *146*, 423–439. [CrossRef]
9.  Mori, M.; Fujishima, M.; Inamasu, Y.; Oda, Y. A study on energy efficiency improvement for machine tools. *CIRP Ann. Manuf. Technol.* **2011**, *60*, 145–148. [CrossRef]
10. Blum, C.; Chiong, R.; Clerc, M.; De Jong, K.; Michalewicz, Z.; Neri, F.; Weise, T. Evolutionary optimization. In *Variants of Evolutionary Algorithms for Real-World Applications*; Chiong, R., Weise, T., Michalewicz, Z., Eds.; Springer: Berlin/Heidelberg, Germany, 2012; pp. 1–29.
11. Liu, Y.; Dong, H.; Lohse, N.; Petrovic, S. Reducing environmental impact of production during a rolling blackout policy—A multi-objective schedule optimisation approach. *J. Clean. Prod.* **2015**, *102*, 418–427. [CrossRef]
12. Nilakantan, J.; Huang, G.; Ponnambalam, S. An investigation on minimizing cycle time and total energy consumption in robotic assembly line systems. *J. Clean. Prod.* **2015**, *90*, 311–325. [CrossRef]
13. Wang, S.; Lu, X.; Li, X.; Li, W. A systematic approach of process planning and scheduling optimization for sustainable machining. *J. Clean. Prod.* **2015**, *87*, 914–929. [CrossRef]
14. Liu, C.; Huang, D. Reduction of power consumption and carbon footprints by applying multi-objective optimisation via genetic algorithms. *Int. J. Prod. Res.* **2014**, *52*, 337–352. [CrossRef]

15. May, G.; Stahl, B.; Taisch, M.; Prabhu, V. Multi-objective genetic algorithm for energy-efficient job shop scheduling. *Int. J. Prod. Res.* **2015**, *53*, 7071–7089. [CrossRef]
16. Zhang, R.; Chiong, R. Solving the energy-efficient job shop scheduling problem: A multi-objective genetic algorithm with enhanced local search for minimizing the total weighted tardiness and total energy consumption. *J. Clean. Prod.* **2016**, *112*, 3361–3375. [CrossRef]
17. Mouzon, G.; Yildirim, M.B.; Twomey, J. Operational methods for minimization of energy consumption of manufacturing equipment. *Int. J. Prod. Res.* **2007**, *45*, 4247–4271. [CrossRef]
18. Dai, M.; Tang, D.; Giret, A.; Salido, M.A.; Li, W.D. Energy-efficient scheduling for a flexible flow shop using an improved genetic-simulated annealing algorithm. *Robot. Comput. Integr. Manuf.* **2013**, *29*, 418–429. [CrossRef]
19. Mansouri, S.A.; Aktas, E.; Besikci, U. Green scheduling of a two-machine flowshop: Trade-off between makespan and energy consumption. *Eur. J. Oper. Res.* **2016**, *248*, 772–788. [CrossRef]
20. Hecker, F.T.; Hussein, W.B.; Paquet-Durand, O.; Hussein, M.A.; Becker, T. A case study on using evolutionary algorithms to optimize bakery production planning. *Expert Syst. Appl.* **2013**, *40*, 6837–6847. [CrossRef]
21. Hecker, F.T.; Stanke, M.; Becker, T.; Hitzmann, B. Application of a modified GA, ACO and a random search procedure to solve the production scheduling of a case study bakery. *Expert Syst. Appl.* **2014**, *41*, 5882–5891. [CrossRef]
22. Fonseca, C.M.; Fleming, P.J. An overview of evolutionary algorithms in multiobjective optimization. *Evol. Comput.* **1995**, *3*, 1–16. [CrossRef]
23. Hosseinabadi, A.A.R.; Siar, H.; Shamshirband, S.; Shojafar, M.; Nasir, M.H.N.M. Using the gravitational emulation local search algorithm to solve the multi-objective flexible dynamic job shop scheduling problem in Small and Medium Enterprises. *Ann. Oper. Res.* **2015**, *229*, 451–474. [CrossRef]
24. Deb, K.; Pratap, A.; Agarwal, S.; Meyarivan, T. A fast and elitist multiobjective genetic algorithm: NSGA-II. *IEEE Trans. Evol. Comput.* **2002**, *6*, 182–197. [CrossRef]
25. Jaimes, A.L.; Coello, C.A.C. Interactive Approaches Applied to Multiobjective Evolutionary Algorithms. In *Multicriteria Decision Aid and Artificial Intelligence*; Doumpos, M., Grigoroudis, E., Eds.; John Wiley & Sons, Ltd.: Chichester, UK, 2013. [CrossRef]
26. Lu, L.; Anderson-Cook, C.; Lin, D. Optimal designed experiments using a pareto front search for focused preference of multiple objectives. *Comput. Stat. Data Anal.* **2014**, *71*, 1178–1192. [CrossRef]
27. Wagner, T.; Trautmann, H. Integration of preferences in hypervolume-based multiobjective evolutionary algorithms by means of desirability functions. *IEEE Trans. Evol. Comput.* **2010**, *14*, 688–701. [CrossRef]
28. Baker, K.R. *Introduction to Sequencing and Scheduling*; John Wiley & Sons, Ltd.: New York, NY, USA, 1974.
29. Pinedo, M. *Scheduling: Theory, Algorithms, and Systems*; Prentice Hall: Upper Saddle River, NJ, USA, 2002; 586p.
30. Abyaneh, S.H.; Zandieh, M. Bi-objective hybrid flow shop scheduling with sequence-dependent setup times and limited buffers. *Int. J. Adv. Manuf. Technol.* **2012**, *58*, 309–325. [CrossRef]
31. Graham, R.L.; Lawler, E.L.; Lenstra, J.K.; Kan, A.H.G.R. Optimization and approximation in deterministic sequencing and scheduling: A survey. *Ann. Discret. Math.* **1979**, *5*, 287–326. [CrossRef]
32. Larrañaga, P.; Kuijpers, C.M.H.; Murga, R.H.; Inza, I.; Dizdarevic, S. Genetic algorithms for the travelling salesman problem: A review of representations and operators. *Artif. Intell. Rev.* **1999**, *13*, 129–170. [CrossRef]
33. Tan, K.C.; Lee, L.H.; Zhu, Q.L.; Ou, K. Heuristic methods for vehicle routing problem with time windows. *Artif. Intell. Eng.* **2001**, *15*, 281–295. [CrossRef]
34. Gog, A.; Chira, C. *Comparative Analysis of Recombination Operators in Genetic Algorithms for the Travelling Salesman Problem*; Springer: Berlin/Heidelberg, Germany, 2011; pp. 10–17.
35. Hwang, H. An improved model for vehicle routing problem with time constraint based on genetic algorithm. *Comput. Ind. Eng.* **2002**, *42*, 361–369. [CrossRef]
36. Prins, C. A simple and effective evolutionary algorithm for the vehicle routing problem. *Comput. Oper. Res.* **2004**, *31*, 1985–2002. [CrossRef]
37. Starkweather, T.; McDaniel, S.; Mathias, K.E.; Whitley, C. A comparison of genetic sequencing operators. In Proceedings of the 4th International Conference on Genetic Algorithms; Belew, R.K., Booker, L.B., Eds.; Morgan Kaufmann: San Francisco, CA, USA, 1991; pp. 69–76.
38. The R Project for Statistical Computing. Available online: https://www.r-project.org/ (accessed on 4 March 2017).

39. Montgomery, D.C.; Runger, G.C. *Applied Statistics and Probability for Engineers*; John Wiley & Sons, Inc.: New York, NY, USA, 1994.

40. Rahimi-Vahed, A.; Dangchi, M.; Rafiei, H.; Salimi, E. A novel hybrid multi-objective shuffled frog-leaping algorithm for a bi-criteria permutation flow shop scheduling problem. *Int. J. Adv. Manuf. Technol.* **2009**, *41*, 1227–1239. [CrossRef]

**MDPI**

*Article*

# A Heuristic Approach to Solving the Train Traffic Re-Scheduling Problem in Real Time

**Omid Gholami \* and Johanna Törnquist Krasemann**

Department of Computer Science and Engineering, Blekinge Institute of Technology, Valhallavägen 1,
371 79 Karlskrona, Sweden; johanna.tornquist.krasemann@bth.se
\* Correspondence: omid.gholami@bth.se; Tel.: +46-(0)455-385-845

Received: 28 February 2018; Accepted: 12 April 2018; Published: 21 April 2018

**Abstract:** Effectiveness in managing disturbances and disruptions in railway traffic networks, when they inevitably do occur, is a significant challenge, both from a practical and theoretical perspective. In this paper, we propose a heuristic approach for solving the real-time train traffic re-scheduling problem. This problem is here interpreted as a blocking job-shop scheduling problem, and a hybrid of the mixed graph and alternative graph is used for modelling the infrastructure and traffic dynamics on a mesoscopic level. A heuristic algorithm is developed and applied to resolve the conflicts by re-timing, re-ordering, and locally re-routing the trains. A part of the Southern Swedish railway network from Karlskrona centre to Malmö city is considered for an experimental performance assessment of the approach. The network consists of 290 block sections, and for a one-hour time horizon with around 80 active trains, the algorithm generates a solution in less than ten seconds. A benchmark with the corresponding mixed-integer program formulation, solved by commercial state-of-the-art solver Gurobi, is also conducted to assess the optimality of the generated solutions.

**Keywords:** railway traffic; disturbance management; real-time re-scheduling; job-shop scheduling; optimization; alternative graph

---

## 1. Introduction

The definitions of "system performance" and "quality of service" in railway traffic and transportation vary, but more recent reports, e.g., from the Boston Consultancy Group [1] and the European Commission [2], indicate that many European countries face challenges with regard to the reliability and punctuality of rail services. Several different factors contribute to these challenges, where the frequency and magnitude of disruptions and disturbances is one factor. The effectiveness in managing the disturbances and disruptions in railway traffic networks when they inevitably do occur is another aspect to consider. In this paper, we focus on the latter, and more specifically, on the problem of real-time train traffic re-scheduling, also referred to as train dispatching [3].

In general, the problem of re-scheduling train traffic in real-time is known to be a hard problem to solve by expert traffic dispatchers in practice, as well as by state-of-the-art scheduling software. There is thus often a trade-off that needs to be made regarding permitted computation time and expected solution quality. Heuristic approaches, or approaches based on, e.g., discrete-event simulation, are often rather quick, since they work in a greedy manner, but sometimes they fail to deliver a reasonable solution or even end up in deadlock. Exact approaches, based on, e.g., mixed-integer programming models solved by branch-and-cut algorithms are, on the other hand, less greedy, but can be very time-consuming, especially if the search space is large, and suffer from significant redundancy. Another aspect concerns the selected level of granularity of the traffic and infrastructure model, and this may indirectly affect the computation time and the quality of the solution as well.

Over the latest three decades, a variety of algorithms and methods have been proposed to solve the train traffic re-scheduling (or, train dispatching) problem, see [3–6] for more recent surveys.

These previously proposed approaches have different strengths and limitations, depending on the intended application and context in mind. In Section 2, we provide a more comprehensive summary of state-of-the-art methods addressing this problem.

In this paper, we propose a heuristic approach to solve the real-time train traffic re-scheduling problem. We view the problem as a blocking/no-wait, parallel-machine job-shop scheduling problem [3]. A job corresponds to a train's path along a pre-determined sequence of track sections (i.e., machines). Each part of that path corresponds to an operation. "blocking/no-wait" constraint refers to that there is no possibility to finish an operation if the next operation of the same job cannot get access to the required machine for processing, i.e. there is no storage space between machines that allows jobs and their current operation to wait for an occupied machine to become available. That is, when an operation of a job is completed, the next operation of that job must immediately start. This corresponds to the train needing at all times to have explicit access to a track section (independent of whether the train is moving or waiting). This problem is here modeled as a graph, where the graph is a hybrid between a mixed graph and an alternative graph. The benefit of using a hybrid graph is the possibility of reducing the number of required computations when constructing the graph. The problem is then solved by a heuristic algorithm. The proposed mixed graph model is discussed in Section 3, and the proposed heuristic algorithm is presented in detail in Section 4. The performance of the proposed approach is assessed in an experimental study, where the approach has been applied to solve a number of disturbance scenarios for a dense part of the Swedish southern railway network system. The performance assessment also includes a benchmark with the commercial optimization software Gurobi (v 6.5.1), using the corresponding mixed integer programming (MIP) model (which is presented in Appendix A). This experimental study is presented in Section 5. Section 6 presents some conclusions from the experimental study and provides pointers to future work.

## 2. Related Work

Szpigel [7] was, in 1973, one of the pioneers, adapting the job-shop scheduling (JSS) problem formulation for the train scheduling problem. This was later developed by other researchers such as D'Ariano et al. [8], Khosravi et al. [9], Liu and Kozan [10], Mascis and Pacciarelli [11], and Oliveira and Smith [12].

The JSS paradigm is also explored in the mixed-integer linear programming (MILP) approach proposed by Törnquist Krasemann and Persson in 2007 [13], where the primary focus was on creating and evaluating strategies for reducing the search space via model re-formulations. Different problem formulations were experimentally evaluated in several disturbance scenarios occurring on a double-tracked network with bi-directional traffic, and solved by commercial software for a time horizon of 60–90 min. The effectiveness of the approach and the different strategies were shown to be highly dependent on the size of the problem and type of disturbance.

A heuristic approach based on a MILP formulation was also proposed in [14]. A time horizon of 30 to 60 min was defined for testing the disturbing scenarios. The commercial solver CPLEX was used to solve the problem, with an allowed computation time of 180 s. Significant effort was made to tune the heuristic algorithm, and the results were compared with exact ones. For up to a 40 min time horizon, the algorithm generated good results.

A MIP model was also applied on a double-track railway corridor, where the focus is on local re-routing of trains to allow for bi-directional traffic and special constraints to respect the safety requirements [15].

As already mentioned, the real-time train traffic re-scheduling problem is a hard problem to solve. This is often an effect of the large solution space, which may be difficult to efficiently navigate in due to significant redundancy. That is, depending on the problem formulation and selected objective function, particularly, optimization solvers may find multiple equally good solutions and also have difficulties finding strong bounds. Finding optimal, or near-optimal, solutions may thus be very time-consuming.

In order to reduce the computational time, one can partition the problem into a set of different sub-problems and solve them in parallel, in a distributed manner with some sort of coordination mechanism. The partitioning can be done in space (i.e., by dividing the infrastructure, or the associated constraints, into separate parts), or in time (i.e., rolling-time horizon). Corman et al. [16–18] proposed and benchmarked a centralized and a distributed re-scheduling approach for the management of a part of the Dutch railway network based on a branch-and-bound approach. These variations come from various static and dynamic priorities that are considered for scheduling. They also worked on a coordination system between multiple dispatching areas. The aim was to achieve a globally optimal solution by combining local solutions. Some constraints are defined for the border area and a branch-and-bound algorithm solves the coordinator problem.

Another approach for reducing the complexity of the problem is to use classical decomposition techniques, which are becoming more frequently used to overcome the complexity (and the associated long computation times) of the train re-scheduling problem when using exact approaches. Lamorgese and Mannino proposed an exact method, which decomposes and solves the problem with a master-slave procedure using column generation [19]. The master is associated with the line traffic control problem and the slaves with the station traffic control.

Decomposition has also been used to reduce the complexity of the associated MIP model of the train re-scheduling problem [20]. An iterative Lagrangian relaxation algorithm is used to solve the problem.

Tormo et al. [21] have investigated different disturbance management methods, for the purpose of supporting railway dispatchers with the assessment of the appropriateness of each method for different problems in the railway operations. They categorized the disturbance management approaches into three subcategories. The first one is to minimize the overall delays at the network level. The second one is the evaluation of disturbances based on the severity, and the third one is to assume the equal importance of each incident. An incremental heuristic method has been compared with a two-phase heuristic approach in [22]. At each processing step, partial resource allocation and partial scheduling are repeated until a complete solution is generated. In a two-phase heuristic approach, the algorithm creates the routes for all the trains with a complete labelling procedure in phase one, and then solves the problem with an approximation algorithm in phase two.

The use of task parallelization to reduce computation time was explored by Bettinelli et al. [23], who proposed a parallel algorithm for the train scheduling problem. The problem is modelled by a graph. For conflict resolution, some dispatching rules are applied. The algorithm solves the problem in an iterative, greedy manner, and to decrease the computational time, the tasks are parallelized. The algorithm enhances the quality of the solutions using neighbourhood search and tabu search. The neighbourhood search has a significant impact on the quality of generated solutions, according to the authors.

Another interesting study of different approaches to variable neighbourhood search and quality assessment is available in [24]. Local search techniques are also used in [25], where the problem is modeled as a hybrid job-shop scheduling problem.

In [10], the train scheduling problem is formulated as a blocking parallel-machine job-shop scheduling problem and modeled as an alternative graph. A heuristic algorithm, referred to as "the feasibility satisfaction procedure", is proposed to resolve the conflicts. The model attempts to consider train length, upgrading the track sections, increasing the trains' speed on the identified tardy section and shortening the local runtime of each train on bottleneck sections. The application was designed for the train scheduling problem and is not considering re-scheduling. In the first step, the problem is solved by a shifting bottleneck algorithm without considering the blocking condition. In the next step, the blocking condition is satisfied by using an alternative graph model.

Khosravi et al. [9] also address the train scheduling and re-scheduling problem using a job-shop scheduling problem formulation, and the problem is solved using a shifting bottleneck approach. Decomposition is used to convert the problem into several single-machine problems. Different variations of the method are considered for solving the single-machine problems. The algorithm benefits from re-timing and re-ordering trains, but no re-routing of trains is performed.

A job-shop scheduling approach is also used to insert additional trains to an existing timetable without introducing significant consecutive delays to the already-scheduled trains [26]. A branch and bound algorithm and an iterative re-ordering strategy are proposed to solve this problem in real-time. Re-timing and re-ordering are conducted in a greedy manner. They have suggested a new bound for the branch and bound algorithm and an iterative re-ordering algorithm, which helps to find solutions in an acceptable computational time.

The main challenge in this study is to—inspired by previous promising research work—develop, apply and evaluate an effective algorithm that, within 10 s, can find sufficiently good solutions to a number of different types of relevant disturbance scenarios for a medium-sized sub-network and a time horizon of 60–90 min. A hybrid of the mixed graph and alternative graph (a derivation of the disjunctive graph, which satisfies the blocking constraint [11]) formulation is used for modelling the train traffic re-scheduling problem. This graph allows us to model the problem with a minimum number of arcs, which is expected to improve the efficiency of the algorithm. The model of the infrastructure and traffic dynamics is on the mesoscopic level, which considers block sections, clear time and headway distance, based on static parameter values. A heuristic algorithm is developed for conflict resolution. This algorithm is an extension of the algorithm developed by Gholami and Sotskov for hybrid job-shop scheduling problems [27]. The extended algorithm provides an effective strategy for visiting the conflicting operations and the algorithm makes use of re-timing, re-ordering, and local re-routing of trains while minimizing train delays.

## 3. Modelling the Job-Shop Scheduling Problem Using a Graph

A well-known and efficient way of modelling the job-shop scheduling problem is to use graphs. Graph theory is a well-studied area in the computational sciences, and it provides efficient algorithms and data structures for implementation. These features make the graph a suitable candidate for modeling the problem addressed in this paper.

In a job-shop scheduling problem, there are $n$ jobs that have to be served by $m$ different types of machines. Each job $j \in J$ has its own sequence of operations $O_j = \{o_{j,1}, o_{j,2}, \ldots, o_{j,m}\}$ to be served by different predefined machines from the set $M$. The jobs have to be served by machines exclusively. The job-shop scheduling problem can be formulated using a mixed graph model $G = (O, C, D)$, or equivalently by a disjunctive graph model. Let $O$ denote the set of all operations to be executed by a set of different machines $M$. The set $C$ represents a predefined sequence of operations for each job $j$ to visit machines from the set $M$. The two operations $o_{j,i}$ and $o_{j',i'}$ which have to be executed on the same machine $m \in M$, cannot be processed simultaneously. This restriction is modelled by an edge $[o_{j,i}, o_{j',i'}] \in D$ (if a mixed graph is used) or equivalently by pairs of disjunctive arcs $\{(o_{j,i}, o_{j',i'})$, and $(o_{j',i'}, o_{j,i})\}$ (if a disjunctive graph is used). In this paper, the mixed graph is used for modelling the primary state of the problem. Two vertices, $o_s$ and $o_d$ as the source and destination vertices, respectively, are added to the model to transform it to a directed acyclic graph.

Using the terms "arc" and "edge" may be confusing. Edges may be directed or undirected; undirected edges are also called "lines", and directed edges are called "arcs" or "arrows". In this paper, the term "edge" is used for undirected edges, and the term "arc" is used for directed edges.

To serve the operations in a job-shop scheduling problem, only one instance of each machine of type $m$ is available. In a hybrid job-shop as a derivation of the job-shop scheduling problem, a set $M_m$ ($|M_m| \geq 1$ ) is available to serve the operations [28]. The notation $\langle m, u \rangle$ is used to indicate a specific machine or resource $u$ from a set $M_m$ if it is necessary. These uniform parallel machine sets increase the throughput and provide more flexibility in the scheduling process.

A train is here synonymous with a job in the job-shop scheduling problem. A train route from the source station to the destination can be considered to be a sequence of operations for a job $j$. Each railroad section is synonymous to a machine $m$. In Table 1, below, we introduce the notations that are used to describe and define the hybrid graph model, algorithmic approach, and corresponding MIP model.

**Table 1.** The notations used for sets, indices, parameters, and variables.

| Sets and Indices | Description |
| --- | --- |
| $M$ | Set of all railway sections (i.e., machines). |
| $m$ | The index used to denote a specific section in the set $M$. |
| $M_m$ | The sub-set of tracks and platforms that belongs to section $m$. (i.e., parallel machines) |
| $m, u$ | A track/platform/resource instance number $u$ from a set $M_m$. |
| $J$ | Set of all trains (i.e., jobs). |
| $j$ | The index used to denote a specific train. |
| $O$ | Set of all train events (i.e., operations). |
| $O_j$ | Set of all events that belong to train $j$. |
| $O^m$ | Set of all events to be scheduled on a section $m$. |
| $O^{m,u}$ | Set of all events to be scheduled on track $u$ of section $m$. |
| $i$ | The index used to denote a specific train event $i$. |
| $o_{j,i}$ | The symbol used to denote the event $i$ which belongs to train $j$. |
| $\langle j, i, m, u \rangle$ | Tuple which refers to train $j$ and its event $i$ which occurs in section $m$ and scheduled for track $u$. When the section is single line $u$ will be ignored. |
| $(\prime)$ | Prime symbol used to distinguish between two instances (i.e., job $j$ and $j'$ ). |

| Parameters | Description |
| --- | --- |
| $c_m$ | The required minimum clear time that must pass after a train has released the assigned track $u$ of section $m$ and before another train may enter track $u$ of section $m$. |
| $h_m$ | The minimum headway time distance that is required between trains (head-head and tail-tail) that run in sequence on the same track $u$ of section $m$. |
| $d_{j,i}$ | The minimum running time, or dwell time, of event $i$ that belongs to train $j$. |
| $b_{j,i}^{initial}$ | This parameter specifies the initial starting time of event $i$ that belongs to train $j$. |
| $e_{j,i}^{initial}$ | This parameter specifies the initial completion time of event $i$ that belongs to train $j$. |
| $ps_{j,i}$ | This parameter indicates if event $i$ includes a planned stop at the associated segment. |

| Variables | Description |
| --- | --- |
| $x_{j,i}^{begin}$ | The re-scheduled starting time of event $i$ that belongs to train $j$. |
| $x_{j,i}^{end}$ | The re-scheduled completion time of event $i$ that belongs to train $j$. |
| $t_{j,i}$ | The tardiness (i.e., delay for train $j$ to complete event $i$ ). |
| $z_{j,i}$ | Delay of event $i, i \in O$, exceeding $\mu$ time units, which is set to three minutes here. |
| $q_{j,i,u}$ | A binary variable which is 1, if the event $i$ uses track $u$. |
| $\gamma_{j,i,j',i'}$ | A binary variable which is 1, if the event $i$ occurs before event $i\prime$. |
| $\lambda_{j,i,j',i'}$ | A binary variable which is 1, if the event $i$ is rescheduled to occur after event $i\prime$. |
| $buf_{j,i}$ | Remaining buffer time for train $j$ to complete an event $\langle j, i \rangle$. |
| $TFD_j$ | The delay for train $j$ once it reaches its final destination, i.e., $t_{j,last}$ which corresponds to the delay when completing its last event. |

To solve the job-shop scheduling problem modelled by a graph, for every two conflicting operations the algorithm is responsible for determining the best execution order to reduce the objective function value. A potential conflict occurs when for two events $o_{j,i}$ and $o_{j',i'}$ belonging to jobs $j$ and $j'$ respectively, the following criteria become true:

$$x_{j,i}^{begin} < x_{j',i'}^{begin} \quad \text{and} \quad x_{j',i'}^{end} \leq x_{j,i}^{begin} + d_{j',i'} \tag{1}$$

The algorithm is in charge of replacing the set of edges $D$ with a subset of arcs $D'$, which defines the order to be served by a resource. As a result, the set of edges $D$ will be substituted by a selected subset of directed arcs $D'$, and the mixed graph $G = (O, C, D)$ will be transformed into a digraph $G' = (O, C \cup D', \varnothing)$.

In this paper, a multi-track railway traffic network will be considered as a job-shop scheduling problem with parallel machines (hybrid job-shop) at each stage.

## 4. A Heuristic Algorithm for Conflict Resolution in the Mixed Graph $G$

When a disturbance occurs, a graph $G$ has to be generated from the ongoing and remaining events for all the trains within the predefined time horizon. Later, the algorithm has to resolve the conflicts between trains affected by disturbance and potential knock-on effects. To resolve all the conflicts in a graph $G$, the algorithm has to visit all conflict pairs, and by replacing the edges with directed arcs, clarify the priorities for using the resources (i.e., the sections and associated tracks). In this research, the objective is to minimize the sum of the total final delay that each train experiences at its final destination (i.e., $TFD_j = Max\left(0, x_{j,last}^{end} - e_{j,last}^{initial}\right), j \in J$ ). We also measure the corresponding when the delay threshold is three minutes, i.e., only delays larger than three minutes are considered, i.e., $\left(\sum TFD_j^{+3}, if \ TFD_j \geq 3 \ min\right)$. The reason why the threshold of three minutes is selected is that this threshold is used to log train delays in Sweden. Delays smaller or equal to three minutes are not registered in the system of the responsible authority, Trafikverket (the Swedish National Transport Administration). Furthermore, delays larger than three minutes may cause train connections to be missed, while below three minutes the consequences for transfers in the studied regional-national train system are not that likely.

To visit the conflicting operations, we followed the strategy presented by Gholami and Sotskov for the hybrid job-shop scheduling problem [27]. In this strategy, a list of potential candidate operations for conflict resolution will be generated. This list is a collection of ongoing operations, or the first operation of the next trains. After any conflict resolution, the list will be updated, some candidates will be added and, if it is necessary, some will be deleted. An operation will be added to the list if the in-degree value of the related vertex becomes zero (the number of arc ends to a vertex is called the in-degree value of the vertex). The in-degree value decreases when the conflict resolution is done for a parent vertex. When the list is regenerated, the shortest release time of all vertices in the list will be calculated or updated again, and the minimum one will be selected for conflict resolution. A re-ordering may cause a vertex to be deleted from the list. By this approach, all the vertices will be visited only once, unless a re-ordering happens. This strategy decreases the computational time of the algorithm. The second benefit of this approach is that adding a new arc $(o_{j,i}, \ o_{j\prime,i\prime})$ does not affect any previously visited vertices unless a re-ordering occurs. Dynamic update of data is possible due to this feature. In Algorithm 1, the pseudocode for conflict resolution is presented.

The algorithm starts from a vertex $o_s$ and finds all the neighbourhood vertices and adds them to a list of candidate vertices. At each iteration, a vertex with a minimum release time will be selected from the candidate list for conflict resolution. If there is a conflict between the candidate event and those events that previously used that track or section (*checkList*), a local re-routing will be applied. For local re-routing, a platform or track with minimum availability time will be selected. Meanwhile, if there is still a conflict, the algorithm tries to use re-timing or re-ordering. For conflict resolution between the candidate operation $o_{j\prime,i\prime}$ and an operation $o_{j,i}$ from the check list, a new arc will be added to the graph $G$. After adding the arc $(o_{j,i}, o_{j\prime,i\prime})$, the start time $(x_{j\prime,i\prime}^{begin})$ of operation $o_{j\prime,i\prime}$ will be postponed to the finishing time $(x_{j,i}^{end})$ of operation $o_{j,i}$ on the conflicting track or platform. If the algorithm adds the opposite arc $(o_{j\prime,i\prime}, o_{j,i})$, then operation $o_{j\prime,i\prime}$ have to be served before the operation $o_{j,i}$, which means that the predefined order of the trains for using this section is changed. This condition occurs when the algorithm tries to prevent deadlocks (unfeasible solutions), or the operation $o_{j,i}$ has a high tardiness (a delayed train). Six dispatching rules are provided for the conflict resolution process. Table 2 is a description of those dispatching rules.

In the train scheduling problem, a train blocks its current section until obtaining the next section or block. Therefore, swapping is not possible in the train scheduling problem. Accordingly, the *addArc* function has to satisfy the blocking condition in the graph $G$ (the alternative graph approach). To fulfil the blocking condition, instead of adding an arc from the operation $o_{j,i}$ to the operation $o_{j\prime,i\prime}$, an arc from the next operation $o_{j,i+1}$ with a zero weight will be added to the operation $o_{j\prime,i\prime}$. This directed arc means that the operation $o_{j\prime,i\prime}$ cannot proceed with its execution on the machine $m$ until the operation $o_{j,i+1}$ starts $(x_{j\prime,i\prime}^{begin} \geq x_{j,i+1}^{begin}$, meanwhile, the operation $o_{j,i}$ is finished). This condition blocks job $j\prime$ on the previous machine, even if it is completed. If the operation $o_{j,i}$ is the last operation of job $j$, then an arc with a weight $d_{j,i}$ has to be added.

---

**Algorithm 1.** The pseudocode for conflict resolution strategy

---

### Heuristic Algorithm for Conflict Resolution in the Mixed Graph G

---

**Require**: The weighted mixed graph $G = (O, C, D)$;

```
candidList = findNeighbours(o_s);
current = findMinimumReleaseTime(candidList);
while (current ≠ o_d);
        checkList = findConflictOperations(current.machineNumber);
        if (conflictExist(current, checkList)) /*local re-routing*/
                vt = minimumVacantTrack(current.machineNumber);
          modifyGraph(G, current, vt);
          checkList = findConflictOperations(vt);
    end_if
    for (node cl: checkList) /*conflict resolution, re-timing, re-ordering*/
        a,b = findBestOrder(current, cl);
        if (not reachable(b, a))
            addArc(a, b);
            updateData(G,a);
        else_if (not reachable(a,b))
                addArc(b, a);
                updateData(G,b);
          else
                checkFeasibility(G);
          end_if
        end_if
        end_for
        candidList += findNeighbours(current);
        candidList -= current;
        current = findMinimumReleaseTime(candidList);
end_while
```

---

**Table 2.** Description of dispatching rules used for conflict resolution.

| Conflict Resolution Strategy | Description |
|---|---|
| 1. Minimum release time goes first | The train with the earliest initial start time ($b_{j,i}^{initial}$) goes first if no deadlock occurs. |
| 2. More delay goes first | If there is a conflict between two trains, the one with the largest tardiness (delay) goes first. The tardiness is calculated as $t_{j,i} = Max\left(0, x_{j,i}^{begin} - b_{j,i}^{initial}\right)$. |
| 3. Less real buffer time goes first | The train with the smallest buffer time goes first. Buffer time is defined as a subtraction of initial ending time and real finishing time $\left(buf_{j,i} = e_{j,i}^{initial} - \left(x_{j,i}^{begin} + d_{j,i}\right)\right)$ for two operations to be scheduled on the same, occupied machine. |
| 4. Less programmed buffer time goes first | The train with smallest buffer time goes first. Buffer time is defined as a subtraction of initial ending time and programmed ending time $\left(buf_{j,i} = e_{j,i}^{initial} - \left(b_{j,i}^{initial} + d_{j,i}\right)\right)$ for two operations to be scheduled on the same, occupied machine. |
| 5. Less total buffer goes first | The train with smallest total buffer time goes first. Total buffer time is defined as a summation of programmed buffer times until the destination point for the trains, i.e., $\left(\sum_{k=i}^{last} buf_{j,k}\right)$. |
| 6. Less total processing time | The train with smallest running time to get to the destination goes first (i.e., the minimum total processing time). The total processing time is defined as a summation of required time to pass each section, for a train, i.e., $\left(\sum_{k=i}^{last} d_{j,k}\right)$. |

Figures 1 and 2 illustrate the difference between the mixed graph and alternative graph models for conflict resolution. There are two conflicting trains in the same path on the sections $m_1$ and $m_2$.

The train $j_1$ is planned on the sections $m_1, m_2$, and $m_3$. The route for the train $j_2$ is $m_1, m_2$, and $m_4$. The events $o_{1,1}$ and $o_{2,1}$ have a conflict on the machines $m_1$ and events $o_{1,2}$ and $o_{2,2}$ on the machine $m_2$. To present a conflict between the two operations in the mixed graph approach only one edge is required ($\left[o_{j,i}, o_{j',i'}\right] \in D$) (See Figure 1a). While in the alternative graph model two arcs are needed, denoted as ($o_{j,i+1}, o_{j',i'}$) and ($o_{j',i'+1} o_{j,i}$) (See Figure 1b).

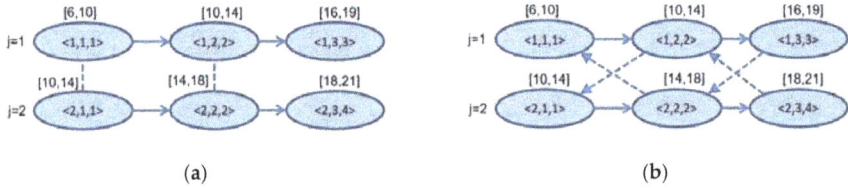

(a)　　　　　　　　　　　　　(b)

**Figure 1.** Presenting a conflict in a graph $G$ : (**a**) using an edge in the mixed graph model to present a conflict between the two operations; (**b**) using two arcs in the alternative graph model to present a conflict between two operations.

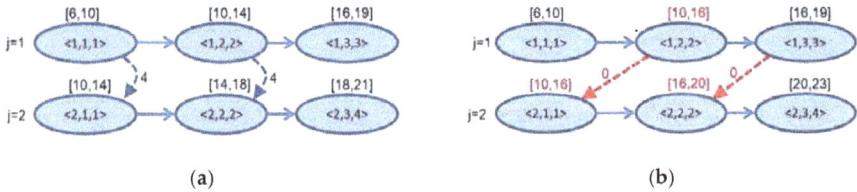

(a)　　　　　　　　　　　　　(b)

**Figure 2.** Resolving the conflicts on the machine $m_1$ and $m_2$ in a graph $G$ : (**a**) conflict resolution using the mixed graph model; (**b**) conflict resolution using the alternative graph model to support the blocking condition.

After conflict resolution in the mixed graph mode (Figure 2a), the operation $o_{2,2}$ could start its processing in minute 14 on the machine $m_2$ (immediately after the completion of the operation $o_{1,2}$). While in the alternative graph case (Figure 2b), the starting time of events $o_{2,2}$ is postponed to minute 16, which job $j_1$ has started the event $o_{1,3}$. The job $j_1$ has waited for 2 min for the machine $m_3$ and is blocked on the machine $m_2$.

In this paper, we propose an approach that is a hybrid between the mixed graph and alternative graph. This hybrid approach makes use of (1) a mixed graph formulation to represent the non-rescheduled timetable in the initial stage of the solution process, and (2) an alternative graph approach when the timetable is to be re-scheduled. The reasons for doing so are as follows:

One way to speed up the algorithm is to reduce the number of edges and arcs in the graph $G$. This reduction leads to a lower number of neighbourhood vertices needing to be handled, less feasibility and constraint checking being required, less computational time to update the data at each stage, and a faster traverse in the graph. As the mixed graph model uses one edge to present a conflict between two vertices and alternative graph needs two arcs, the non-rescheduled timetable in the initial stage uses the mixed graph approach (See Figure 1a). However, after the conflict resolution, the algorithm uses the alternative graph approach (adding an arc from next operation) to satisfy the blocking condition (See Figure 2b). This means that for the unsolved part, the graph is modelled like a mixed graph, and for the solved part, it follows the alternative graph modelling approach.

For safety reasons, the trains have to obey a minimum clear time and headway time distance [23]. The clear time ($c_m$) is the minimum time that a train $j'$ must wait before entering a section $m$, which the train $j$ just left. This time interval between completion of train $j$ and start time of the train $j'$ can be modelled by changing the weight of the priority arc from zero to $c_m$. The headway time distance

$(h_m)$ is the minimum time between two trains $j$ and $j'$ running in the same direction and on the same track of section $m$. The headway distance is the minimum time interval between the start times, and end times respectively, of two consecutive trains, $j$ and $j'$, which can be modelled by adding a new arc with a weight $h_m$ from operation $o_{j,i}$ to $o_{j',i'}$. The *addArc* procedure is responsible for adopting the clear time and headway distance.

Figure 3, is an illustration of conflict resolution, considering the clear time and headway distance for the alternative graph model. In Figure 3a, the starting time of operation $o_{2,1}$ is increased to minute 11, because, the operation $o_{1,1}$ had 4 min of processing time and 1 min of clear time. In Figure 3b, the headway distance is $h_1 = 7$, which means that the start time of operation $o_{2,1}$ must be at least 7 min after the start time of the operation $o_{1,1}$. The starting time of the operation $o_{2,1}$ is increased to minute 13.

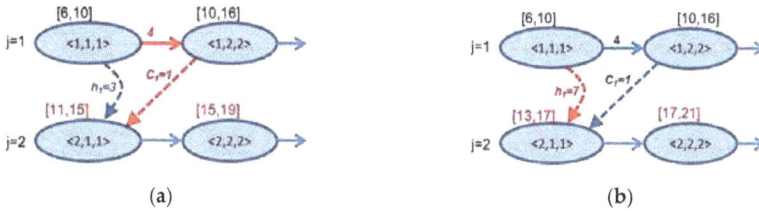

**Figure 3.** Adding the clear time and headway time distance to the alternative graph: (**a**) the starting time of operation $o_{2,1}$ is increased due to the clear time; (**b**) the starting time of operation $o_{2,1}$ is increased due to the headway distance.

For a decision-making procedure, the algorithm needs some data, such as release time, or buffer time. After any re-timing, re-ordering or local re-routing, these data change. The experiments demonstrate that recalculation of these data has a significant effect on the computational time of the algorithm. Our special visiting approach enables a dynamic update of data for those vertices that will be affected.

After adding a new directed arc from the operation $o_{j,i}$ to the operation $o_{j',i'}$, by considering the commercial stop time (passenger trains are not allowed to leave a station before the initial completion time $e_{j,i}^{initial}$) the minimum start time of an operation $o_{j',i'}$ on a track $m.u$ will be

$$x_{j',i'}^{begin} = \begin{cases} \text{Max}(b_{j',i'}^{initial}, x_{j',i'-1}^{end}, \max_{j \in O_{l^{m,u}}}(x_{j,i}^{begin} + d_{j,i})) & \text{if } i'-1 \text{ was a commercial stop,} \\ \text{Max}(x_{j',i'-1}^{end}, \max_{j \in O_{l^{m,u}}}(x_{j,i}^{begin} + d_{j,i})) & \text{if } i'-1 \text{ was not a commercial stop,} \end{cases} \quad (2)$$

where $O_{l^{m,u}}$ is a set of all operations processed by the same resource $u$ from a set $M_m$ until now. Additionally, the ending time can be calculated as follows:

$$x_{j',i'}^{end} = \begin{cases} \max(e_{j',i'}^{initial}, x_{j',i'}^{begin} + d_{j',i'}) & \text{if } i' \text{ is a commercial stop.} \\ x_{j',i'}^{begin} + d_{j',i'} & \text{if } i' \text{ is not a commercial stop.} \end{cases} \quad (3)$$

Considering the clear time restriction, the starting time for an operation $o_{j',i'}$ will be calculated as follows:

$$x_{j',i'}^{begin} = \begin{cases} \text{Max}(b_{j',i'}^{initial}, x_{j',i'-1}^{end}, \max_{j \in O_{l^{m,u}}}(x_{j,i}^{end} + c_m)) & \text{if } i'-1 \text{ was a commercial stop.} \\ \text{Max}(x_{j',i'-1}^{end}, \max_{j \in O_{l^{m,u}}}(x_{j,i}^{end} + c_m)) & \text{if } i'-1 \text{ was not a commercial stop.} \end{cases} \quad (4)$$

The proposed starting time with consideration of headway distance will change to:

$$
x_{j',i'}^{begin} = \begin{cases} \text{Max}(b_{j',i'}^{initial}, x_{j',i'-1}^{end}, \max\limits_{j \in O_{l}m,u}\left(x_{j,i}^{end} + c_m, x_{j,i}^{begin} + h_s\right)) & \text{if } i' - 1 \text{ was a commercial stop.} \\ \text{Max}(x_{j',i'-1}^{end}, \max\limits_{j \in O_{l}m,u}\left(x_{j,i}^{end} + c_m, x_{j,i}^{begin} + h_s\right)) & \text{if } i' - 1 \text{ was not a commercial stop.} \end{cases} \quad (5)
$$

A recursive function is used in the *updateData* function, which updates the release time for all operations that are reachable by the operation $o_{j',i'}$ until $o_d$. The recursive function stops whenever the $x_{j',i'}^{begin}$ value does not change. By viewing the headway distance in the *updateData* function (Equation (5)), it is possible to ignore the extra arc for the headway distance (see Figure 3). Deletion of this extra arc, which was needed for considering the headway time distance, also helps to reduce the computational time.

Before adding a new arc, the algorithm applies the solution feasibility function. If, by adding a new arc, a circuit appears in the graph $G$, the generated solution is not feasible. A circuit in the graph is a sign of infeasible resource allocation. A circuit is a chain of two or more operations that are trying to exchange the machines with the next operations (swapping condition). To avoid a circuit, the following lemma is used:

**Lemma 1.** *In a directed acyclic graph $G$, by adding an arc from vertex a to b, a circuit appears if and only if a is reachable from vertex b in the graph $G$.*

If the reachability test confirms a circuit, the algorithm considers the opposite priority and applies the arc related to the opposite alternative approach (re-ordering). Mostly, by adding the opposite alternative arc, no circuit appears, but in rare cases when the algorithm generates a new circuit, the solution would be rejected.

## 5. Experimental Application and Performance Assessment

### 5.1. Experimental Set-Up

For the experimental evaluation, the approach was applied to a number of disturbance scenarios occurring within a selected sub-network in the south of Sweden, namely the railway stretch between Karlskrona city to Malmö, via Kristianstad and Hässleholm (see Figure 4, below). From Karlskrona to Hässleholm, the railway line is single-track, and from Hässleholm to Malmö, the line is double-track with a small portion having four tracks between Arlöv and Malmö. This stretch consists of approximately 90 segments, with a total of 290 block sections. For the regional trains that operate between Karlskrona and Malmö (and even further, into Denmark via the Öresund Bridge), there is a travel time of 1 h and 31 min between Karlskrona and Kristianstad, and a travel time between Kristianstad and Malmö of 1 h and 5 min.

In line with the categorization suggested in [29], three types of disturbance scenarios are used:

- Category 1 refers to a train suffering from a temporary delay at one particular section, which could occur due to, e.g., delayed train staff, or crowding at platforms resulting in increasing dwell times at stations.
- Category 2 refers to a train having a permanent malfunction, resulting in increased running times on all line sections it is planned to occupy.
- Category 3 refers to an infrastructure failure causing, e.g., a speed reduction on a particular section, which results in increased running times for all trains running through that section.

**Figure 4.** Illustration of the studied railway line Karlskrona-Kristianstad-Malmö. Source: Trafikverket.

All disturbance scenarios occur between 16:00 and 18:00, which is during peak hours. The re-scheduling time horizons are 1 and 1.5 h time windows, respectively, counting from when the disturbance occurs. The scenarios are described in Table 3 below. The experiments were tested on a laptop with 64-bit Windows 10, equipped with an Intel i7-CPU, 2.60 GHz, with 8 Gigabytes of RAM.

**Table 3.** Description of the 30 × 2 scenarios that were used in the experimental study. The first number in the scenario-ID specifies which disturbance category it is. For category 2, the disturbance is a percentage increase of the runtime, e.g., 40%. The two rightmost columns specify the size of the problem expressed in a number of train events that are to be re-scheduled.

| Scenario | Disturbance | | | Problem Size: #Events [1] | |
|---|---|---|---|---|---|
| Category: ID | Location | Initially Disturbed Train | Initially Delay (min) | 1 h Time Window | 1.5 h Time Window |
| 1:1 | Karlshamn-Ångsågsmossen | 1058 (Eastbound) | 10 | 1753 | 2574 |
| 1:2 | Bromölla Sölvesborg | 1064 (Eastbound) | 5 | 1717 | 2441 |
| 1:3 | Kristianstad-Karpalund | 1263 (Southbound) | 8 | 1421 | 2100 |
| 1:4 | Bergåsa-Gullberna | 1097 (Westbound) | 10 | 1739 | 2482 |
| 1:5 | Bräkne Hoby-Ronneby | 1103 (Westbound) | 15 | 1393 | 2056 |
| 1:6 | Flackarp-Hjärup | 491 (Southbound) | 5 | 1467 | 2122 |
| 1:7 | Eslöv-Dammstorp | 533 (Southbound) | 10 | 1759 | 2578 |
| 1:8 | Burlöv-Åkarp | 544 (Northbound) | 7 | 1748 | 2572 |
| 1:9 | Burlöv-Åkarp | 1378 (Northbound) | 4 | 1421 | 2100 |
| 1:10 | Höör-Stehag | 1381 (Southbound) | 10 | 1687 | 2533 |
| 2:1 | Karlshamn-Ångsågsmossen | 1058 (Eastbound) | 40% | 1753 | 2574 |
| 2:2 | Bromölla Sölvesborg | 1064 (Eastbound) | 20% | 1717 | 2441 |
| 2:3 | Kristianstad-Karpalund | 1263 (Southbound) | 20% | 1421 | 2100 |
| 2:4 | Bergåsa-Gullberna | 1097 (Westbound) | 40% | 1739 | 2482 |
| 2:5 | Bräkne Hoby-Ronneby | 1103 (Westbound) | 100% | 1393 | 2056 |
| 2:6 | Flackarp-Hjärup | 491 (Southbound) | 100% | 1467 | 2122 |
| 2:7 | Eslöv-Dammstorp | 533 (Southbound) | 50% | 1759 | 2578 |
| 2:8 | Burlöv-Åkarp | 544 (Northbound) | 80% | 1748 | 2572 |
| 2:9 | Burlöv-Åkarp | 1378 (Northbound) | 40% | 1421 | 2100 |
| 2:10 | Höör-Stehag | 1381 (Southbound) | 40% | 1687 | 2533 |
| 3:1 | Karlshamn-Ångsågsmossen | All trains passing through | 4 | 1753 | 2574 |
| 3:2 | Bromölla Sölvesborg | All trains passing through | 2 | 1717 | 2441 |
| 3:3 | Kristianstad-Karpalund | All trains passing through | 3 | 1421 | 2100 |
| 3:4 | Bergåsa-Gullberna | All trains passing through | 6 | 1739 | 2482 |
| 3:5 | Bräkne Hoby-Ronneby | All trains passing through | 5 | 1393 | 2056 |
| 3:6 | Flackarp-Hjärup | All trains passing through | 3 | 1467 | 2122 |
| 3:7 | Eslöv-Dammstorp | All trains passing through | 4 | 1759 | 2578 |
| 3:8 | Burlöv-Åkarp | All trains passing through | 2 | 1748 | 2572 |
| 3:9 | Burlöv-Åkarp | All trains passing through | 2 | 1421 | 2100 |
| 3:10 | Höör-Stehag | All trains passing through | 2 | 1687 | 2533 |

[1] The size of the generated graph $G$ is the squared size of number of events.

## 5.2. Results and Analysis

In Tables 4 and 5, the results are summarized. The grey cells include the best solution values found by different dispatching rules (DR-1 to DR-6) configurations. These solution values can be compared with the optimal values generated by MIP model. The two rightmost columns compare the computational time for the MIP model and heuristic algorithm. Since the computation time of the algorithm is not significantly affected by the choice of dispatching rule, the computation time is very similar for all six different configurations, and therefore only one computation time value per scenario is presented in Tables 4 and 5.

**Table 4.** Computational results for 1 h time horizon.

| Scenario | | $TFD_j^{+3}$ Objective Function (hh:mm:ss) | | | | | | Computational Time (hh:mm:ss) | |
|---|---|---|---|---|---|---|---|---|---|
| Category: ID | Optimal Results | Dispatching Rules (DR) | | | | | | MIP Model | Heuristic Algorithm |
| | | 1 | 2 | 3 | 4 | 5 | 6 | | |
| 1:1 | 0:01:03 | 00:01:14 | 00:01:14 | 00:01:14 | 00:15:22 | 00:15:22 | 00:24:05 | 00:00:04 | 00:00:06 |
| 1:2 | 0:00:00 | 00:00:00 | 00:00:00 | 00:00:00 | 00:00:00 | 00:00:00 | 01:16:47 | 00:00:04 | 00:00:05 |
| 1:3 | 0:00:00 | 00:02:09 | 00:02:09 | 00:02:09 | 00:36:33 | 00:36:33 | 00:26:32 | 00:00:04 | 00:00:03 |
| 1:4 | 0:01:02 | 00:02:28 | 00:02:28 | 00:02:28 | 00:12:40 | 00:12:40 | 00:08:11 | 00:00:06 | 00:00:06 |
| 1:5 | 0:07:01 | 00:16:15 | 00:16:15 | 00:16:15 | 00:19:49 | 00:19:49 | 00:13:14 | 00:00:05 | 00:00:03 |
| 1:6 | 0:00:23 | 00:05:46 | 00:05:46 | 00:05:46 | 00:05:46 | 00:05:46 | 00:05:46 | 00:00:04 | 00:00:03 |
| 1:7 | 0:05:05 | 00:06:24 | 00:06:24 | 00:06:24 | 00:06:24 | 00:06:24 | 00:14:45 | 00:00:04 | 00:00:05 |
| 1:8 | 0:01:34 | 00:12:01 | 00:12:01 | 00:12:01 | 00:12:01 | 00:12:01 | 00:13:40 | 00:00:04 | 00:00:06 |
| 1:9 | 0:00:00 | 00:14:01 | 00:14:01 | 00:14:01 | 00:13:02 | 00:13:02 | 00:14:01 | 00:00:04 | 00:00:03 |
| 1:10 | 0:00:00 | 00:01:18 | 00:01:18 | 00:00:00 | 00:00:00 | 00:00:00 | 00:01:33 | 00:00:04 | 00:00:05 |
| 2:1 | 0:05:24 | 00:45:37 | 00:45:37 | 00:06:42 | 00:06:42 | 00:06:42 | 00:08:16 | 00:00:04 | 00:00:06 |
| 2:2 | 0:02:43 | 00:03:29 | 00:39:53 | 00:03:29 | 00:03:29 | 00:03:29 | 01:06:56 | 00:00:05 | 00:00:05 |
| 2:3 | 0:01:01 | 00:01:47 | 00:20:06 | 00:20:06 | 00:01:47 | 00:01:47 | 00:20:14 | 00:00:03 | 00:00:03 |
| 2:4 | 0:15:12 | 00:22:12 | 00:22:50 | 00:22:50 | 00:22:12 | 00:22:12 | 00:55:54 | 00:00:05 | 00:00:05 |
| 2:5 | 0:42:09 | 00:43:24 | 00:58:03 | 00:58:03 | 01:05:08 | 01:05:08 | 00:59:16 | 00:00:04 | 00:00:03 |
| 2:6 | 0:01:24 | 00:02:44 | 00:02:44 | 00:02:44 | 00:02:44 | 00:02:44 | 00:02:44 | 00:00:06 | 00:00:04 |
| 2:7 | 0:05:27 | 00:08:09 | 00:08:09 | 00:08:09 | 00:08:09 | 00:08:09 | 00:09:43 | 00:00:04 | 00:00:10 |
| 2:8 | 0:21:12 | 00:43:37 | 00:38:42 | 00:38:42 | 00:43:37 | 00:43:37 | 00:40:16 | 00:00:06 | 00:00:06 |
| 2:9 | 0:00:00 | 00:00:00 | 00:00:00 | 00:00:00 | 00:00:00 | 00:00:00 | 00:00:00 | 00:00:05 | 00:00:03 |
| 2:10 | 0:00:08 | 00:05:09 | 00:05:09 | 00:05:09 | 00:05:09 | 00:05:09 | 00:06:43 | 00:00:04 | 00:00:06 |
| 3:1 | 0:01:00 | 00:01:00 | 00:01:00 | 00:01:00 | 00:01:00 | 00:01:00 | 00:25:05 | 00:00:04 | 00:00:06 |
| 3:2 | 0:00:00 | 00:00:00 | 00:00:00 | 00:00:00 | 00:00:19 | 00:00:19 | 00:25:54 | 00:00:04 | 00:00:06 |
| 3:3 | 0:00:00 | 00:03:49 | 00:21:38 | 01:05:52 | 00:01:17 | 00:15:40 | 00:54:14 | 00:00:05 | 00:00:03 |
| 3:4 | 0:12:21 | 00:16:59 | 00:20:13 | 00:20:13 | 00:16:59 | 00:16:59 | 00:20:22 | 00:00:07 | 00:00:06 |
| 3:5 | 0:05:20 | 00:05:25 | 00:57:34 | 00:15:58 | 00:05:25 | 00:05:25 | 00:16:12 | 00:00:03 | 00:00:03 |
| 3:6 | 0:00:00 | 00:21:04 | 00:21:04 | 00:21:04 | 00:23:12 | 00:23:12 | 00:28:42 | 00:00:05 | 00:00:03 |
| 3:7 | 0:00:09 | 00:07:42 | 00:14:11 | 00:14:11 | 00:14:11 | 00:14:11 | 00:15:45 | 00:00:10 | 00:00:05 |
| 3:8 | 0:00:00 | 00:00:00 | 00:05:06 | 00:05:06 | 00:00:00 | 00:00:00 | 00:07:54 | 00:00:04 | 00:00:05 |
| 3:9 | 0:00:00 | 00:00:00 | 00:00:41 | 00:00:41 | 00:00:00 | 00:00:00 | 00:00:00 | 00:00:04 | 00:00:04 |
| 3:10 | 0:00:00 | 00:04:53 | 00:04:53 | 00:03:35 | 00:01:04 | 00:01:04 | 00:02:38 | 00:00:03 | 00:00:05 |

**Table 5.** The computational results for 1.5 h time horizon (hh:mm:ss).

| Scenario | | $TFD_j^{+3}$ Objective Function | | | | | | Computational Time | |
|---|---|---|---|---|---|---|---|---|---|
| Category: ID | Optimal Results | Dispatching Rules (DR) | | | | | | MIP Model | Heuristic Algorithm |
| | | 1 | 2 | 3 | 4 | 5 | 6 | | |
| 1:1 | 0:01:03 | 00:01:14 | 00:01:14 | 00:01:14 | 00:45:13 | 00:45:13 | 00:46:56 | 00:00:12 | 00:00:20 |
| 1:2 | 0:00:00 | 00:00:00 | 00:00:00 | 00:00:00 | 00:12:09 | 00:12:09 | 05:09:54 | 00:00:10 | 00:00:15 |
| 1:3 | 0:00:00 | 00:02:09 | 00:02:09 | 00:02:09 | 01:34:59 | 01:34:59 | 00:25:21 | 00:00:11 | 00:00:10 |
| 1:4 | 0:00:00 | 00:01:27 | 00:01:27 | 00:01:27 | 01:38:16 | 01:38:16 | 00:51:24 | 00:00:13 | 00:00:16 |
| 1:5 | 0:02:08 | 00:10:23 | 00:10:23 | 00:10:23 | 00:24:07 | 00:24:07 | 00:12:14 | 00:00:07 | 00:00:09 |
| 1:6 | 0:00:23 | 00:05:46 | 00:05:46 | 00:05:46 | 00:05:46 | 00:05:46 | 00:05:46 | 00:00:12 | 00:00:10 |
| 1:7 | 0:05:05 | 00:06:24 | 00:06:24 | 00:06:24 | 00:06:24 | 00:06:24 | 00:14:45 | 00:00:11 | 00:00:18 |
| 1:8 | 0:01:34 | 00:12:01 | 00:12:01 | 00:12:01 | 00:12:01 | 00:12:01 | 00:13:40 | 00:00:12 | 00:00:17 |
| 1:9 | 0:00:00 | 00:13:00 | 00:13:00 | 00:13:00 | 00:13:16 | 00:13:16 | 00:38:13 | 00:00:07 | 00:00:10 |
| 1:10 | 0:00:00 | 00:01:18 | 00:01:18 | 00:00:00 | 00:00:00 | 00:00:00 | 00:01:33 | 00:00:11 | 00:00:16 |
| 2:1 | 0:03:36 | 00:45:37 | 00:45:37 | 00:04:31 | 00:17:11 | 00:17:11 | 00:14:16 | 00:00:11 | 00:00:17 |
| 2:2 | 0:00:00 | 00:00:23 | 01:21:00 | 00:00:23 | 00:00:23 | 00:00:23 | 04:36:13 | 00:00:12 | 00:00:15 |
| 2:3 | 0:02:40 | 00:29:31 | 00:32:58 | 00:32:58 | 00:39:25 | 00:39:25 | 00:33:46 | 00:00:07 | 00:00:09 |
| 2:4 | 0:26:34 | 00:43:50 | 01:11:40 | 01:11:40 | 00:43:50 | 00:43:50 | 01:57:35 | 00:00:16 | 00:00:15 |
| 2:5 | 1:11:44 | 01:17:47 | 01:40:09 | 01:40:09 | 01:41:32 | 01:41:32 | 01:35:29 | 00:00:12 | 00:00:08 |
| 2:6 | 0:01:24 | 00:02:44 | 00:02:44 | 00:02:44 | 00:02:44 | 00:02:44 | 00:02:44 | 00:00:11 | 00:00:09 |

Table 5. *Cont.*

| Scenario | | $TFD_j^{+3}$ Objective Function | | | | | | Computational Time | |
|---|---|---|---|---|---|---|---|---|---|
| Category: ID | Optimal Results | Dispatching Rules (DR) | | | | | | MIP Model | Heuristic Algorithm |
| | | 1 | 2 | 3 | 4 | 5 | 6 | | |
| 2:7 | 0:05:27 | 00:08:09 | 00:08:09 | 00:08:09 | 00:08:09 | 00:08:09 | 00:09:43 | 00:00:11 | 00:00:17 |
| 2:8 | 0:21:12 | 00:44:42 | 00:48:13 | 00:48:13 | 00:44:54 | 00:44:54 | 01:08:14 | 00:00:36 | 00:00:17 |
| 2:9 | 0:00:00 | 00:00:00 | 00:00:00 | 00:00:00 | 00:00:13 | 00:00:13 | 00:00:00 | 00:00:07 | 00:00:10 |
| 2:10 | 0:00:08 | 00:05:09 | 00:05:09 | 00:05:09 | 00:05:09 | 00:05:09 | 00:06:43 | 00:00:12 | 00:00:16 |
| 3:1 | 0:00:00 | 00:00:00 | 00:22:42 | 00:00:00 | 00:00:00 | 00:00:00 | 00:46:56 | 00:00:12 | 00:00:17 |
| 3:2 | 0:00:00 | 00:00:17 | 00:00:17 | 00:00:17 | 00:00:36 | 00:00:36 | 01:36:46 | 00:00:11 | 00:00:14 |
| 3:3 | 0:01:16 | 00:09:36 | 00:57:38 | 01:22:08 | 00:07:17 | 00:14:39 | 02:07:54 | 00:00:11 | 00:00:10 |
| 3:4 | 0:16:37 | 00:31:06 | 00:32:34 | 00:32:34 | 00:23:26 | 00:23:26 | 01:10:03 | 00:00:35 | 00:00:15 |
| 3:5 | 0:04:44 | 00:04:58 | 00:55:43 | 00:13:08 | 00:10:16 | 00:10:16 | 00:19:59 | 00:00:10 | 00:00:09 |
| 3:6 | 0:00:00 | 00:50:13 | 00:50:13 | 00:50:13 | 00:27:54 | 00:27:54 | 01:07:29 | 00:00:17 | 00:00:15 |
| 3:7 | 0:00:41 | 00:06:48 | 00:15:45 | 00:15:45 | 00:15:45 | 00:15:45 | 00:17:19 | 00:00:18 | 00:00:19 |
| 3:8 | 0:00:00 | 00:00:00 | 00:01:22 | 00:01:22 | 00:00:00 | 00:00:00 | 00:03:58 | 00:00:11 | 00:00:17 |
| 3:9 | 0:00:00 | 00:02:06 | 00:03:32 | 00:02:48 | 00:02:06 | 00:02:06 | 00:02:06 | 00:00:14 | 00:00:10 |
| 3:10 | 0:00:00 | 00:08:07 | 00:08:31 | 00:07:13 | 00:01:28 | 00:01:28 | 00:03:02 | 00:00:11 | 00:00:16 |

The *minimum release time goes first* (DR-1) was the best dispatching rule, considering the $\sum TFD_j^{+3}$ objective function. The average $\sum TFD_j^{+3}$ for the MIP model was 259 s for the 1 h time window and 332 s for the 1.5 h time window. The average $\sum TFD_j^{+3}$ for the *minimum release time goes first* (DR-1) was 597 and 849 s, respectively.

Statistical analyses of the results belonging to disturbance scenario type 1 revealed that the *less real buffer time goes first* (DR-3) dispatching rule, with an average delay of 361 s for a 1 h time window and 314 s for a 1.5 h time window, works better than the other dispatching rules. Additionally, an inefficient solution is not able to absorb the delays (i.e., the delays after a temporary route blocking may remain in the system until midnight). The analysis also shows that for all dispatching rules in scenario 1, the $\sum TFD_j^{+3}$ objective function values of the 1.5 h time window are lower than the values for the 1 h time window, which confirms that the algorithm successfully attempts to make the timetable absorb delays when possible.

For scenario type 2, the *minimum release time goes first* (DR-1), *less programmed buffer time goes first* (DR-4), and *less total buffer goes first* (DR-5) worked somewhat the same, and better than the others. The average $\sum TFD_j^{+3}$ objectives were 1056, 953, and 953 s for a 1 h time window, and 1547, 1581, and 1581 for a 1.5 h time window. The optimal values are 568 s for a 1 h time window and 796 s for a 1.5 h time window. In scenario type 3, the *minimum release time goes first* worked better than the others with an average of 365 s delay, but for a 1.5 h time window the *less total buffer goes first* (DR-5) with an average of 532 s was the best. The optimal values were 113 and 139 s, respectively.

With the help of a visualization software, the resulting, revised timetables can be analysed beyond aggregated numbers. The *more delay goes first* (DR-2) dispatching rule gives priority to the trains with the largest tardiness. We observed in the visualization of the solutions that, when the conflict is between two tardy trains, this strategy works well and reduces the delay. However, for conflicts between an on-time train and a tardy train, this dispatching rule gives priority to the tardy train, which causes a delay for an on-time train. In other words, when the tardy train reaches the destination, e.g., *Karlskrona* or *Malmö*, this strategy causes a delay for new trains that have recently started the journey. A more effective decision would potentially be to prioritize the on-time train, because the late train is near to its final destination.

The *less real buffer time goes first* (DR-3) dispatching rule, gives priority to the train with least buffer time. This strategy helps the algorithm to reduce the delay for tardy trains. When the conflict is between two tardy trains, this policy is fair. The *less programmed buffer time goes first* (DR-5) considers the sum of buffer time for a train to its destination. In a disturbance area, this strategy works well. The algorithm gives priority to a train with less programmed buffer time, which seems to be fair between two tardy trains. However, when a tardy train has a conflict with an on-time train, this dispatching rule gives priority to the tardy one, which is not effective if the on-time train is at the beginning of its itinerary and thus may cause knock-on delays if it is delayed. The *less total processing time* (DR-6) dispatching rule, tries to give priority to trains with less remaining events to leave the tracks as soon as possible. The experimental results demonstrate that this strategy does not work well compared to other dispatching rules.

The choice of dispatching rule does not affect the computational time, but the number of events in the re-scheduling time window and selected sub-network has a significant effect on the computational time since the size of the graph $G$ increases quadratically. Figure 5 illustrates the increase of computational time against increase of the time horizon and number of events. Both the size of graph $G$ and the computational time increase quadratically.

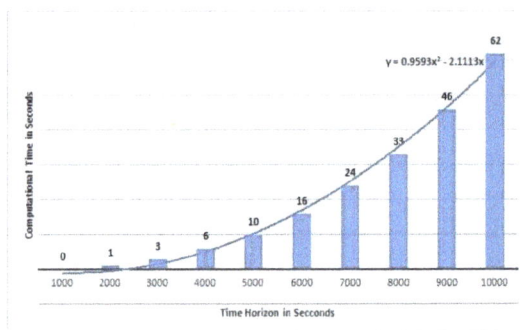

**Figure 5.** The computational time increases quadratically, based on the number of events.

## 6. Conclusions and Future Work

This paper addresses the real-time train traffic re-scheduling problem. A mixed graph is used for modeling the problem as a blocking job-shop scheduling problem. A heuristic algorithm is proposed that benefits from re-timing, re-ordering, and local re-routing. The algorithm benefits also from a dynamic update of data, which accelerates the computations.

The response time for such a real-time computational scheduling tool is a vital factor. In the proposed solution approach, the problem for a 1 h time window is solved in less than 10 s, and for a 1.5 h time window, the computational time is less than 20 s. It is also unknown what time horizon is necessary to consider in different situations and what role this uncertainty would play. Interviews with dispatchers suggest that it differs a lot depending on the situation, context and associated working load.

The $TFD_j^{+3}$ objective function is a relevant criterion to control the lateness of trains. However, based on the situation and type of disturbance scenario, the dispatchers also have other concerns and objectives. The investigation of other objective functions and useful solution quality metrics is necessary to investigate in future research. The graph $G$ is represented by an adjacency matrix in the current implementation. Using alternative data structures such as adjacency lists, can be an option to investigate the possibility to reduce computational time further.

**Acknowledgments:** The presented research work was conducted within the research projects FLOAT (FLexibel Omplanering Av Tåglägen i drift) and BLIXTEN (Beslutstöd för trafikLedare: approxImativa och eXakta opTimEraNde metoder), funded by the Swedish Transport Administration (Trafikverket) with the KAJT (KApacitet i JärnvägsTrafiken) research programme. We are thankful for the financial support, data, and significant guidance provided by Trafikverket.

**Author Contributions:** Omid has developed the graph model and heuristic algorithms. The algorithmic approach has been implemented in Java by Omid. Johanna has provided all input data required to build and perform the experiments, and she is the main developer of the visualization tool used for visual inspection of re-scheduling solutions. Johanna has also managed the associated research project, handled all contacts with the industry partner, Trafikverket, and acquired the associated research funding. Omid has generated the disturbance scenarios, performed the experiments with the algorithmic approach and the associated performance assessment including the analysis of the resulting data and conclusions. Johanna has formulated and implemented the corresponding MIP formulation in java, which was applied to solve the same disturbance scenarios in order to evaluate the level of optimality and efficiency of the algorithmic approach. Both Johanna and Omid contributed significantly to writing the manuscript. Both authors have read and approved the final manuscript.

**Conflicts of Interest:** The authors declare no conflict of interest.

## Appendix A

The MIP formulation is based on the model developed by Törnquist and Persson [13]. The model was implemented in Java and solved by Gurobi 6.5.1. Let $J$ be the set of trains, $M$ the set of segments, defining the rail infrastructure, and $O$ the set of events. An event can be seen as a time slot request by a train for a specific segment. The index $j$ is associated with a specific train service, and index $m$ with a specific infrastructure segment, and $i$ with the event. An event is connected both to an infrastructure segment and a train. The sets $O_j \subseteq O$ are ordered sets of events for each train $j$, while $O^m \subseteq O$ are ordered sets of events for each segment $m$.

Each event has a point of origin, $m_{j,i}$, which is used for determining a change in the direction of traffic on a specific track. Further, for each event, there is a scheduled start time and an end time, denoted by $b_{j,i}^{inital}$ and $e_{j,i}^{inital}$, which are given by the initial timetable. The disturbance is modelled using parameters $b_{j,i}^{static}$ and $e_{j,i}^{static}$, denoting the pre-assigned start and end time of the already active event.

There are two types of segments, modeling the infrastructure between stations and within stations. Each segment $m$ has a number of parallel tracks, indicated by the sets $M_m$ and each track requires a separation in time between its events (one train leaving the track and the next enters the same track). The minimum time separation between trains on a segment is denoted by $\Delta_m^{Meeting}$ for trains that travel in opposite directions, and $\Delta_m^{Following}$ for trains that follow each other; the separation is only required if the trains use the same track on that specific segment.

The parameter $ps_{j,i}$ indicates if event $i$ includes a planned stop at the associated segment (i.e., it is then normally a station). The parameter $d_{j,i}$ represents the minimum running time, pre-computed from the initial schedule, if event $i$ occurs on a line segment between stations. For station segments, $d_{j,i}$ corresponds to the minimum dwell time of commercial stops, where transfers may be scheduled.

The variables in the model are either binary or continuous. The continuous variables describe the timing of the events and the delay, and the binary variables describe the discrete decisions to take on the model concerning the selection of a track on segments with multiple tracks or platforms, and the order of trains that want to occupy the same track and/or platform. The continuous, non-negative, variables are $x_{j,i}^{begin}$, $x_{j,i}^{end}$, and $z_{j,i}$ (delay of the event $i, i \in O$, exceeding $\mu$ time units, which is set to three minutes here).

The variables $x_{j,i}^{end}$ and $x_{j,i}^{begin}$ are modelling the resource allocation, where a resource is a specific track segment. The arrival time at a specific segment is given by $x_{j,i}^{begin}$ and departure from a specific segment is given by $x_{j,i}^{end}$ for a specific train. The binary variables are defined as:

$$q_{j,i,u} = \begin{cases} 1, \text{if event } i \text{ uses track } u, i \in O^m, u \in M_m, m \in M, j \in J \\ 0, \text{otherwise} \end{cases}$$

$$\gamma_{j,i,j\prime,i\prime} = \begin{cases} 1, \text{if event } i \text{ occurs before event } i', i \in O^m, m \in M : i < i', j \text{ and } j\prime \in J \\ 0, \text{otherwise} \end{cases}$$

$$\lambda_{j,i,j\prime,i\prime} = \begin{cases} 1, \text{if event } i \text{ is rescheduled to occur after event } i', i \in O^m, m \in M : i < i', j \text{ and } j\prime \in J \\ 0, \text{otherwise} \end{cases}$$

With the objective to minimize the sum of all delays for all trains reaching their final destination with a delay larger than three minutes, the objective function can be formulated as follows, where the parameter $n_j$ for each train $j \in J$ holds the final event of train $j$:

$$\min_z f := \sum_{j \in J} z_{n_j} \tag{A1}$$

We also have the following three blocks of constraints. The first block concerns the timing of the events belonging to each train $j$ and its sequence of events, defined by the event list $O_j \subseteq O$. These constraints define the relation between the initial schedule and revised schedule, as an effect of the disturbance. Equation (A7) is used to compute the delay of each event exceeding $\mu$ time units, where $\mu$ is set to three minutes in this context.

$$x_{j,i}^{end} = x_{j,i+1}^{begin}, i \in O_j, j \in J : i \neq n_j, j \in J \tag{A2}$$

$$x_{j,i}^{begin} = b_{j,i}^{static}, i \in O : b_{j,i}^{static} > 0, j \in J \tag{A3}$$

$$x_{j,i}^{end} = e_{j,i}^{static}, i \in O : e_{j,i}^{static} > 0, j \in J \tag{A4}$$

$$x_{j,i}^{end} \geq x_{j,i}^{begin} + d_{j,i}, i \in O, j \in J \tag{A5}$$

$$x_{j,i}^{begin} \geq b_{j,i}^{initial}, i \in O : ps_{j,i} = 1, j \in J \tag{A6}$$

$$x_{j,i}^{end} - e_{j,i}^{initial} - u \leq z_{j,i}, i \in O, j \in J \tag{A7}$$

In the following part, $N$ is a large constant. The second block of constraints concerns the permitted interaction between trains, given the capacity limitations of the infrastructure (including safety restrictions):

$$\sum_{u \in M_m} q_{j,i,u} = 1, i \in O^m, m \in M, j \in J \tag{A8}$$

$$q_{j,i,u} + q_{j',i',u} - 1 \leq \lambda_{j,i,j',i'} + \gamma_{j,i,j',i'}, i, i' \in O^m, u \in M_m, m \in M : i < i', j \neq j' \in J \tag{A9}$$

$$x_{j',i'}^{begin} - x_{j,i}^{end} \geq \Delta_m^{Meeting} \gamma_{j,i,j',i'} - N\left(1 - \gamma_{j,i,j',i'}\right), i, i' \in O^m, m \in M : i < i', m_{i'} \neq m_i, j \neq j' \in J \tag{A10}$$

$$x_{j',i'}^{begin} - x_{j,i}^{end} \geq \Delta_m^{Following} \gamma_{j,i,j',i'} - N\left(1 - \gamma_{j,i,j',i'}\right), i, i' \in O^m, m \in M : i < i', m_{i'} \neq m_i, j \neq j' \in J \tag{A11}$$

$$x_{j,i}^{begin} - x_{j',i'}^{end} \geq \Delta_m^{Meeting} \lambda_{j,i,j',i'} - N\left(1 - \lambda_{j,i,j',i'}\right), i, i' \in O^m, m \in M : i < i', m_{i'} \neq m_i, j \neq j' \in J \tag{A12}$$

$$x_{j,i}^{begin} - x_{j',i'}^{end} \geq \Delta_m^{Following} \lambda_{j,i,j',i'} - N\left(1 - \lambda_{j,i,j',i'}\right), i, i' \in O^m, m \in M : i < i', m_{i'} \neq m_i, j \neq j' \in J \tag{A13}$$

$$\lambda_{j,i,j',i'} + \gamma_{i,i'} \leq 1, i, i' \in O^m, m \in M : i < i', j \neq j' \in J \tag{A14}$$

$$x_{j,i}^{begin}, x_{j,i}^{end}, z_{j,i} \geq 0, i \in O, j \in J \tag{A15}$$

$$\gamma_{j,i,j',i'}, \lambda_{j,i,j',i'} \in \{0,1\}, i' \in O^m, i \in M : i < i', j \neq j' \in J \tag{A16}$$

$$q_{j,i,u} \in \{0,1\}, i \in O^m, u \in M_m, m \in M, j \in J \tag{A17}$$

## References

1. Boston Consultancy Group. *The 2017 European Railway Performance Index;* Boston Consultancy Group: Boston, MA, USA, 18 April 2017.
2. European Commission Directorate General for Mobility and Transport. *Study on the Prices and Quality of Rail Passenger Services;* Report Reference: MOVE/B2/2015-126; European Commission Directorate General for Mobility and Transport: Brussel, Belgium, April 2016.

3.   Lamorgese, L.; Mannino, C.; Pacciarelli, D.; Törnquist Krasemann, J. Train Dispatching. In *Handbook of Optimization in the Railway Industry, International Series in Operations Research & Management Science*; Borndörfer, R., Klug, T., Lamorgese, L., Mannino, C., Reuther, M., Schlechte, T., Eds.; Springer: Cham, Switzerland, 2018; Volume 268. [CrossRef]

4.   Cacchiani, V.; Huisman, D.; Kidd, M.; Kroon, L.; Toth, P.; Veelenturf, L.; Wagenaar, J. An overview of recovery models and algorithms for real-time railway rescheduling. *Transp. Res. B Methodol.* **2010**, *63*, 15–37. [CrossRef]

5.   Fang, W.; Yang, S.; Yao, X. A Survey on Problem Models and Solution Approaches to Rescheduling in Railway Networks. *IEEE Trans. Intell. Trans. Syst.* **2015**, *16*, 2997–3016. [CrossRef]

6.   Josyula, S.; Törnquist Krasemann, J. Passenger-oriented Railway Traffic Re-scheduling: A Review of Alternative Strategies utilizing Passenger Flow Data. In Proceedings of the 7th International Conference on Railway Operations Modelling and Analysis, Lille, France, 4–7 April 2017.

7.   Szpigel, B. Optimal train scheduling on a single track railway. *Oper. Res.* **1973**, *72*, 343–352.

8.   D'Ariano, A.; Pacciarelli, D.; Pranzo, M. A branch and bound algorithm for scheduling trains in a railway network. *Eur. J. Oper. Res.* **2017**, *183*, 643–657. [CrossRef]

9.   Khosravi, B.; Bennell, J.A.; Potts, C.N. Train Scheduling and Rescheduling in the UK with a Modified Shifting Bottleneck Procedure. In Proceedings of the 12th Workshop on Algorithmic Approaches for Transportation Modelling, Optimization, and Systems 2012, Ljubljana, Slovenia, 13 September 2012; pp. 120–131.

10.  Liu, S.; Kozan, E. Scheduling trains as a blocking parallel-machine job shop scheduling problem. *Comput. Oper. Res.* **2009**, *36*, 2840–2852. [CrossRef]

11.  Mascis, A.; Pacciarelli, D. Job-shop scheduling with blocking and no-wait constraints. *Eur. J. Oper. Res.* **2002**, *143*, 498–517. [CrossRef]

12.  Oliveira, E.; Smith, B.M. *A Job-Shop Scheduling Model for the Single-Track Railway Scheduling Problem*; Research Report Series 21; School of Computing, University of Leeds: Leeds, UK, 2000.

13.  Törnquist, J.; Persson, J.A. N-tracked railway traffic re-scheduling during disturbances. *Transp. Res. Part B Methodol.* **2007**, *41*, 342–362. [CrossRef]

14.  Pellegrini, P.; Douchet, G.; Marliere, G.; Rodriguez, J. Real-time train routing and scheduling through mixed integer linear programming: Heuristic approach. In Proceedings of the 2013 International Conference on Industrial Engineering and Systems Management (IESM), Rabat, Morocco, 28–30 October 2013.

15.  Xu, Y.; Jia, B.; Ghiasib, A.; Li, X. Train routing and timetabling problem for heterogeneous train traffic with switchable scheduling rules. *Transp. Res. Part C Emerg. Technol.* **2017**, *84*, 196–218. [CrossRef]

16.  Corman, F.; D'Ariano, A.; Pacciarelli, D.; Pranzo, M. Centralized versus distributed systems to reschedule trains in two dispatching areas. *Public Trans. Plan. Oper.* **2010**, *2*, 219–247. [CrossRef]

17.  Corman, F.; D'Ariano, A.; Pacciarelli, D.; Pranzo, M. Optimal inter-area coordination of train rescheduling decisions. *Trans. Res. Part E* **2012**, *48*, 71–88.

18.  Corman, F.; Pacciarelli, D.; D'Ariano, A.; Samá, M. Rescheduling Railway Traffic Taking into Account Minimization of Passengers' Discomfort. In Proceedings of the International Conference on Computational Logistics, ICCL 2015, Delft, The Netherlands, 23–25 September 2015; pp. 602–616.

19.  Lamorgese, L.; Mannino, C. An Exact Decomposition Approach for the Real-Time Train Dispatching Problem. *Oper. Res.* **2015**, *63*, 48–64. [CrossRef]

20.  Meng, L.; Zhou, X. Simultaneous train rerouting and rescheduling on an N-track network: A model reformulation with network-based cumulative flow variables. *Trans. Res. Part B Methodol.* **2014**, *67*, 208–234. [CrossRef]

21.  Tormo, J.; Panou, K.; Tzierpoulos, P. Evaluation and Comparative Analysis of Railway Perturbation Management Methods. In Proceedings of the Conférence Mondiale sur la Recherche Dans les Transports (13th WCTR), Rio de Janeiro, Brazil, 15–18 July 2013.

22.  Rodriguez, J. An incremental decision algorithm for railway traffic optimisation in a complex station. Eleventh. In Proceedings of the International Conference on Computer System Design and Operation in the Railway and Other Transit Systems (COMPRAIL08), Toledo, Spain, 15–17 September 2008; pp. 495–504.

23.  Bettinelli, A.; Santini, A.; Vigo, D. A real-time conflict solution algorithm for the train rescheduling problem. *Trans. Res. Part B Methodol.* **2017**, *106*, 237–265. [CrossRef]

24.  Samà, M.; D'Ariano, A.; Corman, F.; Pacciarelli, D. A variable neighbourhood search for fast train scheduling and routing during disturbed railway traffic situations. *Comput. Oper. Res.* **2017**, *78*, 480–499. [CrossRef]

25. Burdett, R.L.; Kozan, E. A sequencing approach for creating new train timetables. *OR Spectr.* **2010**, *32*, 163–193. [CrossRef]
26. Tan, Y.; Jiang, Z. A Branch and Bound Algorithm and Iterative Reordering Strategies for Inserting Additional Trains in Real Time: A Case Study in Germany. *Math. Probl. Eng.* **2015**, *2015*, 289072. [CrossRef]
27. Gholami, O.; Sotskov, Y.N. A fast heuristic algorithm for solving parallel-machine job-shop scheduling problems. *Int. J. Adv. Manuf. Technol.* **2014**, *70*, 531–546. [CrossRef]
28. Sotskov, Y.; Gholami, O. Mixed graph model and algorithms for parallel-machine job-shop scheduling problems. *Int. J. Prod. Res.* **2017**, *55*, 1549–1564. [CrossRef]
29. Krasemann, J.T. Design of an effective algorithm for fast response to the re-scheduling of railway traffic during disturbances. *Transp. Res. Part C Emerg. Technol.* **2012**, *20*, 62–78. [CrossRef]

*algorithms*

MDPI

*Article*

# Evaluating Typical Algorithms of Combinatorial Optimization to Solve Continuous-Time Based Scheduling Problem

**Alexander A. Lazarev \*, Ivan Nekrasov \*,† and Nikolay Pravdivets †**

Institute of Control Sciences, 65 Profsoyuznaya Street, 117997 Moscow, Russia; pravdivets@ya.ru
* Correspondence: jobmath@mail.ru (A.A.L.); ivannekr@mail.ru (I.N.); Tel.: +7-495-334-87-51 (A.A.L.)
† These authors contributed equally to this work.

Received: 22 February 2018; Accepted: 12 April 2018; Published: 17 April 2018

**Abstract:** We consider one approach to formalize the Resource-Constrained Project Scheduling Problem (RCPSP) in terms of combinatorial optimization theory. The transformation of the original problem into combinatorial setting is based on interpreting each operation as an atomic entity that has a defined duration and has to be resided on the continuous time axis meeting additional restrictions. The simplest case of continuous-time scheduling assumes one-to-one correspondence of resources and operations and corresponds to the linear programming problem setting. However, real scheduling problems include many-to-one relations which leads to the additional combinatorial component in the formulation due to operations competition. We research how to apply several typical algorithms to solve the resulted combinatorial optimization problem: enumeration including branch-and-bound method, gradient algorithm, random search technique.

**Keywords:** RCPSP; combinatorial optimization; scheduling; linear programming; MES; Job Shop

## 1. Introduction

The Resource-Constrained Project Scheduling Problem (RCPSP) has many practical applications. One of the most obvious and direct applications of RCPSP is planning the fulfilment of planned orders at the manufacturing enterprise [1] that is also sometimes named Job Shop. The Job Shop scheduling process traditionally resides inside the Manufacturing Execution Systems scope [2] and belongs to principle basic management tasks of any industrial enterprise. Historically the Job Shop scheduling problem has two formal mathematical approaches [3]: continuous and discrete time problem settings. In this paper, we research the continuous-time problem setting, analyze its bottlenecks, and evaluate effectiveness of several typical algorithms to find an optimal solution.

The continuous-time Job Shop scheduling approach has been extensively researched and applied in different industrial spheres throughout the past 50 years. One of the most popular classical problem settings was formulated in [4] by Manne as a disjunctive model. This problem setting forms a basic system of restrictions evaluated by different computational algorithms depending on the particular practical features of the model used. A wide overview of different computational approaches to the scheduling problem is conducted in [5,6]. The article [5] considers 69 papers dating back to the XX century, revealing the following main trends in Job Shop scheduling:

- Enumerating techniques
- Different kinds of relaxation
- Artificial intelligence techniques (neural networks, genetic algorithms, agents, etc.)

Artificial intelligence (AI) techniques have become mainstream nowadays. The paper [6] gives a detailed list of AI techniques and methods used for scheduling.

From the conceptual point of view, this paper deals with a mixed integer non-linear (MINLP) scheduling problem [5] that is relaxed to a combinatorial set of linear programming problems due to the linear "makespan" objective function. As a basic approach we take the disjunctive model [4]. A similar approach was demonstrated in [7] where the authors deployed the concept of parallel dedicated machines scheduling subject to precedence constraints and implemented a heuristic algorithm to generate a solution.

## 2. Materials and Methods

In order to formalize the continuous-time scheduling problem we will need to introduce several basic definitions and variables. After that, we will form a system of constraints that reflect different physical, logical and economic restrictions that are in place for real industrial processes. Finally, introducing the objective function will finish the formulation of RCPSP as an optimization problem suitable for solving.

### 2.1. Notions and Base Data for Scheduling

Sticking to the industrial scheduling also known as the Job Shop problem, let us define the main notions that we will use in the formulation:

- The enterprise functioning process utilizes resources of different types (for instance machines, personnel, riggings, etc.). The set of the resources is indicated by a variable $R = \{R_r\}$, $r = 1, \ldots, |R|$.
- The manufacturing procedure of the enterprise is formalized as a set of operations $J$ tied with each other via precedence relations. Precedence relations are brought to a matrix $G = \langle g_{ij} \rangle$, $i = 1, \ldots, |J|$, $j = 1, \ldots, |J|$. Each element of the matrix $g_{ij} = 1$ iff the operation $j$ follows the operation $i$ and zero otherwise $g_{ij} = 0$.
- Each operation $i$ is described by duration $\tau_i$. Elements $\tau_i$, $i = 1, \ldots, |J|$, form a vector of operations' durations $\longrightarrow \tau$.
- Each operation $i$ has a list of resources it uses while running. Necessity for resources for each operation is represented by the matrix $Op = \langle op_{ir} \rangle$, $i = 1, \ldots, |J|$, $r = 1, \ldots, |R|$. Each element of the matrix $op_{ir} = 1$ iff the operation $i$ of the manufacturing process allocates the resource $r$. All other cases bring the element to zero value $op_{ir} = 0$.
- The input orders of the enterprise are considered as manufacturing tasks for the certain amount of end product and are organized into a set $F$. Each order is characterized by the end product amount $v_f$ and the deadline $d_f$, $f = 1, \ldots, |F|$. Elements inside $F$ are sorted in the deadline ascending order.

Using the definitions introduced above, we can now formalize the scheduling process as residing all $|J|$ operations of all $|F|$ orders on the set of resources $R$. Mathematically this means defining the start time of each operation $i = 1, \ldots, |J|$ of each order $f = 1, \ldots, |F|$.

### 2.2. Continuous-Time Problem Setting

The continuous-time case is formulated around the variables that stand for the start moments of each operation $i$ of each order $f$: $x_{if} \geq 0$, $i = 1, \ldots, |J|$, $f = 1, \ldots, |F|$ [3]. The variables $x_{if} \geq 0$ can be combined into $|F|$ vectors $\overrightarrow{x_f} \geq 0$. The main constraints of the optimization problem in that case are:

- Precedence graph of the manufacturing procedure:

$$G\overrightarrow{x_f} \geq (\overrightarrow{x_f} + \overrightarrow{\tau}), \qquad f = 1, \ldots, |F|.$$

- Meeting deadlines for all orders

$$x_{|J|f} + \tau_{|J|} \leq d_f, \qquad f = 1, \ldots, |F|.$$

The constraints above do not take into consideration the competition of operations that try to allocate the same resource. We can distinguish two types of competition here: competition of operations within one order and competition of operations of different orders. The competition of operations on parallel branches of one order is considered separately by modelling the scheduling problem for a single order (this is out of the scope of this paper). The main scope of this research is to address the competition of operations from different orders. Assuming we have the rule (constraint) that considers the competition of operations within one and the same order let us formalize the competition between the operations of different orders. Consider we have a resource, which is being allocated by $K$ different operations of $|F|$ different orders.

For each resource $r = 1, \ldots , |R|$ let us distinguish only those operations that allocate it during their run. The set of such operations can be found from the corresponding column of the resource allocation matrix $Op = < op_{ir} >, \ i = 1, \ldots , |J|, \ r = 1, \ldots , |R|$.

$$x^r = x | col_r(Op) = 1.$$

The example for two competing operations of two different orders is shown in Figure 1.

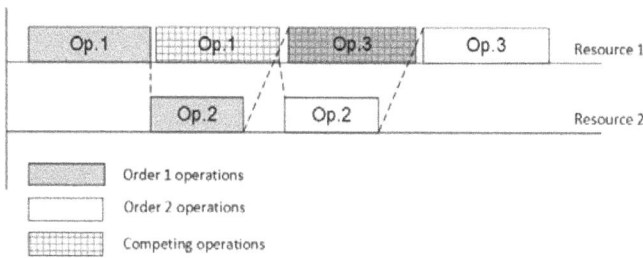

**Figure 1.** Competing operations on one resource.

For each resource $r = 1, \ldots , |R|$ each competing operation $i = index(x^r) = 1, \ldots , K$ of order $f = 1, \ldots , |F|$ will compete with all other competing operations of all other orders, i.e., going back to the example depicted in Figure 1 we will have the following constraint for each pair of operations:

$$x_{i\varphi} \geq x_{j\psi} + \tau_j \ XOR \ x_{j\psi} \geq x_{i\varphi} + \tau_i,$$

where indexes are as follows $i, j = 1, \ldots , K; \ \varphi, \psi = 1, \ldots , |F|; \ \varphi \neq \psi$. Implementing an additional Boolean variable $c_k \in \{0, 1\}$ will convert each pair of constraints into one single formula:

$$c_k \cdot x_{i,\varphi} + (1 - c_k) \cdot x_{j\psi} \geq c_k \cdot (x_{j\psi} + \tau_j) + (1 - c_k) \cdot (x_{i\varphi} + \tau_i).$$

From the above precedence constraints, we can form the following set of constraints for the optimization problem:

$$g_{i,j} \cdot x_{i,f} \geq x_{i,f} + \tau_i, \qquad\qquad f = 1, \ldots , |F|, \quad i = 1, \ldots , |J|, \quad j = 1, \ldots , |J|, \quad (1)$$

$$x_{|J|f} + \tau_{|J|} \leq d_f, \qquad\qquad f = 1, \ldots , |F|, \quad (2)$$

$$c_k \cdot x_{i,\varphi} + (1 - c_k) \cdot x_{j\psi} \geq c_k \cdot (x_{j\psi} + \tau_j) + (1 - c_k) \cdot (x_{i\varphi} + \tau_i), \qquad \varphi, \psi = 1, \ldots , |F|, \quad \varphi \neq \psi, \quad (3)$$

$$x_{if} > 0, \qquad\qquad f = 1, \ldots , |F|,$$

$$c_k \in \{0, 1\}, \qquad\qquad k = 1, \ldots , K.$$

where the variables $x_{if} > 0$, $f = 1,\ldots,|F|$, $i = 1,\ldots,|J|$ are continuous and represent the start times of each operation of each order; and variables $c_k \in \{0,1\}$, $k = 1,\ldots,K$ are Boolean and represent the position of each competing operation among other competitors for each resource.

### 2.3. Combinatorial Optimization Techniques

From the mathematical point of view, the resulting basic problem setting belongs to linear programming class. The variables $c_k \in \{0,1\}$, $k = 1,\ldots,K$ form a combinatorial set of subordinate optimization problems [8]. However, we must suppress that the combinatorial set $c_k \in \{0,1\}$, $k = 1,\ldots,K$, is not full as the sequence of operations within each order is dictated by manufacturing procedure. Consequently, the number of combinations is restricted only to those variants when the operations of different orders trade places without changing their queue within one order. Let us research possible ways to solve the given combinatorial set of problems. In this article, the optimization will be conducted with "makespan" criterion formalized with the formula:

$$\sum_{f=1}^{|F|} x_{|J|f} \to min.$$

### 2.3.1. Enumeration and Branch-and-Bound Approach

Enumerating the full set of combinations for all competing operations is an exercise of exponential complexity [9]. However, as we have mentioned above the number of combinations in constraint (3) is restricted due to other constraints (2) and (1). The precedence graph constraint (1) excludes the permutations of competing operations within one order making the linear problem setting non-feasible for corresponding combinations. The way we enumerate the combinations and choose the starting point of enumeration also influences the flow of computations significantly. The common approach to reducing the number of iterations in enumeration is to detect the forbidden areas where the optimal solution problem does not exist (e.g., the constraints form a non-feasible problem or multiple solutions exist). The most widely spread variant of the technique described is branch-and-bound method [10] and its different variations.

Let us choose the strategy of enumerating permutations of operations in such a way that it explicitly and easily affords to apply branch-and-bound method. The starting point of the enumeration will be the most obvious positioning of orders in a row as shown in Figure 2. We start from the point where the problem is guaranteed to be solvable—when all orders are planned in a row—which means that the first operation of the next order starts only after the last operation of the previous order is finished. If the preceding graph is correct and the resources are enough that means the single order can be placed on the resources that are totally free (the problem is solvable—we do not consider the unpractical case when the graph and/or resources are not enough to fulfill a single order). Infeasibility here can be reached only in the case when we start mixing the orders. This guarantees that the linear programming problem is solvable, however, this combination is far from being optimal/reasonable.

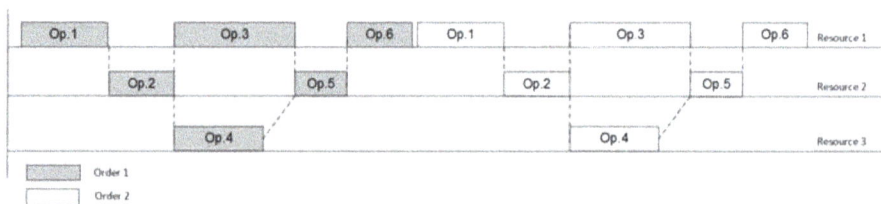

**Figure 2.** Initial state: all orders are placed in a row.

The enumeration is organized by shifting the operations of the last order to the left in such a way that they trade their places with previous competing operations of preceding orders (see Figure 3). As soon as the operation reaches its leftmost feasible position and the current combination corresponds to a linear programming problem with no solution we roll back to the previous combination, stop shifting operations of the last order and switch to the operations of the previous order. After the enumeration for orders on the current resource is over, we proceed the same iterations with the next resource. The formal presentation of the designed branch-and-bound algorithm is described by Procedure 1.

### Procedure 1: branch-and-bound

1. **Find the initial solution** of the **LP problem** (1)–(3) for the combination $c_k \in \{0,1\}$, $k = 1,\dots,K$ that corresponds to the case when all orders are in a row (first operation of the following order starts only after the last operation of preceding order is completed).

2. **Remember** the solution and keep the value of objective function $\Phi$ as temporarily best result $Opt = \Phi$.

3. **Select the last** order $f = |F|$.

4. **Select the resource** $r = |R|$ that is used by last operation $i = |J|$ of the precedence graph $G$ of the manufacturing process.

5. **The branch** $< i, r, f >$ is now formulated.

6. **Condition**: Does the selected resource $r$ have more than one operation in the manufacturing procedure that allocates it?

    6.1. **If yes then**

        **Begin** evaluating branch $< i, r, f >$

        **Shift** the operation $i$ one competing position to the left

        **Find** the solution of the **LP problem** (1)–(3) for current combination and calculate the objective function $\Phi$

        **Condition**: is the solution feasible?

        **If feasible then**

            **Condition**: Is the objective function value $\Phi$ better than temporarily best result $Opt$?

            **If yes then**

                **save current solution** as the new best result $Opt = \Phi$.

        **End of condition**

        **Switch to the preceding operation** $i = i - 1$ of currently selected order $f$ for the currently selected resource $r$.

        Go to the p. 5 and **start evaluating the new branch**

        **If not feasible then**

        Stop evaluating branch $< i, r, f >$

        Switch to the preceding resource $r = r - 1$

        Go to the p. 5 and **start evaluating the new branch**

        **End of condition**: is the solution feasible?

    **End of condition**: p. 6

7. **Switch to the preceding order** $f = f - 1$

8. **Repeat** pp. 4–7 **until** no more branches $< i, r, f >$ are available

9. **Repeat** pp. 3–8 **until** no more feasible shifts are possible for all operations in all branches.

Iterating combinations for one order on one resource can be considered as evaluating one branch of the algorithm [10]. Finding the leftmost position for an operation is similar to detecting a bound. Switching between branches and stopping on bounds is formalized with a general branch-and-bound procedure.

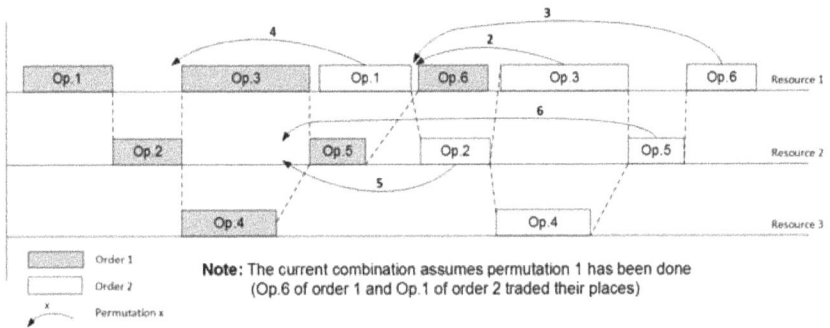

**Figure 3.** Permutations of competing operations in branch-and-bound algorithm.

### 2.3.2. Gradient-Alike Algorithm

The opposite of the mentioned above approach would be searching for a local minimum among all combinations with a gradient algorithm [11]. Here, we will compute an analogue of the derivative that is used to define the search direction at each point [12]. 'Derivative' for each competing operation is calculated by trying to shift it by one position to the left. In case the objective function reduces (or increases for maximization problems) we shift the operation to the left with the defined step (the step ratio is measured in a number of position operation skips moving to the left). The formal presentation of the designed gradient-alike algorithm is described by Procedure 2.

**Procedure 2: gradient algorithm**

1. **Find the initial solution** of the **LP problem** (1)–(3) for the combination $c_k \in \{0,1\}, k = 1, \ldots, K$ that corresponds to the case when all orders are in a row (first operation of the following order starts only after the last operation of preceding order is completed).

2. **Remember** the solution and keep the value of objective function $\Phi$ as temporarily best result $Opt = \Phi$.

3. **Select the last** order $f = |F|$.

4. **Select the resource** $r = |R|$ that is used by first operation $i = 1$ of the sequence graph of the manufacturing process.

5. **The current optimization variable for gradient optimization is selected**—position of operation $i$ of the order $f$ on the resource $r < i, r, f >$.

6. **Condition**: Does the selected resource $r$ have more than one operation in the manufacturing procedure that allocates it?

6.1. **If yes then**

**Begin** optimizing position $< i, r, f >$

6.1.1. **Set the step of shifting** the operation to maximum (shifting to the leftmost position). $D(i, r, f) = \max$

6.1.2. **Find the 'derivative' of shifting** the operation $i$ to the left

**Shift** the operation $i$ one position to the left

**Find** the solution of the **LP problem** (1)–(3) for current combination and calculate the objective function $\Phi$

**Condition A**: is the solution feasible?

**If feasible then**

**Condition B**: Is the objective function value $\Phi$ better than temporarily best result $Opt$?

**If yes then**

We found the optimization direction for position $< i, r, f >$, proceed to p. 6.1.3

**If not then**

No optimization direction for the current position $< i, r, f >$

**stop** optimizing position $< i, r, f >$

**switch** to the next operation $i = i + 1$

**go to p 6 and repeat search for position** $< i + 1, r, f >$

**End of condition B**

**If** not feasible **then**

No optimization direction for the current position $< i, r, f >$

**stop** optimizing position $< i, r, f >$

**switch** to the next operation $i = i + 1$

**go to p 6 and repeat search for position** $< i + 1, r, f >$

**End of condition A**

6.1.3. **Define the maximum possible optimization step** for the current position $< i, r, f >$, initial step value $D(i, r, f) = \max$

**Shift** the operation $i$ left using the step $D(i, r, f)$.

**Find** the solution of the **LP problem** (1)–(3) for current combination and calculate the objective function $\Phi$

**Condition C**: Is the solution feasible and objective function value $\Phi$ better than temporarily best result $Opt$?

**If** yes **then**

**save current solution** as the new best result $Opt = \Phi$

**stop** optimizing position $< i, r, f >$

**switch** to the next operation $i = i + 1$

**go to p 6 and repeat search for position** $< i + 1, r, f >$

**If** not **then**

reduce the step twice $D(i, r, f) = \frac{D(i,r,f)}{2}$ and repeat operations starting from p. 6.1.3

**End of condition C**

**Switch** to the next operation $i = i + 1$, go to p. 6 and optimize position $< i, r, f >$

7. **Switch** to the preceding resource $r = r - 1$

8. **Repeat** pp. 5–7 for currently selected resource $< i, r - 1, f >$

9. **Switch** to the preceding order $f = f - 1$

10. **Repeat** pp. 4–9 for currently selected order $< i, r, f - 1 >$.

11. **Repeat** pp. 3–10 **until** no improvements and/or no more feasible solutions exist.

The optimization variables in gradient algorithm represent positions of all competing operations relative to each other. The maximum optimization step for each iteration is detected on-the-fly by trying to shift the current operation to the leftmost position (as it is shown in Figure 4) that shows the LP problem (1)–(3) is solvable and the objective function is improved.

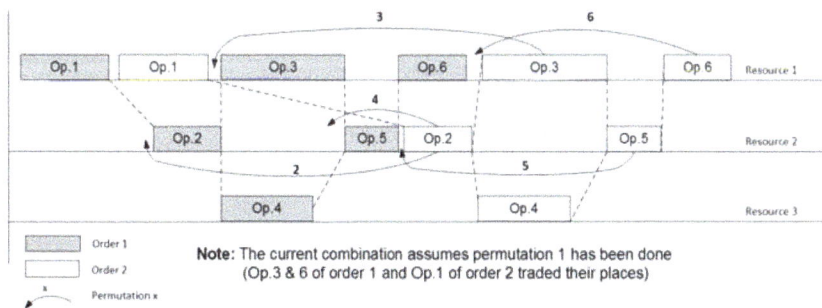

**Figure 4.** Permutations of competing operations in gradient algorithm.

## 3. Results of the Computational Experiment

As a reference trial example let us take the following scheduling problem. The manufacturing procedure contains 13 operations that require 6 different resources. The schedule of the manufacturing procedure for a single order is shown in Figure 5. In this example, we want to place two orders at the depicted resources and start evaluating from the point 'all orders in a row' (see Figure 6).

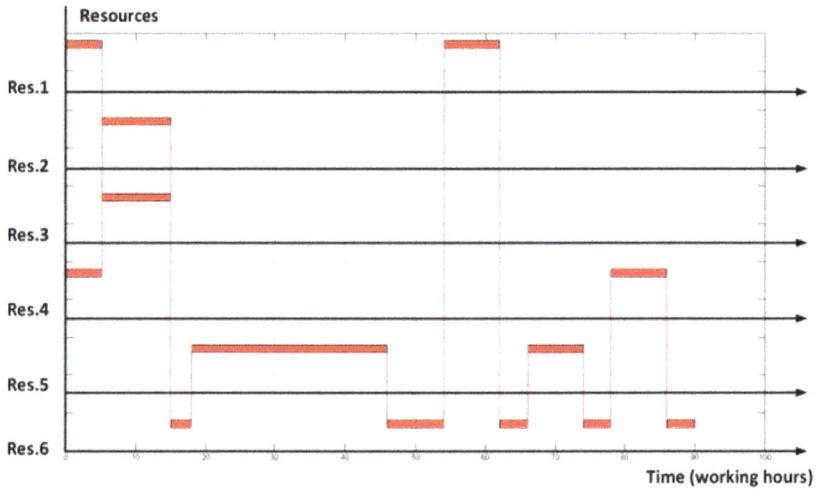

**Figure 5.** Gannt diagram for single order manufacturing procedure.

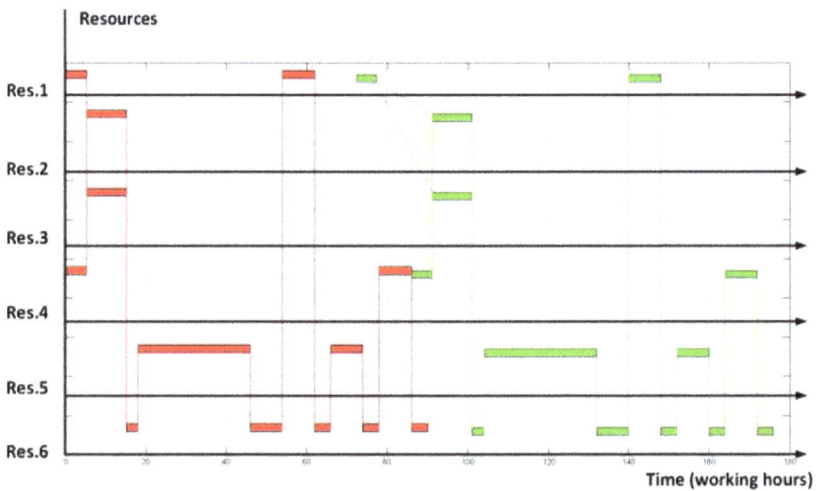

**Figure 6.** Gannt diagram for starting point of two orders schedule optimization.

The results of evaluating branch-and-bound and gradient algorithm are placed in the following Table 1. The resulting schedules are depicted in Figure 7 (for branch-and-bound algorithm) and Figure 8 (for gradient algorithm).

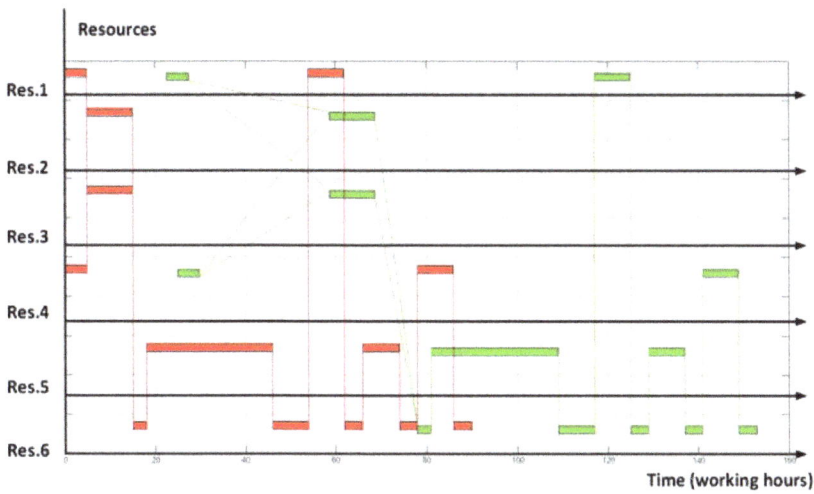

**Figure 7.** Branch-and-bound optimization result.

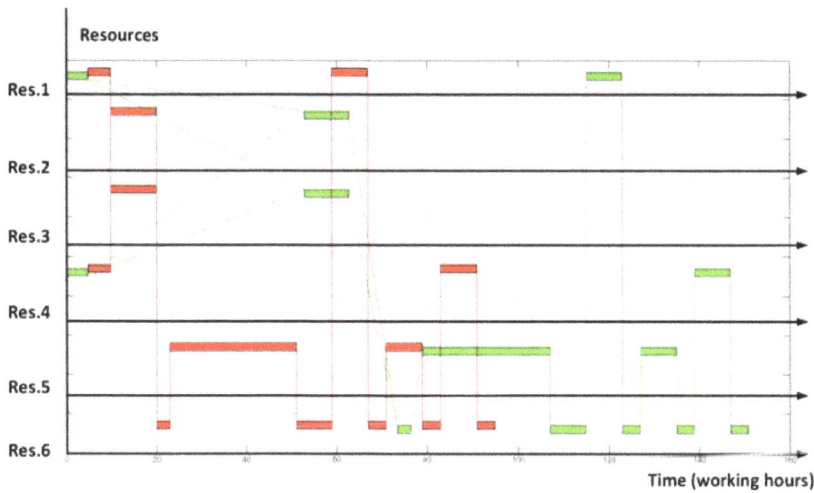

**Figure 8.** Gradient algorithm optimization result.

Results of analysis gives us understanding that for a modest dimensionality scheduling problem both considered approaches are applicable. However, as the number of operations in manufacturing procedure grows and the number of orders increases, we will experience an enormous growth of algorithm iterations for enumeration (branch-and-bound) technique and the gradient-alike algorithm will obviously detect local optimum as the best achievable solution (which means the optimization results will differ more and more compared to the digits in Table 1). Rising difference between the results is shown in Table 2.

Expanding the previous problem for a procedure of 500 operations in 2 orders we will get the results presented in Table 3.

**Table 1.** Algorithms test results: small number of operations (13 operations).

| Algorithm | Resulting "Makespan" Objective Function | Times of Orders Finished | | Full Number of Operations Permutations (Including Non-Feasible) | Number of Iterated Permutations | Number of Iterated Non-Feasible Permutations | Calculation Time, Seconds |
|---|---|---|---|---|---|---|---|
| | | Order 1 | Order 2 | | | | |
| B&B | 235 | 86 | 149 | 318 | 25 | 11 | 3.6 |
| gradient | 238 | 91 | 147 | 318 | 18 | 4 | 2.8 |

**Table 2.** Results for 2 orders with number of operations increasing.

| Number of Operations | Algorithm | Resulting "Makespan" Objective Function | Times of Orders Finished | | Number of Iterated Permutations | Number of Iterated Non-Feasible Permutations | Calculation Time, Seconds |
|---|---|---|---|---|---|---|---|
| | | | Order 1 | Order 2 | | | |
| 25 | B&B | 348 | 134 | 214 | 71 | 29 | 5.4 |
| | gradient | 358 | 139 | 219 | 36 | 9 | 3.1 |
| 50 | B&B | 913 | 386 | 527 | 201 | 59 | 25 |
| | gradient | 944 | 386 | 558 | 84 | 22 | 3.7 |
| 65 | B&B | 1234 | 488 | 746 | 490 | 70 | 112 |
| | gradient | 1296 | 469 | 826 | 126 | 66 | 11.2 |
| 100 | B&B | 1735 | 656 | 1079 | 809 | 228 | 288 |
| | gradient | 1761 | 669 | 1092 | 677 | 83 | 237 |

**Table 3.** Algorithms test results: increased number of operations (500 operations).

| Algorithm | Resulting "Makespan" Objective Function | Times of Orders Finished | | Full Number of Operations Permutations (Including Non-Feasible) | Number of Iterated Permutations | Number of Iterated Non-Feasible Permutations | Calculation Time, Seconds |
|---|---|---|---|---|---|---|---|
| | | Order 1 | Order 2 | | | | |
| B&B | 4909 | 1886 | 3023 | Unknown | ≈1200 (Manually stopped) | Unknown (≈10% of total) | Manually stopped after ≈7 h |
| Gradient | 4944 | 1888 | 3056 | Unknown | ≈550 | Unknown (≈30% of total) | ≈3 h |
| Random (gradient extension) | Trying implementing randomized algorithm led to high computation load of PC with no reasonable estimation of calculation time. | | | | | | |

Time to start first gradient iteration from feasible point took enumeration of more than 1000 variants.

## 4. Discussion

Analyzing the computational experiment result, we come to a classical research outcome that to solve the global optimization problem effectively we need to find a more computationally cheap algorithm that gives a solution closer to global optimum. Let us try several standard approaches to improve the solution with non-exponential algorithm extensions.

### 4.1. Random Search as an Effort to Find Global Optimum

As mentioned above, the gradient algorithm, being more effective from the computational complexity point of view, affords to find only local suboptimal solutions [12]. A typical extension to overcome this restriction would be random search procedure. The main idea of this modification is to iterate gradient search multiple times going out from different starting points [13]. In case the starting points are generated randomly we can assume that the more repeating gradient searches we do the higher the probability of finding a global optimum we achieve. There was some research conducted in this area whose outcome recommends how many starting points to generate in order to cover the problem's acceptance region with high value of probability [14]. According to [14] the acceptable number of starting points is calculated as

$$N \cdot \dim(\Phi),$$

where $N = 5 \ldots 20$, and $\dim(\Phi)$ is the dimensionality of optimization problem being solved, i.e., for our case this is the number of all competing operations of all orders on all resources. The result of applying random search approach is represented in Table 2.

The main barrier for implementing the random search procedure for the combinatorial scheduling problem is generating enough feasible optimization starting points. As we can see from the results in Table 2, number of failures to generate feasible starting point is much higher than the quantity of successful trials. Leaning upon the results of enumeration algorithm in Table 1 we can assume that the tolerance regions for optimization problem (1)–(3) are very narrow. Even from the trial example in Table 1 we see that the number of feasible iterations (25) collect less than 10 percent of all possible permutations (318) which leaves us very low probability of getting a feasible initial point for further gradient optimization. Thus, we can make a conclusion that 'pure' implementation of random search procedure will not give a huge effect but should be accompanied with some analytic process of choosing feasible initial points of optimization. Such a procedure may be based on non-mathematical knowledge such as industry or management expertise. From our understanding, this question should be investigated separately.

## 5. Conclusions

Research of a continuous-time scheduling problem is conducted. We formalized the scheduling problem as a combinatorial set [8] of linear programming sub problems and evaluated typical computational procedures with it. In addition to the classical and estimated resulting conflict between the "complexity" and "locality" of optimization algorithms we came to the conclusion that the classical approach of randomization is unhelpful in terms of improving the locally found suboptimal solution. The reason here is that the scheduling problem in setting (1)–(3) has a very narrow feasibility area which makes it difficult to randomly detect a sufficient number of starting points for further local optimization. The efficiency of random search might be increased by introducing a martial rule or procedure of finding typical feasible starting points. The other effective global optimization procedures are mentioned in a very short form and are left for further authors' research. They are:

- Genetic algorithms. From the first glance evolutionary algorithms [15] should have a good application case for the scheduling problem (1)–(3). The combinatorial vector of permutations $c_k \in \{0,1\}$, $k = 1, \dots, K$, seems to be naturally and easily represented as a binary crossover [15] while the narrow tolerance region of the optimization problem will contribute to the fast convergence of the breeding procedure. Authors of this paper leave this question for further research and discussion.
- Dynamic programming. A huge implementation area in global optimization (and particularly in RCPSP) is left for dynamic programming algorithms [16]. Having severe limitations in amount and time we do not cover this approach but will come back to it in future papers.

The computation speed of the high dimension problem using an average PC is not satisfactory. This fact forces authors to investigate parallel computing technologies. Future research assumes adoption of created algorithms to a parallel paradigm, for instance, implementing map-reduce technology [17].

**Acknowledgments:** This work was supported by the Russian Science Foundation (grant 17-19-01665).

**Author Contributions:** A.A.L. conceived conceptual and scientific problem setting; I.N. adopted the problem setting for manufacturing case and designed the optimization algorithms; N.P. implemented the algorithms, performed the experiments and analyzed the data.

**Conflicts of Interest:** The authors declare no conflict of interest. The founding sponsors had no role in the design of the study; in the collection, analyses, or interpretation of data; in the writing of the manuscript, and in the decision to publish the results.

## References

1. Artigues, C.; Demassey, S.; Néron, E.; Sourd, F. *Resource-Constrained Project Scheduling Models, Algorithms, Extensions and Applications*; Wiley-Interscience: Hoboken, NJ, USA, 2008.
2. Meyer, H.; Fuchs, F.; Thiel, K. *Manufacturing Execution Systems. Optimal Design, Planning, and Deployment*; McGraw-Hill: New York, NY, USA, 2009.
3. Jozefowska, J.; Weglarz, J. *Perspectives in Modern Project Scheduling*; Springer: New York, NY, USA, 2006.
4. Manne, A.S. On the Job-Shop Scheduling Problem. *Oper. Res.* **1960**, *8*, 219–223, doi:10.1287/opre.8.2.219.
5. Jones, A.; Rabelo, L.C. Survey of Job Shop Scheduling Techniques. In *Wiley Encyclopedia of Electrical and Electronics Engineering*; National Institute of Standards and Technology: Gaithersburg, ML, USA, 1999. Available online: http://citeseerx.ist.psu.edu/viewdoc/download?doi=10.1.1.37.1262&rep=rep1&type=pdf (accessed on 10 April 2017).
6. Taravatsadat, N.; Napsiah, I. Application of Artificial Intelligent in Production Scheduling: A critical evaluation and comparison of key approaches. In Proceedings of the 2011 International Conference on Industrial Engineering and Operations Management, Kuala Lumpur, Malaysia, 22–24 January 2011; pp. 28–33.
7. Hao, P.C.; Lin, K.T.; Hsieh, T.J.; Hong, H.C.; Lin, B.M.T. Approaches to simplification of job shop models. In Proceedings of the 20th Working Seminar of Production Economics, Innsbruck, Austria, 19–23 February 2018.
8. Trevisan, L. *Combinatorial Optimization: Exact and Approximate Algorithms*; Stanford University: Stanford, CA, USA, 2011.
9. Wilf, H.S. *Algorithms and Complexity*; University of Pennsylvania: Philadelphia, PA, USA, 1994.
10. Jacobson, J. *Branch and Bound Algorithms—Principles and Examples*; University of Copenhagen: Copenhagen, Denmark, 1999.
11. Erickson, J. *Models of Computation*; University of Illinois: Champaign, IL, USA, 2014.
12. Ruder, S. *An Overview of Gradient Descent Optimization Algorithms*; NUI Galway: Dublin, Ireland, 2016.
13. Cormen, T.H.; Leiserson, C.E.; Rivest, R.L.; Stein, C. *Introduction to Algorithms*, 3rd ed.; Massachusetts Institute of Technology: London, UK, 2009.
14. Kalitkyn, N.N. *Numerical Methods*; Chislennye Metody; Nauka: Moscow, Russia, 1978. (In Russian)
15. Haupt, R.L.; Haupt, S.E. *Practical Genetic Algorithms*, 2nd ed.; Wiley-Interscience: Hoboken, NJ, USA, 2004.

16. Mitchell, I. *Dynamic Programming Algorithms for Planning and Robotics in Continuous Domains and the Hamilton-Jacobi Equation*; University of British Columbia: Vancouver, BC, Canada, 2008.

17. Miner, D.; Shook, A. *MapReduce Design Patterns: Building Effective Algorithms and Analytics for Hadoop and Other Systems*; O'Reilly Media: Sebastopol, CA, USA, 2013.

*algorithms*

MDPI

*Article*

# Dual Market Facility Network Design under Bounded Rationality

**D. G. Mogale [1], Geet Lahoti [1], Shashi Bhushan Jha [1], Manish Shukla [2], Narasimha Kamath [3] and Manoj Kumar Tiwari [1,*]** 

1   Department of Industrial & Systems Engineering, Indian Institute of Technology, Kharagpur 721302, India; dgmogle@gmail.com (D.G.M.); geetlahoti2454@gmail.com (G.L.); sbjhakdk2009@gmail.com (S.B.J.)
2   Durham University Business School, Durham University, Durham DH1 3LB, UK; scholarmanish@gmail.com
3   O9 Solutions, Bangalore 560037, India; narasimha.kamath@gmail.com
*   Correspondence: mktiwari9@iem.iitkgp.ernet.in; Tel.: +91-9734444693

Received: 18 February 2018; Accepted: 16 April 2018; Published: 20 April 2018

**Abstract:** A number of markets, geographically separated, with different demand characteristics for different products that share a common component, are analyzed. This common component can either be manufactured locally in each of the markets or transported between the markets to fulfill the demand. However, final assemblies are localized to the respective markets. The decision making challenge is whether to manufacture the common component centrally or locally. To formulate the underlying setting, a newsvendor modeling based approach is considered. The developed model is solved using Frank-Wolfe linearization technique along with Benders' decomposition method. Further, the propensity of decision makers in each market to make suboptimal decisions leading to bounded rationality is considered. The results obtained for both the cases are compared.

**Keywords:** facility network design; newsvendor model; bounded rationality; non-linear programming; decomposition method; linearization technique

## 1. Introduction

Globalization has brought about large scale integration of regional economies and culture through effective transportation and communication. International trade has been the central attention of major firms in last couple of decades. Economic globalization has led to major changes in how the companies deal with the market. Any attempt towards a competitive advantage at an international level has far reaching benefits to the company. Even supporters of globalization agree that the benefits of globalization are not without risks—such as those arising from volatile capital movements. With the recent global meltdown, capital flows reversing and international trade shrinking, getting into a non-imitable market advantage that has assumed the prime importance [1].

In this work, the focus is on firms that are present in many markets across the globe. With Asian countries like India and China gaining importance in the international market, there are issues that globalization still has to iron out before we can really see homogeneity across the globe. Low income countries are slowly gaining ground in the recent globalized world. On one hand, these countries offer huge market opportunities, with large gain in the labor cost, but it has itself associated with increased logistics cost and foreign trade barriers. Considering these paradoxical factors simultaneously, this paper works on quantifying the decision of off-shoring production facilities, modeled by taking into account the real life uncertainties and response variances. In particular, we consider the case of two economies served by a firm manufacturing two products at two localized final assemblies, using a common component which can either be produced locally or transported from a central setup which would also cause the company to incur respective transportation costs. Each of the markets has respective demand characteristics, selling price, and assembly costs. We intend to maximize the total profit of the firm, by optimally deciding where to locate the common component manufacturing setup, and also the individual transportation quantities. This situation is analyzed using the Newsvendor Model.

The notion of bounded rationality in operation management was taken from the economics literature. Initially, the "bounded rationality" term was introduced by Simon [2] to denote the decision-making behaviors where the decision maker finds the alternatives. Later, scholars constructed diverse model frameworks to cover agents' bounded rationality by the quantal response model. To account for the limitations in the normative approach adopted by a traditional Newsvendor Model, which assumes that decision makers are perfect optimizers, the descriptive approach of bounded rationality is incorporated. The decision model of bounded rationality is basically based on classical quantal choice theory [3]. Considering all possible utility generating alternatives to be the candidates for selection, decision makers try to choose much captivating alternatives (obtaining greater utility) because their probability of selection is high. For this purpose, the analytically convenient logit choice model can be used where selection probability of alternate $j$ is proportional to $\exp(u_j)$ [4]. The quantal choice models take into account that decision makers may commit errors. It is not necessary that one always makes the best decisions; better decisions are made frequently.

Thus, we formulate the problem with the framework of bounded rationality for the multi-market network design. The resulting formulation leads to a quadratic objective function with linear constraints. Since the constraints are not all equality constraints, it is not simple to reduce the problem into linear system of equations. The problem is solved through a novel approach of Frank-Wolfe linearization method along with Generalized Benders' Decomposition method. The results are compared with standard gradient search method and the superior convergence is empirically proved.

The remaining part of the text is organized as follows. In Section 2, the literature is reviewed. The model and methodology are described in Section 3. Section 4 discusses the test results and comparative analysis. Some concluding remarks along with future work are given in Section 5.

## 2. Literature Review

A broad review has been presented by Snyder [5] on stochastic and robust facility location models, illustrating the optimization techniques that have evolved in this area. Hayes and Wheelwright [6] describe several approaches such as geographical network analysis, physical facilities analysis, functional analysis and organizational mission analysis, and product-process analysis for the formulation of multi-plant facility strategy. Skinner [7] proposes the concept of operational focus using the product-process approach, and Van Mieghem [8] models the approach of product-process focus for a multi-market scenario. Cohen and Lee [9] are the pioneers towards formulating global manufacturing strategy as a mathematical programming problem accounting factors such as local demand, sourcing constraints and taxations. Organizations may select for developing their facility with respect to product, volumes, or process. Lu et al. [1] analyze such scenarios under known demand and incorporating scale economies. Their model includes two products with distinct demand features, two geographically disjoint markets with diverse economic features, and two processes with distinct goals comprises of common component manufacturing and dedicated assembly. Multi-product firms frequently usage the commonality of foster flexibility for their process networks. In another study, the trade-offs between risk pooling and logistics cost with respect to two extreme configurations such as process and product has been investigated for commonality in a multi-plant network [10]. Melkote et al. [11] develops an optimization model for facility location and transportation network design problem.

Rationality refers to perfect optimization for the given scenario. On the contrary, the concept of bounded rationality identifies the intrinsic imperfections in human decision making. Chen et al. [12] described utility functions for the decision maker and inferred stochastic probabilities. Su [13] modeled the bounded rationality including stochastic elements in the decision process. Despite selecting the utility-maximizing alternative constantly, decision makers embrace a probabilistic opting rule such that much captivating alternatives are selected frequently. The resulting formulation led to a quadratic optimization with linear constraints.

Exact methods like Benders' decomposition method (Benders, 1962) are frequently found in applications of operations management. Bahl et al. [14] were amongst the first to use multi-item production scheduling. Benders' decomposition methods have also found applications in discretely constrained mathematical programs [15]. Even stochastic optimization problems have found applications of Benders' decomposition methods. Archibald et al. [16] compared the performance of Dynamic programming versus the nested Benders' decomposition method for optimization of hydro-electric generation problems. Velarde et al. [17] implemented a heuristic based Benders' decomposition method for the robust international sourcing model. The methodology involved generation of cuts via Tabu search, using the dual variables from the sub-problem to obtain the neighborhoods. Benders' decomposition method has been employed to solve the strategic and operations management problems, especially in networks based problems, which as such can be solved by linear programming methods. Ali et al. utilized it for solving multi-commodity network problems [18]. Dogan et al. [19] used Benders decomposition method for solving a mixed integer programming problem for the strategic production-distribution allocation problem for the supply chain network. Benders' decomposition method is also utilized in project time compression problems in Critical Path Method (CPM)/Programme Evaluation Review Technique (PERT) networks, by approximating the convex or concave activity cost-duration functions to piecewise linear time cost curves [20]. The Frank-Wolfe algorithm is used to solve quadratic programming problems with linear constraints [21].

Lee et al. [22] investigated a plant allocation and inventory level decisions for serving global supply chains and revealed that the importing firm escalates its inventory level if the transportation cost increases or the exchange rate of the inventory country lessen. The result of this study has been empirically confirmed using data of Korean firms yielded from the Export-Import Bank of Korea. Jean et al. [23] studied the relationship-based product innovation in global supply chains where this research offered a context-based explanation for the contradictory and conflicting empirical indication with respect to relational capital innovation links. In another study, a single period inventory model has been proposed to encapsulate the trade-off between inventory policies and disruption risks considering the scenario of dual-sourcing supply chain [24]. Tang et al. [25] examined multiple-attribute decision making using Pythagorean 2-tuple linguistic numbers, and proposed two operators, namely, Pythagorean 2-tuple linguistic weighted Bonferroni mean (WBM) and Pythagorean 2-tuple linguistic weighted geometric Bonferroni mean (WGBM). Further, the effectiveness of this approach is verified considering the green supplier selection problem. Zhang et al. [26] developed a greedy insertion heuristic algorithm using a multi-stage filtering mechanism comprising coarse granularity and fine granularity filtering for ameliorating the energy efficiency of single machine scheduling problems. A two-product, multi-period newsvendor problem has been formulated by Zhang and Yang [27] considering fixed demand. In addition, this research used the online learning method for performing the experiment, and also proposed real and integer valued online ordering policies. Egri and Vancza [28] presented an extended version of the newsvendor model to fulfil all demands of the customer by the supplier. This model particularly minimizes the total cost comprising setup, obsolete inventory and inventory holding costs. Furthermore, the model has been developed considering the decentralized setting using asymmetric information of customer and supplier. Ren and Huang [29] summarized several methods of modeling customer bounded rationality and also surveyed implementation of approaches with respect to appropriate operations management settings. In a dyadic supply chain, Du et al. [30] examined fairness preferences especially individuals' psychological understanding. Moreover, to formulate the fairness concerns, they used Nash bargaining solution as fairness reference. Di et al. [31] developed a systematic methodology to obtain boundedly rational user equilibria (BRUE) solutions for networks with fixed demands that assist in predicting BRUE link traffic flow patterns in a given network to guide planners for making network design decisions accordingly when travellers behave boundedly rational. Di and Liu [32] presented a comprehensive review of models as well as methodologies of bounded rationality route choice behavior. The models are mainly divided into two parts: substantive and procedural bounded rationality models. While methodologies applied on these two models are game theory and interactive congestion game for substantive bounded rationality, and random utility and non- or semi-compensatory for procedural bounded rationality.

### 3. Mathematical Model Formulation

The present research investigates the operations policy of a multi-national firm which makes goods for serving geographically disjointed markets with localized final assemblies and joint parts. The joint part can be shipped from the markets with diverse financial and demand features. The cost required for transferring the joint components (intermediate goods between markets), and for foreign trade barriers like tariffs and duties provides the total transportation cost. The generalized operations strategy structure is illustrated in Figure 1 for multiple-product and multiple-market model with commonality.

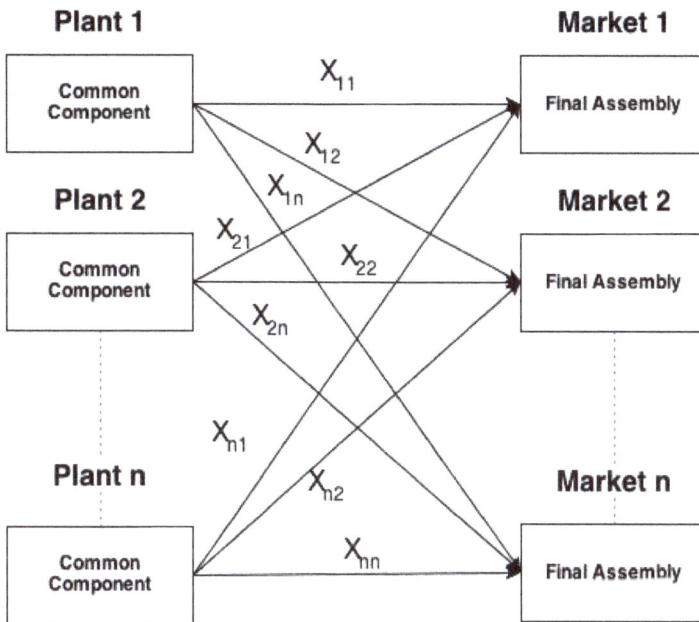

**Figure 1.** Generalized multi-product multi-market network design.

This study dealt with the strategic facility location decision of engine production by taking into account the financial and demand features. The newsvendor network approach is used to evaluate the model. Firstly, we analyze the scenario under deterministic demand. The sensitivity of the trade-off between the manufacturing costs and the transportation costs are examined. Secondly, the situation under demand uncertainty is evaluated. Here, we assume uniformly distributed demand. The profit function is accordingly formulated and similar analysis is carried out. The function is solved using Frank Wolfe's sequential linearization technique with Bender's Decomposition method. Thirdly, the results of optimal decision scenario is compared with the scenario of bounded rationality considering uniformly distributed demand. When we consider the bounded rationality, the best option is not always picked by individuals, however, they opt for better options often. Finally, we intend to give a statistical proof of the superiority of the algorithm used over the traditional gradient search method by Rosen. Thus, a three-fold comprehensive comparison is made for the multi-market network design. The canonical setting at each market includes a newsvendor setup who has to obtain how much of the commonality to order. Every item costs $c$ but can be sold at price $p$, where $p > c$. For a given demand, the profit is given as the product of price and minimum of demand and ordering quantity minus the procurement cost, which is cost per item multiplied by ordering quantity. Suppose we have a random demand, $D$ having density $f$ and cumulative distribution $F$. The expected profit in any of the selling points in the network under newsvendor model with demand uncertainty is given by:

$$\pi(x) = p\, E(\min(D, x)) - cx \qquad (1)$$

where,

    $x$: Number of items ordered
    $p$: Selling price of the item
    $c$: Cost of item, where $c < p$
    $D$: Demand
    $E(y)$: Expected value of $y$

Considering that the demand $D$ is uniformly distributed between $[a, b]$. Then the profit function in Equation (1) can be expressed as [13]:

$$\pi(x) = Ax^2 + Bx + C \qquad (2)$$

where,

$$A = \frac{-p}{2(b-a)},\ B = \frac{pb}{b-a} - c,\ C = \frac{-pa^2}{2(b-a)}$$

Now, we apply the above model to the multi-market scenario. Each of the markets can be considered as a newsvendor model. Consider the multi-market model as depicted in Figure 2.

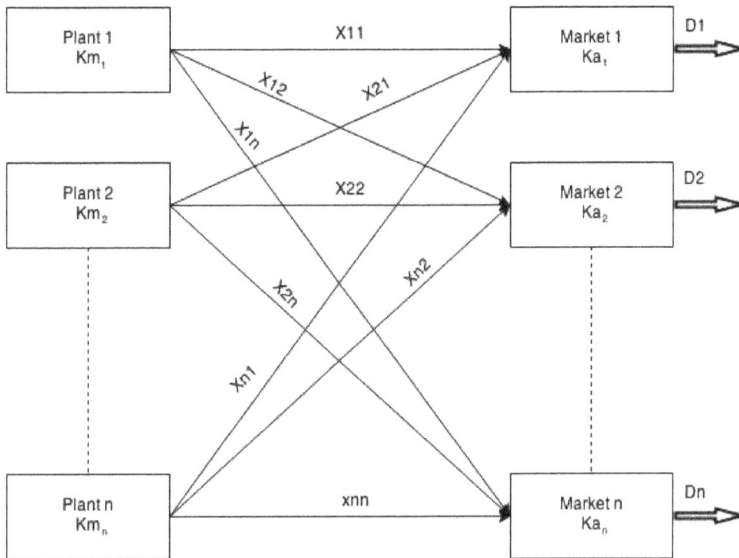

**Figure 2.** Generalized multi-market design.

With reference to Figure 2, we use the following notations:

Notations

$x_{ij}$: Quantity of common component manufactured in plant $i$, and assembled at market $j$.

$p_j$: Selling price of final assembly at market $j$.

$Ct_{ij}$: Cost of transportation of common component from plant $i$ to market $j$. ($Ct_{ii} = 0$)

$Cm_i$: Cost of manufacturing the common component in plant $i$

$Km_i$: Capacity limit on the common component manufacturing at plant $i$

$Ka_j$: Capacity limit on the final assembly at market $j$.

$D_j$: Demand at the market $j$, following a uniform distribution, $Dj \sim U[aj,bj]$

$a_j$: Lower bound on the uniform demand distribution in market $j$

$b_j$: Upper bound on the uniform demand distribution in market $j$

$Ca_j$: Average cost per item incurred at market $j$ due to various sources from which the items are transported

$v_{ij}$: Value addition in terms of per unit profit obtained due to a common component transportation from plant $i$ and sold at market $j$

$A_j$: Coefficient of $x^2$ term in the Equation (2) for profit function in market $j$

$B_j$: Coefficient of $x$ term in the Equation (2) for profit function in market $j$

$C_j$: Constant term in the Equation (2) for profit function in market $j$

$\beta_j$: Bounded rationality parameter for market $j$

$\pi_j$: Profit in market $j$

$n$: Total number of countries in the multi-market network.

Plant, $i = 1, 2 \ldots n$

Market $j = 1, 2 \ldots n$

To apply the newsvendor's model to each market, we need to determine the average cost per item at each market. This is taken as the weighted average of the common component manufacturing cost per item from the various source plants from which the items are transported.

$$Ca_j = \frac{\sum\limits_{j=1}^{a} x_{ij}(Cm_i + Ct_{ij})}{\sum\limits_{i=1}^{a} x_{ij}} \tag{3}$$

Note here that $Ct_{ii} = 0$, as there is no transportation cost from plant $i$ to market $i$. The value addition in each path in terms of per unit profit obtained due to a common component transportation from plant $i$ and sold at market $j$ is given by,

$$V_{ij} = p_j - Cm_i - Ct_{ij}$$

Thus, from Equation (2), the profit in each market $j$, under uniform demand distribution $Dj \sim U[aj,bj]$ would be given as,

$$\pi_j = A_j \left(\sum_{i=1}^{a} x_{ij}\right)^2 + B_j \left(\sum_{i=1}^{a} x_{ij}\right) + C_j \tag{4}$$

where,

$$A_j = \frac{-p_j}{2(b_j - a_j)}, \ B_j = \frac{p_j b_j}{(b_j - a_j)} - Ca_j, \ C_j = \frac{-p_j a_j^2}{2(b_j - a_j)}$$

Note here that $Caj$ is not a constant term. Thus the equations is expanded and taking $BB_j = \frac{p_j b_j}{(b_j - a_j)}$, we have the expression for the total profit $\pi$ as,

$$\pi = \sum_{i=1}^{n} \pi_j = \sum_{j=1}^{n} \sum_{i=1}^{n} A_j x_{ij}^2 + 2 \sum_{j=1}^{n} \sum_{i=1}^{n} \sum_{k=i+1}^{n} A_j x_{ij} x_{kj} + \sum_{j=1}^{n} \sum_{i=1}^{n} x_{ij} (Cm_i + Ct_{ij} + BB_j) \tag{5}$$

Thus, we have the objective function to maximize the total profit: max $\pi$

This is subject to capacity constraints in each plant as well as each market for the final assembly,

Plant capacity constraint: $\sum_{j=1}^{n} x_{ij} \leq Km_i \quad \forall i = 1 \ldots n$

Final Assembly capacity constraint: $\sum_{j=1}^{n} x_{ij} \leq Ka_j \quad \forall j = 1 \ldots n$

The formulation in Equation (5) is having a quadratic objective function and linear constraints. We apply Frank Wolfe's sequential linearization method, and Bender's Decomposition method is used for decomposing the linear problem within the Frank Wolfe's method. Stepwise procedure for the integrated methodology is given as follows:

Consider the problem to,

$$\text{Minimize } \pi(x)$$
$$\text{subject to } x \in S;$$
$$\text{where } S \subset R^n$$

Let, $f$ is a continuously differentiable function.

***Step 0.*** Choose an initial solution, $x^0 \in S$. Let $k = 0$. Here an arbitrary basic feasible solution is chosen, that is, an extreme point.

***Step 1.*** Determine a search direction, $p^k$. In the Frank Wolfe algorithm $p^k$ is determined through the solution of the approximation of the problem (1) that is obtained by replacing the function $f$ with its first-order Taylor expansion around $x_k$. Therefore, solve the problem to minimize:

$$z_k(x) = \pi(x_k) + \nabla \pi(x_k)^T (x - x_k)$$

Subject to $x \in S$

This is a Linear Programming (LP) problem. In large scale problems this would be computationally complex, and would require to be decomposed into smaller problems which can be solved easily. Hence we apply the Bender's decomposition method at this point to solve the LP.

***Step 2.*** Thus by Bender's decomposition method $x^*$ is obtained as an optimal solution to gradient equation. The search direction is $p_k = x^{*-} - x_k$, that is, the direction vector from the feasible point $x_k$ towards the extreme point.

***Step 3.*** A step length is determined and represented as $\alpha_k$, such that $f(x_k + \alpha_k p_k) < f(x_k)$.

***Step 4.*** New iteration point is found out using $x_{k+1} = x_k + \alpha_k p_k$.

***Step 5.*** Check if stopping criteria is reached, else go to Step 1.

*3.1. Illustration of Methodology on the Problem*

The formulation in Equation (5) can be rewritten in matrix format as:

$$\text{Max } x^T Q x + b^T x + c$$
$$\text{s.t } Ax \leq K$$

where $Q$, $b$, $c$, $A$ and $K$ are appropriately defined depending on the size of the problem. With this notation on the formulation, the application of integrated Frank Wolfe's and Bender's decomposition method as applied to the multi-market network design problem under uniform demand uncertainty is depicted in the flowchart, in Figure 3.

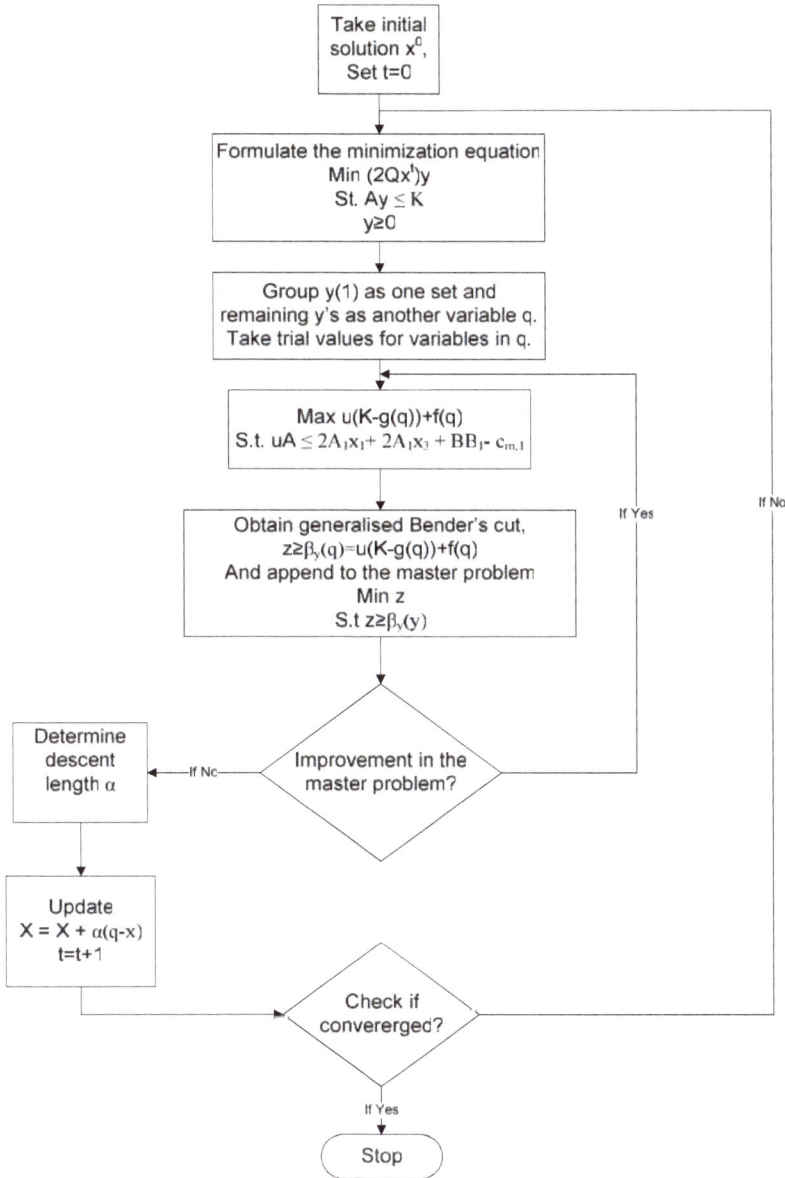

**Figure 3.** Flow-chart of the algorithm.

## 3.2. Newsvendor Model under Bounded Rationality

The assumption of perfect rationality for decision makers is a typical tactic in a majority of normative analysis. In order to obtain bounded rationality, we followed the approach of [13], where multi-nomial logit choice model is applied and assumed that decision makers selects alterative $j$, with probability.

$$\phi_j = \frac{e^{\frac{u_j}{\beta}}}{\sum_j e^{\frac{u_j}{\beta}}} \tag{6}$$

Correspondingly, the following density function gives the logit choice probabilities over a continuous domain:

$$\phi(y) = \frac{e^{\frac{u(y)}{\beta}}}{\int_y e^{\frac{u(y)}{\beta}}} \tag{7}$$

Thus, the selection of agents is a random variable. The superior options are selected frequently using the above logit structure. The best alternative depends on the mode of the selection distribution. The magnitude of cognitive and computational restrictions suffered by the decision makers is elucidated by $\beta$ parameter. This can be observed when selection distribution approximates the uniform distribution over the options in $\beta \rightarrow \infty$ limit. In some utmost instances, decision makers randomize the options with equal probabilities when they failed to make any informed choices. In the other scenario, when $\beta \rightarrow \infty$ the choice distribution in (1) completely focused on utility maximizing options. The selection of perfect rational decision maker corresponds with this choice. Therefore, we can consider the magnitude of $\beta$ as the extent of bounded rationality. The logit choice framework to the newsvendor problem is employed here. The profit function is as stated in Equation (1) is

$$\pi(x) = p\,E(\min(D,x)) - cx$$

This is uniquely maximized at,

$$x^* = F^{-1}(1 - c/p) \tag{8}$$

where, $F$ is the cumulative distribution function of demand D. All other symbols are as used before. This solution is selected when the decision maker is perfectly rational, although the newsvendor's ordering amount is dependent on noise and it called the behavioral solution in bounded rationality. The bounded rational newsvendor can order any amount of product within the range of minimum possible and maximum possible demand realizations in the domain S. Then, the Equation (9) gives the probability density function associated with behavioral solution:

$$\phi(y) = \frac{e^{\frac{\pi(x)}{\beta}}}{\int_s e^{\frac{\pi(v)}{\beta}}dv} = \frac{e^{\frac{pE(\min(D,x))-cx}{\beta}}}{\int_s e^{\frac{pE(\min(D,v))-cv}{\beta}}dv} \tag{9}$$

Now, if the demand $D$ is uniformly distributed in in the interval of $[a, b]$, then the behavioral solution would follow truncated normal distribution over $[a, b]$ with mean $\mu$ and standard deviation $\sigma$ [13].

$$\mu = b - \frac{c}{p}(b - a) \tag{10}$$

$$\sigma^2 = \beta\frac{b - a}{p} \tag{11}$$

The expected value of this truncated normal distribution would be,

$$E(x) = \mu - \sigma\frac{\phi((b-\mu)/\sigma) - \phi((a-\mu)/\sigma)}{\varphi((b-\mu)/\sigma) - \varphi((a-\mu)/\sigma)} \tag{12}$$

where, $\phi(.)$ denotes cumulative normal distribution and $\varphi(.)$ denotes normal density function.

As an example in the dual market scenario, each market decision maker would be subjected to bounded rationality. Thus, the expected ordering quantities in each market, with uniformly distributed demand between $[a_i, b_i]$ $(i = 1 \ldots n)$ and with bounded rationality parameters as $\beta_1$ and $\beta_2$ would be,

$$E(x_1 + x_3) = \mu_1 - \sigma_1 \frac{\phi((b_1 - \mu_1)/\sigma_1) - \phi((a_1 - \mu_1)/\sigma_1)}{\varphi((b_1 - \mu_1)/\sigma_1) - \varphi((a_1 - \mu_1)/\sigma_1)} \tag{13}$$

where,

$$\mu_1 = b_1 - \frac{c_1}{p_1}(b_1 - a_1) \qquad\qquad \sigma_1^2 = \beta_1 - \frac{b_1 - a_1}{p_1}$$

and,

$$E(x_2 + x_4) = \mu_1 - \sigma_1 \frac{\phi((b_1 - \mu_1)/\sigma_1) - \phi((a_1 - \mu_1)/\sigma_1)}{\varphi((b_1 - \mu_1)/\sigma_1) - \varphi((a_1 - \mu_1)/\sigma_1)} \tag{14}$$

The concept of bounded rationality can be reconciled with original problem of multi-market network design in way that initially under uniform demand distribution, the production allocation strategy is decided by the company and then ordering decisions are made under bounded rationality, giving suboptimal. Similarly, the model can be applied to multi-market scenario with any number of plants and markets. Thus, under bounded rational conditions, we can find the expected profit as given in (3). Hence, a comprehensive comparison is made amongst the profits under various conditions in the following section.

## 4. Test Results and Comparative Analysis

A comprehensive analytical study was made for the different test examples of varying problem sizes. Each case depicts the results in different scenarios. We illustrate the optimal production allocation in the network design for the uniform demand distribution. Finally, we make a comparison of this with profits in the scenario of bounded rational decision maker.

All data sets used for the test examples are provided in Appendix A. These include the pricing details, the manufacturing and transportation costs and the demand distribution parameters.

X: matrix production allocation, where $x_{ij}$ represents the optimal production allocation between plant $i$ and market $j$.

### 4.1. I. Test Case 1

**Problem size n = 2.**

*1. Scenario of Uniform Demand Distribution*

X:

The production allocation values for test case 1 are given in Table 1.

**Table 1.** Production allocation values for test case 1.

|         | Market 1 | Market 2 |
|---------|----------|----------|
| Plant 1 | 1636.36  | 0        |
| Plant 2 | 0        | 1187.5   |

Objective function value = 108,678.

We observe that the solution indicates a market focused strategy for the company as depicted in Figure 4.

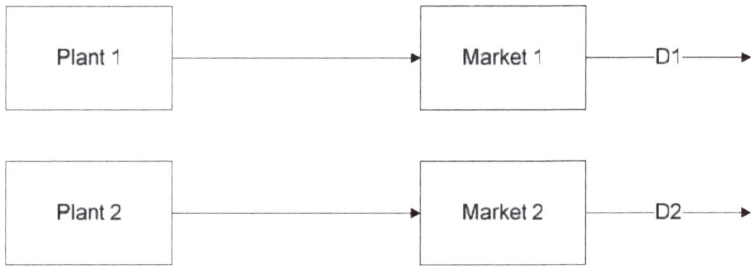

**Figure 4.** Market focused strategy.

*2. Scenario of Bounded Rational Decision Maker*

Table 2 depicts the ordering quantities and profits under bounded rationality for test case 1

**Table 2.** Ordering quantities and profits under bounded rationality for test case 1.

| Cases for Bounded Rationality Conditions with Varying β Parameter | | | | | |
|---|---|---|---|---|---|
| $\beta_1$ | $\beta_2$ | Ordering Quantity in Market 1 | Ordering Quantity in Market 2 | Total Profits | % Deviation from Achievable Profits |
| 0 | 0 | 1409 | 1156 | 108,640 | 0.12 |
| 10 | 10 | 1409 | 1187 | 105,802 | 1.34 |
| 100 | 100 | 1491 | 1482 | 104,054 | 1.88 |
| 1000 | 1000 | 1499 | 1499 | 103,753 | 1.97 |
| 10,000 | 10,000 | 1499 | 1499 | 103,752 | 2.03 |
| 100,000 | 100,000 | 1500 | 1500 | 103,750 | 2.45 |
| 1,000,000 | 1,000,000 | 1500 | 1500 | 103,750 | 2.56 |

*4.2. II. Test Case 2*

**Problem size n = 4.**

*1. Scenario of Uniform Demand Distribution*: Table 3 illustrates the Production allocation values.

X:

**Table 3.** Production allocation values for test case 2.

| | Market 1 | Market 2 | Market 3 | Market 4 |
|---|---|---|---|---|
| Plant 1 | 1518 | 0 | 482 | 0 |
| Plant 2 | 0 | 1187.5 | 0 | 0 |
| Plant 3 | 0 | 0 | 968 | 0 |
| Plant 4 | 0 | 0 | 0 | 1608 |

Objective function value = 286,553.59.

Strategic allocation of common component is shown in Figure 5.

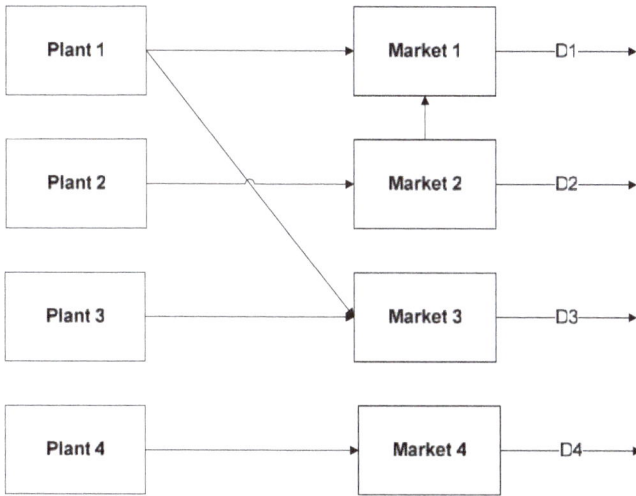

**Figure 5.** Strategic allocation of common component in Test Case 2.

*2. Scenario of Bounded Rational Decision Maker*

The ordering quantities and profits under bounded rationality for test case 2 are reported in Table 4.

**Table 4.** Ordering quantities and profits under bounded rationality for test case 2.

| | Cases for Bounded Rationality Conditions with Varying $\beta$ Parameter | | | | | |
|---|---|---|---|---|---|---|
| $\beta$ | Ordering Quantity in Market 1 | Ordering Quantity in Market 2 | Ordering Quantity in Market 3 | Ordering Quantity in Market 4 | Total Profits | % Deviation from Achievable Profits |
| 0 | 1225 | 1263 | 923 | 1234 | 286,550 | 0.45 |
| 10 | 1232 | 1305 | 940 | 1258 | 262,834 | 1.53 |
| 100 | 1242 | 1360 | 1051 | 1264 | 213,850 | 2.24 |
| 1000 | 1335 | 1374 | 1066 | 1395 | 204,467 | 2.2 |
| 10,000 | 1409 | 1444 | 1066 | 1395 | 216,800 | 2.57 |
| 100,000 | 1462 | 1471 | 1066 | 1402 | 283,890 | 2.83 |
| 1,000,000 | 1500 | 1498 | 1075 | 1428 | 278,777 | 3.44 |

*4.3. III. Test Case 3*

**Problem size n = 10**

*1. Scenario of Uniform Demand Distribution*

X:

Table 5 shows the production allocation values of test case 3.

**Table 5.** Production allocation values for test case 3.

| | Market 1 | Market 2 | Market 3 | Market 4 | Market 1 | Market 2 | Market 3 | Market 4 | Market 9 | Market 10 |
|---|---|---|---|---|---|---|---|---|---|---|
| Plant 1 | 1153 | 0 | 0 | 780 | 0 | 0 | 0 | 0 | 0 | 0 |
| Plant 2 | 0 | 593 | 782 | 0 | 0 | 0 | 0 | 543 | 0 | 0 |
| Plant 3 | 0 | 0 | 0 | 0 | 0 | 0 | 1472 | 0 | 0 | 0 |
| Plant 4 | 1673 | 0 | 0 | 0 | 0 | 0 | 0 | 0 | 0 | 0 |
| Plant 5 | 0 | 630 | 0 | 226 | 0 | 0 | 0 | 0 | 0 | 476 |
| Plant 6 | 1885 | 0 | 0 | 0 | 0 | 0 | 0 | 0 | 0 | 0 |
| Plant 7 | 0 | 252 | 378 | 150 | 0 | 0 | 0 | 474 | 0 | 720 |

**Table 5.** *Cont.*

| | Market 1 | Market 2 | Market 3 | Market 4 | Market 1 | Market 2 | Market 3 | Market 4 | Market 9 | Market 10 |
|---|---|---|---|---|---|---|---|---|---|---|
| Plant 8 | 0 | 0 | 0 | 0 | 0 | 1836 | 0 | 0 | 0 | 0 |
| Plant 9 | 0 | 0 | 0 | 1566 | 0 | 0 | 0 | 0 | 0 | 0 |
| Plant 10 | 1724 | 0 | 0 | 0 | 0 | 0 | 0 | 0 | 0 | 0 |

Objective function value = 723,977

2. *Scenario of Bounded Rational Decision Maker*

The Profits under bounded rationality for test case 3 are presented in Table 6. Bounded rationality factor β = 100.

**Table 6.** Profits under bounded rationality for test case 3.

| Cases for Bounded Rationality Conditions with Varying β Parameter | | |
|---|---|---|
| β | Total profits | % Deviation from Achievable Profits |
| 0 | 723,908 | 0.53 |
| 10 | 615,322 | 2.19 |
| 100 | 553,790 | 2.85 |
| 1000 | 498,411 | 2.68 |
| 10,000 | 448,570 | 2.61 |
| 100,000 | 723,908 | 3.45 |

*4.4. Comparative Analysis*

For a problem size of 2 (dual market scenario), there are various possible cases and scenarios that might arise due to the tradeoffs between the prices and the costs. In order to evaluate the financial desirability of the options processing actions for every product, the investment decisions declined. For the analysis of various possible cases in the dual market scenario, we take the value addition into consideration, which represents the per-unit profit of the company, due to the particular path taken in the production allocation. Let $v1$ denote the value addition due to path from plant 1 to market 1, similarly, $v2$ be for plant 1 to market 2, $v3$ for plant 2 to market 1 and v4 for plant 2 to market 2. The probable orderings of the four net values are 24 (=4!) without considering any assumption, but their definitions suggest an interdependence, $v4 \geq v1 \Rightarrow v2 \geq v3$, which removes 6 orderings (where, as mentioned earlier $vi$ represents the per unit profit value addition in any path). The outstanding 18 are distributed into two clusters, one of which is a replica of the other. The following study emphasizes the 9 orderings because of symmetry in results. These result in 4 possible allocation strategies of market focused, centralized with two possible centralization and hybrid type of model as indicated in Figure 6.

Table 7 depicts 1 of the 9 possible scenarios which is analyzed under cases of deterministic demand, uniform demand and uniform demand under bounded rationality, and the total profits (T.P) in each case is found. For the case of uniform distribution, the Frank Wolfe's Bender decomposition method is compared with the Rosen's gradient search method which is depicted in the Figure 7.

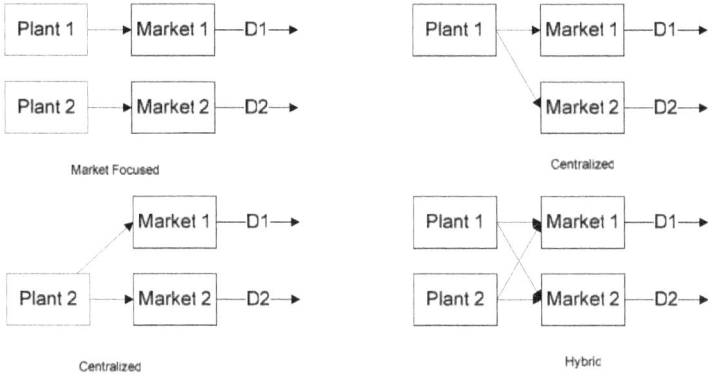

**Figure 6.** Different possible scenarios of production allocation for dual market case.

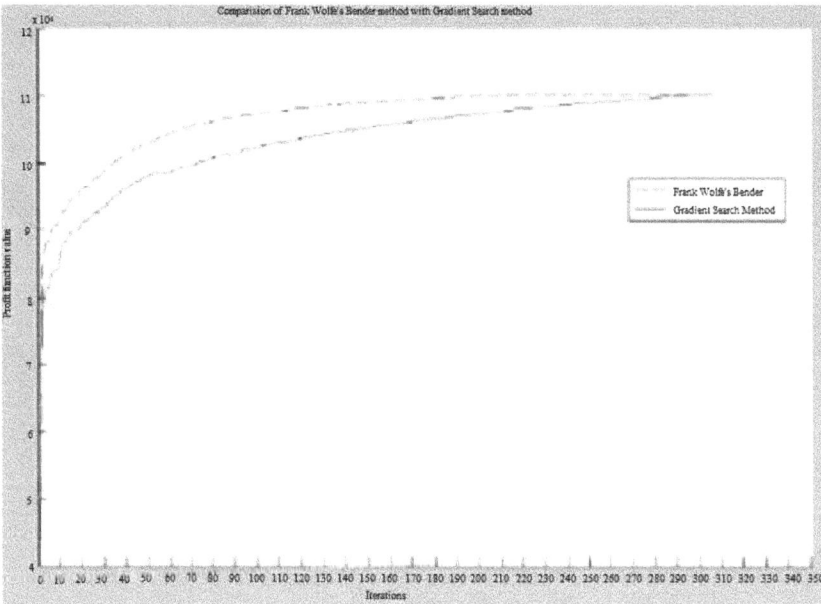

**Figure 7.** Graphical comparison of the two algorithms for scenario 1.

**Table 7.** Profit results of Scenario 1.

| Scenario 1 | | Case of Deterministic Demand | | Case of Uniform Demand Distribution D ~ [1000, 200] |
|---|---|---|---|---|
| | $p_1 = 100$ | $x_1 = D_1$ | | $x_1 = 1565$ |
| | $p_2 = 80$ | $x_2 = 0$ | | $x_2 = 0$ |
| | $c_{m,1} = 40$ | $x_3 = 0$ | | $x_3 = 0$ |
| $v_1 > v_2 > v_3 > v_4$ | $c_{m,2} = 65$ | $x_4 = D_2$ | | $x_4 = 1142$ |
| $c_{t,1} > \Delta c_m \quad c_{t,2} > \Delta c_m$ | $c_{t,1} = 30$ | $D = D_1 = D_2$ | T.P | |
| | | 1000 | 85,000 | |
| | $c_{t,2} = 25$ | 2000 | 170,000 | T.P. = 105,951.9 |
| | | 1500 | 127,500 | |

## 5. Conclusions and Future Work

The decision regarding the manufacturing of the common component has been analytically modeled and solved for a multi-market scenario. For the case of deterministic demand, the linear formulation was solved for the 3 distinct scenarios. Then, the case of uncertainty has been analyzed, where the demand is uniformly distributed. This was formulated as a quadratic problem and solved using a proposed integrated methodology exploiting the Frank-Wolfe linearization technique, along with Benders' decomposition method. To establish the superiority of the proposed solution methodology, in terms of convergence, we compared it with Rosen's gradient search technique. In case of uniform demand with bounded rationality, the decision maker ordering a specific number of units for selling in the respective market cannot make the optimal choice, but his choice is guided by a probabilistic logit choice model, with higher probability of making the right choice. We have demonstrated (as expected) that bounded rational decisions are less profitable than the rational decision making and get reinforced when the lack of clarity of decision makers becomes more prominent. Scope for future studies can include, but is not limited to, incorporating some other behavioral characteristics along with bounded rationality.

**Author Contributions:** Manoj Kumar Tiwari and Narasimha Kamath contributed to the overall idea and manuscript drafting. D. G. Mogale and Geet Lahoti defined the problem and formulated the mathematical model. Shashi Bhushan Jha and Manish Shukla written the overall manuscript. Finally, all the authors have read and approved the final manuscript.

**Conflicts of Interest:** The authors declare no conflict of interest.

## Appendix A

The notations used have following implications:

$n$ — Number of countries where the markets are present.
$p$ — Vector where $p_j$ is the selling price of final assembly at market $j$.
$Ct$ — Matrix for the transportation cost where, $Ct_{ij}$ is the cost of transportation of common component from plant i to market $j$. ($Ct_{ii} = 0$).
$Cm$ — Vector where $Cm_i$ is cost of manufacturing the common component in plant $i$.
$Km$ — Vector where $Km_i$ is the capacity limit on the common component manufacturing at plant $i$.
$Ka$ — Vector where $Ka_j$ is the capacity limit on the final assembly at market $j$.
$a$ — Vector where $a_i$ is lower bound on the uniform demand distribution.
$b$ — Vector where $b_i$ is upper bound on the uniform demand distribution.

I.    *Test Case 1*

1. $n = 2$
2. $p$

| 110 | 80 |
|---|---|

3. $Ct$

| 0 | 30 |
|---|---|
| 0 | 25 |

4. $Cm$

| 40 | 65 |
|---|---|

5. $Km$

| 2000 | 2000 |
|---|---|

6.  *Ka*

| 2000 | 2000 |
|------|------|

7.  *A*

| 1000 | 1000 |
|------|------|

8.  *b*

| 2000 | 2000 |
|------|------|

## II.  *Test Case 2*

1.  *n = 4*
2.  *p*

| 110 | 80 | 140 | 120 |
|-----|----|----|-----|

3.  *Ct*

| 0  | 30 | 24 | 45 |
|----|----|----|----|
| 25 | 0  | 32 | 56 |
| 21 | 28 | 0  | 58 |
| 56 | 68 | 43 | 0  |

4.  *Cm*

| 40 | 65 | 77 | 47 |
|----|----|----|----|

5.  *Km*

| 2000 | 2000 | 2000 | 2000 |
|------|------|------|------|

6.  *Ka*

| 2000 | 2000 | 2000 | 2000 |
|------|------|------|------|

7.  *a*

| 1000 | 1000 | 1000 | 1000 |
|------|------|------|------|

8.  *b*

| 2000 | 2000 | 2000 | 2000 |
|------|------|------|------|

## III.  *Test Case 3*

1.  *n = 10*
2.  *p*

| 110 | 80 | 140 | 120 | 110 | 80 | 140 | 120 | 110 | 80 |
|-----|----|----|-----|-----|----|----|-----|-----|----|

3.  *Ct*

| | | | | | | | | | |
|---|---|---|---|---|---|---|---|---|---|
| 0 | 48 | 20 | 42 | 47 | 56 | 15 | 16 | 31 | 20 |
| 31 | 0 | 13 | 49 | 29 | 58 | 31 | 49 | 40 | 52 |
| 13 | 13 | 0 | 57 | 22 | 30 | 58 | 28 | 31 | 29 |
| 29 | 32 | 29 | 0 | 53 | 13 | 54 | 36 | 57 | 20 |
| 21 | 34 | 24 | 59 | 0 | 15 | 46 | 52 | 55 | 40 |
| 23 | 35 | 27 | 29 | 41 | 0 | 58 | 40 | 46 | 21 |
| 39 | 30 | 45 | 59 | 20 | 42 | 0 | 23 | 37 | 26 |
| 41 | 36 | 28 | 15 | 48 | 15 | 54 | 0 | 13 | 37 |
| 20 | 30 | 23 | 46 | 32 | 56 | 14 | 26 | 0 | 19 |
| 29 | 56 | 37 | 33 | 29 | 23 | 38 | 60 | 29 | 0 |

4.   *Cm*

| | | | | | | | | | |
|---|---|---|---|---|---|---|---|---|---|
| 40 | 65 | 77 | 47 | 40 | 65 | 77 | 47 | 40 | 65 |

5.   *Km*

| | | | | | | | | | |
|---|---|---|---|---|---|---|---|---|---|
| 2000 | 2000 | 2000 | 2000 | 2000 | 2000 | 2000 | 2000 | 2000 | 2000 |

6.   *Ka*

| | | | | | | | | | |
|---|---|---|---|---|---|---|---|---|---|
| 2000 | 2000 | 2000 | 2000 | 2000 | 2000 | 2000 | 2000 | 2000 | 2000 |

7.   *a*

| | | | | | | | | | |
|---|---|---|---|---|---|---|---|---|---|
| 1000 | 1000 | 1000 | 1000 | 1000 | 1000 | 1000 | 1000 | 1000 | 1000 |

8.   *b*

| | | | | | | | | | |
|---|---|---|---|---|---|---|---|---|---|
| 2000 | 2000 | 2000 | 2000 | 2000 | 2000 | 2000 | 2000 | 2000 | 2000 |

## References

1.   Lu, L.X.; Van Mieghem, J.A. Multi-market facility network design with offshoring applications. *Manuf. Serv. Oper. Manag.* **2009**, *11*, 90–108. [CrossRef]
2.   Simon, H.A. A behavioral model of rational choice. *Q. J. Econ.* **1955**, *69*, 99–118. [CrossRef]
3.   Luce, R.D. *Individual Choice Behavior: A Theoretical Analysis*; Courier Corporation: North Chelmsford, MA, USA, 2005.
4.   McFadden, D. *Econometric models of probabilistic choice. Structure Analysis Discrete Data Economic Application*; MIT Press Cambridge: Cambridge, MA, USA, 1981; pp. 198–272.
5.   Snyder, L.V. Facility location under uncertainty: A review. *IIE Trans.* **2006**, *38*, 547–564. [CrossRef]
6.   Hayes, R.H.; Wheelwright, S.C. *Restoring Our Competitive Edge: Competing through Manufacturing*; John Wiley & Sons: New York, NY, USA, 1984; Volume 8.
7.   Skinner, W. *The Focused Factory*; Harvard Business Review: Brighton, UK, 1974; Volume 52.
8.   Van Miegham, J. *Operations Strategy*; Dynamic Ideas: Belmont, MA, USA, 2008.
9.   Cohen, M.A.; Lee, H.L. Resource deployment analysis of global manufacturing and distribution networks. *J. Manuf. Oper. Manag.* **1989**, *2*, 81–104.
10.   Kulkarni, S.S.; Magazine, M.J.; Raturi, A.S. Risk pooling advantages of manufacturing network configuration. *Prod. Oper. Manag.* **2004**, *13*, 186–199. [CrossRef]
11.   Melkote, S.; Daskin, M.S. An integrated model of facility location and transportation network design. *Transp. Res. Part A Policy Pract.* **2001**, *35*, 515–538. [CrossRef]

12. Chen, H.-C.; Friedman, J.W.; Thisse, J.-F. Boundedly rational Nash equilibrium: A probabilistic choice approach. *Games Econ. Behav.* **1997**, *18*, 32–54. [CrossRef]

13. Su, X. Bounded rationality in newsvendor models. *Manuf. Serv. Oper. Manag.* **2008**, *10*, 566–589. [CrossRef]

14. Bahl, H.C.; Zionts, S. Multi-item scheduling by Benders' decomposition. *J. Oper. Res. Soc.* **1987**, *38*, 1141–1148. [CrossRef]

15. Gabriel, S.A.; Shim, Y.; Conejo, A.J.; de la Torre, S.; García-Bertrand, R. A Benders decomposition method for discretely-constrained mathematical programs with equilibrium constraints. *J. Oper. Res. Soc.* **2010**, *61*, 1404–1419. [CrossRef]

16. Archibald, T.W.; Buchanan, C.S.; McKinnon, K.I.M.; Thomas, L.C. Nested Benders decomposition and dynamic programming for reservoir optimisation. *J. Oper. Res. Soc.* **1999**, *50*, 468–479. [CrossRef]

17. Velarde, J.L.G.; Laguna, M. A Benders-based heuristic for the robust capacitated international sourcing problem. *IIE Trans.* **2004**, *36*, 1125–1133. [CrossRef]

18. Ali, I.; Barnett, D.; Farhangian, K.; Kennington, J.; Patty, B.; Shetty, B.; McCarl, B.; Wong, P. Multicommodity network problems: Applications and computations. *IIE Trans.* **1984**, *16*, 127–134. [CrossRef]

19. Dogan, K.; Goetschalckx, M. A primal decomposition method for the integrated design of multi-period production—Distribution systems. *IIE Trans.* **1999**, *31*, 1027–1036. [CrossRef]

20. Kuyumcu, A.; Garcia-Diaz, A. A decomposition approach to project compression with concave activity cost functions. *IIE Trans.* **1994**, *26*, 63–73. [CrossRef]

21. Frank, M.; Wolfe, P. An algorithm for quadratic programming. *Nav. Res. Logist.* **1956**, *3*, 95–110. [CrossRef]

22. Lee, S.; Park, S.J.; Seshadri, S. Plant location and inventory level decisions in global supply chains: Evidence from Korean firms. *Eur. J. Oper. Res.* **2017**, *262*, 163–179. [CrossRef]

23. Jean, R.J.; Kim, D.; Bello, D.C. Relationship-based product innovations: Evidence from the global supply chain. *J. Bus. Res.* **2017**, *80*, 127–140. [CrossRef]

24. Xanthopoulos, A.; Vlachos, D.; Iakovou, E. Optimal newsvendor policies for dual-sourcing supply chains: A disruption risk management framework. *Comput. Oper. Res.* **2012**, *39*, 350–357. [CrossRef]

25. Tang, X.; Huang, Y.; Wei, G. Approaches to Multiple-Attribute Decision-Making Based on Pythagorean 2-Tuple Linguistic Bonferroni Mean Operators. *Algorithms* **2018**, *11*, 5. [CrossRef]

26. Zhang, H.; Fang, Y.; Pan, R.; Ge, C. A New Greedy Insertion Heuristic Algorithm with a Multi-Stage Filtering Mechanism for Energy-Efficient Single Machine Scheduling Problems. *Algorithms* **2018**, *11*, 18. [CrossRef]

27. Zhang, Y.; Yang, X. Online ordering policies for a two-product, multi-period stationary newsvendor problem. *Comput. Oper. Res.* **2016**, *74*, 143–151. [CrossRef]

28. Egri, P.; Váncza, J. Channel coordination with the newsvendor model using asymmetric information. *Int. J. Prod. Econ.* **2012**, *135*, 491–499. [CrossRef]

29. Ren, H.; Huang, T. Modeling customer bounded rationality in operations management: A review and research opportunities. *Comput. Oper. Res.* **2018**, *91*, 48–58. [CrossRef]

30. Du, S.; Nie, T.; Chu, C.; Yu, Y. Newsvendor model for a dyadic supply chain with Nash bargaining fairness concerns. *Int. J. Prod. Res.* **2014**, *52*, 5070–5085. [CrossRef]

31. Di, X.; Liu, H.X.; Pang, J.-S.; Ban, X.J. Boundedly rational user equilibria (BRUE): Mathematical formulation and solution sets. *Transp. Res. Part B Methodol.* **2013**, *57*, 300–313. [CrossRef]

32. Di, X.; Liu, H.X. Boundedly rational route choice behavior: A review of models and methodologies. *Transp. Res. Part B Methodol.* **2016**, *85*, 142–179. [CrossRef]

![algorithms logo] *algorithms*

MDPI

*Article*

# Optimal Control Algorithms and Their Analysis for Short-Term Scheduling in Manufacturing Systems

**Boris Sokolov** [1], **Alexandre Dolgui** [2] and **Dmitry Ivanov** [3,*]

[1]   Saint Petersburg Institute for Informatics and Automation of the RAS (SPIIRAS), V.O. 14 line, 39, 199178 St. Petersburg, Russia; sokol@iias.spb.su

[2]   Department of Automation, Production and Computer Sciences, IMT Atlantique, LS2N—CNRS UMR 6004, La Chantrerie, 4 rue Alfred Kastler, 44300 Nantes, France; alexandre.dolgui@imt-atlantique.fr

[3]   Department of Business Administration, Berlin School of Economics and Law, 10825 Berlin, Germany

*   Correspondence: divanov@hwr-berlin.de; Tel.: +49-30-30877-1155

Received: 18 February 2018; Accepted: 16 April 2018; Published: 3 May 2018

**Abstract:** Current literature presents optimal control computational algorithms with regard to state, control, and conjunctive variable spaces. This paper first analyses the advantages and limitations of different optimal control computational methods and algorithms which can be used for short-term scheduling. Second, it develops an optimal control computational algorithm that allows for the solution of short-term scheduling in an optimal manner. Moreover, qualitative and quantitative analysis of the manufacturing system scheduling problem is presented. Results highlight computer experiments with a scheduling software prototype as well as potential future research avenues.

**Keywords:** scheduling; optimal control; manufacturing; algorithm; attainable sets

## 1. Introduction

Short-term scheduling in manufacturing systems (MS) considers jobs that contain operation chains with equal (i.e., flow shop) or different (i.e., job shop) machine sequences and different processing times. Operations which need to be scheduled for machines with different processing power are subject to various criteria including makespan, lead time, and due dates ([1–4]).

Over the last several decades, various studies have investigated scheduling problems from different perspectives. A rich variety of methods and applications can be observed in the development of rigorous theoretical models and efficient solution techniques. [5–12] and [13] have demonstrated that specific large-scale scheduling problems with complex hybrid logical and terminal constraints, process execution non-stationary (i.e., interruptions in machine availability), complex interrelations between process dynamics, capacity evolution and setups (i.e., intensity-dependent processing times for machine work) require further investigation in terms of a broad range of methodical approaches. One of these is optimal control.

Optimal control approaches differ from mathematical programming methods and represent schedules as trajectories. The various applications of optimal control to scheduling problems are encountered in production systems with single machines [14], job sequencing in two-stage production systems [15], and multi-stage machine structures with alternatives in job assignments and intensity-dependent processing rates. Specifically, such multi-stage machine structures include flexible MSs ([16–18]), supply chain multi-stage networks ([19,20]), and Industry 4.0 systems that allow data interchange between the product and stations, flexible stations dedicated to various technological operations, and real-time capacity utilization control [13].

A diversity of knowledge and findings in optimal control applications exists which pertains to scheduling. However, these approaches typically pertain to trajectories which are assumed to be optimal but are subject to some specific constraint system and process model forms such as finite dimensionality, convexity, etc. In general, optimal control algorithms only provide the necessary conditions for optimal solution existence. The maximum principle provides both optimality and necessary conditions only for some specific cases, i.e., linear control systems. As such, further investigations are required in each concrete application case.

This paper seeks to bring the discussion forward by carefully elaborating on the optimality issues described above and providing some ideas and implementation guidance on how to confront these challenges. The purpose of the present study is to contribute to existing works by providing some closed forms of algorithmic optimality proven in the rich optimal control axiomatic.

The rest of this paper is organized as follows. In Section 2, we propose general elements of the multiple-model description of industrial production scheduling and its dynamic interpretation. In Section 3, optimal control computational algorithms and their analyses are proposed. Section 4 presents a combined method and algorithm for short-term scheduling in MS. In Section 5, qualitative and quantitative analysis of the MS scheduling problem is suggested. The paper concludes in Section 6 by summarizing the results of this study.

## 2. Optimal Control Applications to Scheduling in Manufacturing Systems

Optimal control problems belong to the class of extremum optimization theory, i.e., the analysis of the minimization or maximization of some functions ([21–28]. This theory evolved on the basis of calculus variation principles developed by Fermat, Lagrange, Bernulli, Newton, Hamilton, and Jacobi. In the 20th century, two computational fundamentals of optimal control theory, maximum principle [21] and dynamic programming method [29], were developed. These methods extend the classical calculus variation theory that is based on control variations of a continuous trajectory and the observation of the performance impact of these variations at the trajectory's terminus. Since control systems in the middle of the 20th century were increasingly characterized by continuous functions (such as 0–1 switch automats), the development of both the maximum principle and dynamic programming was necessary in order to solve the problem with complex constraints on control variables.

Manufacturing managers are always interested in non-deterministic approaches to scheduling, particularly where scheduling is interconnected with the control function [30]. Studies by [31] and [11] demonstrated a wide range of advantages regarding the application of control theoretic models in combination with other techniques for production and logistics. Among others, they include a non-stationary process view and accuracy of continuous time. In addition, a wide range of analysis tools from control theory regarding stability, controllability, adaptability, etc. may be used if a schedule is described in terms of control. However, the calculation of the optimal program control (OPC) with the direct methods of the continuous maximum principle has not been proved efficient [32]. Accordingly, the application of OPC to scheduling is important for two reasons. First, a conceptual problem consists of the continuous values of the control variables. Second, a computational problem with a direct method of the maximum principle exists. These shortcomings set limitations on the application of OPC to purely combinatorial problems.

To date, various dynamic models, methods and algorithms have been proposed to solve planning and scheduling problems in different application areas ([13,19,33–38]). In these papers, transformation procedures from classical scheduling models to their dynamic analogue have been developed. For example, the following models of OPC were proposed:

$M_g$—dynamic model of MS elements and subsystems motion control;

$M_k$—dynamic model of MS channel control;

$M_o$—dynamic model of MS operations control;

$M_f$—dynamic model of MS flow control;

$M_p$—dynamic model of MS resource control;

$M_e$—dynamic model of MS operation parameters control;

$M_c$—dynamic model of MS structure dynamic control;

$M_v$—dynamic model of MS auxiliary operation control.

The detailed mathematical formulation of these models was presented by [36,37] as well as [38]. We provide the generalized dynamic model of MS control processes ($M$ model) below:

$$M = \left\{ \mathbf{u}(t) \,\middle|\, \dot{\mathbf{x}} = \mathbf{f}(\mathbf{x}, \mathbf{u}, t); \mathbf{h}_0(\mathbf{x}(T_0)) \leq \mathbf{O}, \right.$$
$$\left. \mathbf{h}_1(\mathbf{x}(T_f)) \leq \mathbf{O}, \mathbf{q}^{(1)}(\mathbf{x}, \mathbf{u}) = \mathbf{O}, \mathbf{q}^{(2)}(\mathbf{x}, \mathbf{u}) < \mathbf{O} \right\}, \tag{1}$$

$$J_\vartheta = J_\vartheta(\mathbf{x}(t), \mathbf{u}(t), t) = \varphi_\vartheta\left(\mathbf{x}(t_f)\right) + \int_{T_0}^{T_f} f_\vartheta(\mathbf{x}(\tau), \mathbf{u}(\tau), \tau) d\tau,$$

$$\vartheta \in \{g, k, o, f, p, e, c, v\},$$

where $\vartheta \in \{g, k, o, f, p, e, c, v\}$—lower index which correspond to the motion control model; channel control model; operations control model; flow control; $M_p$—resource control; operation parameters control; structure dynamic control model; auxiliary operation control model; $\mathbf{x} = \left\| \mathbf{x}^{(g)T}, \mathbf{x}^{(k)T}, \mathbf{x}^{(o)T}, \mathbf{x}^{(p)T}, \mathbf{x}^{(f)T}, \mathbf{x}^{(e)T}, \mathbf{x}^{(c)T}, \mathbf{x}^{(v)T} \right\|^T$ is a vector of the MS generalized state, $\mathbf{u} = \left\| \mathbf{u}^{(g)T}, \mathbf{u}^{(k)T}, \mathbf{u}^{(o)T}, \mathbf{u}^{(p)T}, \mathbf{u}^{(f)T}, \mathbf{u}^{(e)T}, \mathbf{u}^{(c)T}, \mathbf{u}^{(v)T} \right\|^T$ is a vector of generalized control, $\mathbf{h}_0$, $\mathbf{h}_1$ are known vector functions that are used for the state $\mathbf{x}$ end conditions at the time points $t = T_0$ and $t = T_f$, and the vector functions $\mathbf{q}^{(1)}$, $\mathbf{q}^{(2)}$ define the main spatio–temporal, technical, and technological conditions for MS execution; $J_\vartheta$ are indicators characterizing the different aspects of MS schedule quality.

Overall, the constructed model $M$ (1) is a deterministic, non-linear, non-stationary, finite-dimensional differential system with a reconfigurable structure. Figure 1 shows the interconnection of models $M_g$, $M_k$, $M_o$, $M_p$, $M_f$, $M_e$, $M_c$, and $M_v$ embedded in the generalized model.

In Figure 1 the additional vector function $\boldsymbol{\xi} = \left\| \boldsymbol{\xi}^{(g)T}, \boldsymbol{\xi}^{(k)T}, \boldsymbol{\xi}^{(o)T}, \boldsymbol{\xi}^{(p)T}, \boldsymbol{\xi}^{(f)T}, \boldsymbol{\xi}^{(e)T}, \boldsymbol{\xi}^{(c)T}, \boldsymbol{\xi}^{(v)T} \right\|^T$ of perturbation influences is introduced.

The solutions obtained in the presented multi-model complex are coordinated by the control inputs vector $\mathbf{u}^{(o)}(t)$ of the model Mo. This vector determines the sequence of interaction operations and fixes the MS resource allocation. The applied procedure of solution adjustment refers to resource coordination.

The model complex $M$ evolves and generalizes the dynamic models of scheduling theory. The predominant distinctive feature of the complex is that non-linear technological constraints are actualized in the convex domain of allowable control inputs rather than in differential equations [36,37].

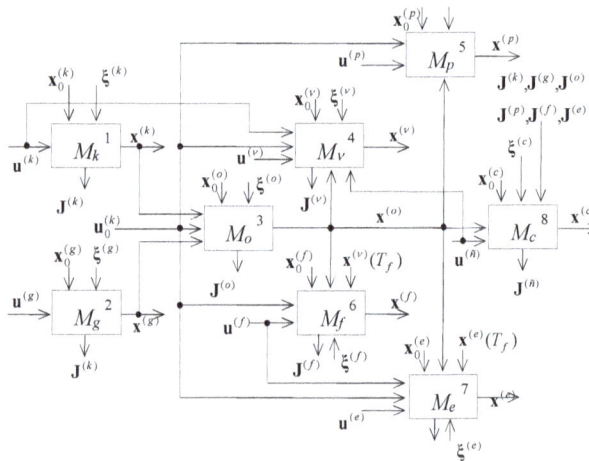

**Figure 1.** The scheme of optimal program control model interconnection.

In this case, the MS job shop scheduling problem can be formulated as the following problem of OPC: it is necessary to find an allowable control $\mathbf{u}(t)$, $t \in (T_0, T_f]$ that ensures for the model (1) meeting of vector constraint functions $\mathbf{q}^{(1)}(\mathbf{x}, \mathbf{u}) = 0$, $\mathbf{q}^{(2)}(\mathbf{x}, \mathbf{u}) \leq 0$ and guides the dynamic system (i.e., MS job shop schedule) $\dot{\mathbf{x}}(t) = f(t, \mathbf{x}(t), \mathbf{u}(t))$, from the initial state to the specified final state. If there are several allowable controls (schedules), then the best one (optimal) should be selected to maximize (minimize) $J_\theta$. The formulated model is a linear, non-stationary, finite-dimensional, controlled differential system with the convex area of admissible control. Note that the *boundary problem* is a standard OPC problem ([21,36,37]). This model is linear in the state and control variables, and the objective is linear. The transfer of non-linearity to the constraint ensures convexity and allows use of interval constraints.

In this case, the adjoint system can be written as follows:

$$\dot{\psi}_l = -\frac{\partial H}{\partial x_l} + \sum_{\alpha=1}^{I_1} \lambda_\alpha(t) \frac{\partial \mathbf{q}_\alpha^{(1)}(\mathbf{x}(t), \mathbf{u}(t))}{\partial x_l} + \sum_{\beta=1}^{I_2} \rho_\beta(t) \frac{\partial \mathbf{q}_\beta^{(2)}(\mathbf{x}(t), \mathbf{u}(t))}{\partial x_l} \tag{2}$$

The coefficients $\lambda_\alpha(t)$, $\rho_\beta(t)$ can be determined through the following expressions:

$$\rho_\beta(t)\mathbf{q}_\beta^{(2)}(\mathbf{x}(t), \mathbf{u}(t)) \equiv 0, \ \beta \in \{1, \dots, I_2\} \tag{3}$$

$$\operatorname{grad}_{\mathbf{u}} H(\mathbf{x}(t), \mathbf{u}(t), \boldsymbol{\psi}(t)) = \sum_{\alpha=1}^{I_1} \lambda_\alpha(t)\operatorname{grad}_{\mathbf{u}}\mathbf{q}_\alpha^{(1)}(\mathbf{x}(t), \mathbf{u}(t)) + \sum_{\beta=1}^{I_2} \rho_\beta(t)\operatorname{grad}_{\mathbf{u}}\mathbf{q}_\beta^{(2)}(\mathbf{x}(t), \mathbf{u}(t)) \tag{4}$$

Here, $x_l$ are elements of a general state vector, $\psi_l$ are elements of an adjoint vector. Additional transversality conditions for the two ends of the state trajectory should be added for a general case:

$$\psi_l(T_0) = -\frac{\partial J_{ob}}{\partial x_l}\bigg|_{x_l(T_0)=x_{l0}}, \ \psi_l(T_f) = -\frac{\partial J_{ob}}{\partial x_l}\bigg|_{x_l(T_f)=x_{lf}} \tag{5}$$

Let us consider the algorithmic realization of the maximum principle. In accordance with this principle, two systems of differential equations should be solved: the main system (1) and the adjoint system (2). This will provide the optimal program control vector $\mathbf{u}^*(t)$ (the indices «pl» are omitted) and the state trajectory $\mathbf{x}^*(t)$. The vector $\mathbf{u}^*(t)$ at time $t = T_0$ under the conditions $\mathbf{h}_0(\mathbf{x}(T_0)) \leq 0$ and for the given value of $\boldsymbol{\psi}(T_0)$ should return the maximum to the Hamilton's function:

$$H(\mathbf{x}(t), \mathbf{u}(t), \boldsymbol{\psi}(t)) = \boldsymbol{\Psi}^{\mathsf{T}} \mathbf{f}(\mathbf{x}, \mathbf{u}, t) \tag{6}$$

Here we assume that general functional of MS schedule quality is transformed to Mayer's form [21].

The received control is used in the right members of (1), (2), and the first integration step for the main and for the adjoint system is made: $t_1 = T_0 + \tilde{\delta}$ ($\tilde{\delta}$ is a step of integration). The process of integration is continued until the end conditions $\mathbf{h}_1\left(\mathbf{x}(T_f)\right) \leq \vec{O}$ are satisfied and the convergence accuracy for the functional and for the alternatives is adequate. This terminates the construction of the optimal program control $\mathbf{u}^*(t)$ and of the corresponding state trajectory $\mathbf{x}^*(t)$.

However, the only question that is not addressed within the described procedure is how to determine $\boldsymbol{\psi}(T_0)$ for a given state vector $\mathbf{x}(T_0)$.

The value of $\boldsymbol{\psi}(T_0)$ depends on the end conditions of the MS schedule problem. Therefore, the maximum principle allows the transformation of a problem of non-classical calculus of variations to a boundary problem. In other words, the problem of MS OPC (in other words, the MS schedule) construction is reduced to the following problem of transcendental equations solving:

$$\Phi = \Phi(\boldsymbol{\psi}(T_0)) = 0 \tag{7}$$

At this stage, we consider the most complicated form of the MS OPC construction with MS initial state at time $t = T_0$. The end point of the trajectory and the time interval are fixed:

$$\mathbf{x}(T_f) = \mathbf{a} \tag{8}$$

where $\mathbf{a} = ||a_1, a_2, \ldots, a_{\tilde{n}}||^T$ is a given vector; $\tilde{n}$ is the dimensionality of the general MS state vector. Hence, the problem of optimal MS OPC construction is reduced to a search for the vector $\boldsymbol{\psi}(T_0)$, such that:

$$\boldsymbol{\Phi} = \boldsymbol{\Phi}(\boldsymbol{\psi}(T_0)) = \mathbf{x}(T_f) - \mathbf{a} \tag{9}$$

The problem of solving Equations (9) is equivalent to a minimization of the function:

$$\Delta_u(\boldsymbol{\psi}(T_0)) = \frac{1}{2}\left(\mathbf{a} - \mathbf{x}(T_f)\right)^T \left(\mathbf{a} - \mathbf{x}(T_f)\right) \tag{10}$$

The main feature of the problems (9), (10) is that the interdependency of $\boldsymbol{\psi}(T_0)$ and $\boldsymbol{\rho}(T_f) = \left(\mathbf{a} - \mathbf{x}(T_f)\right)$ is defined indirectly via the system of differential Equation (1). This feature leads to the use of iterative methods.

## 3. Classification and Analyses of Optimal Control Computational Algorithms for Short-Term Scheduling in MSs

Real-world optimal control problems are characterized by high dimensionality, complex control constraints, and essential non-linearity. Prior to introducing particular algorithms and methods of optimal control computation, let us consider their general classification (Figure 2).

Two main groups are distinguished: the group of general optimal control algorithms and methods and the group of specialized algorithms. The first group includes direct and indirect methods and algorithms. The direct methods imply search processes in spaces of variables. The first and most widely used methods of this group are the methods of reduction to finite-dimensional choice problems. Another subgroup includes direct gradient methods of search in functional spaces of control. To implement complex constraints for control and state variables, the methods of penalty functions and of gradient projection were developed. The third subgroup represents methods based on Bellman's optimality principle and methods of variation in state spaces. The methods of this subgroup help to consider various state constraints.

Unlike the methods of the first group, the indirect methods and algorithms are aimed at the acquisition of controls that obey the necessary or/and sufficient optimality conditions. Firstly, the methods and algorithms for two-point boundary problems can be referenced. Particular methods are used here: Newton's method (and its modifications), Krylov and Chernousko's methods of successive approximations, and the methods of perturbation theory.

Methods based on sufficient conditions of optimality are preferable for irregular problems such as optimal zero-overshoot response choice. The third group includes specialized methods and algorithms for linear optimal control problems and approximate analytical methods and algorithms. The latter methods imply that the nonlinear components of models are not essential.

Let us limit a detailed analysis of computational procedures to two-point boundary problems with fixed ends of the state trajectory $\mathbf{x}(t)$ and a fixed time interval $(T_0, T_f)$. For this system class, the following methods can be considered ([23,24,33]): Newton's method and its modifications, methods of penalty functionals, gradient methods, and the Krylov–Chernousko method.

Let us consider the possible implementation of the above-mentioned methods to the stated problem.

Methods and algorithms of optimal control

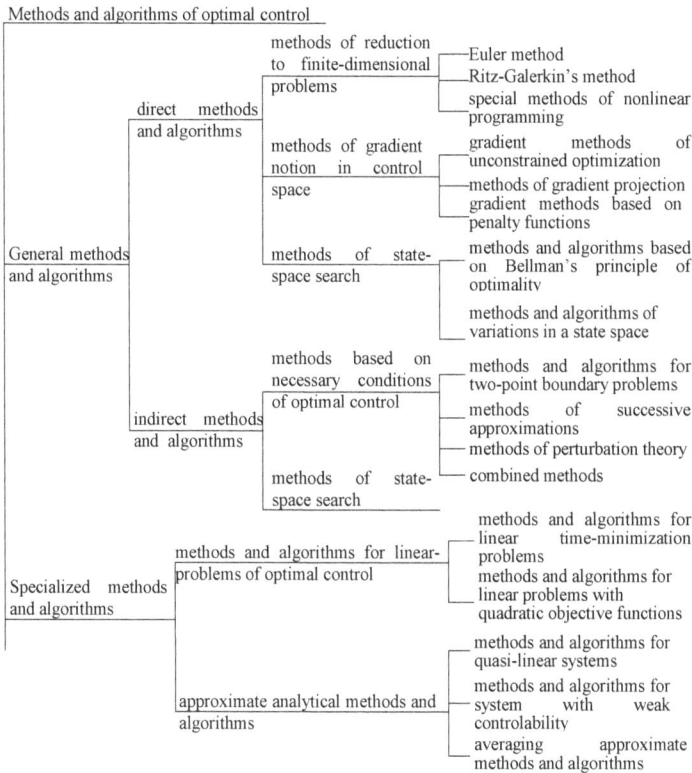

**Figure 2.** Optimal control computational methods.

Newton's method and its modifications. For the first iteration of the method, an initial approximation $\psi(T_0) = \psi_{(0)}(T_0)$ is set. Then, the successive approximations are received from the formula:

$$\psi_{(r)}(T_0) = \psi_{(r-1)}(T_0) - \left[\frac{\partial \boldsymbol{\rho}_{(r-1)}(T_f)}{\partial \psi_{(r\ 1)}(T_0)}\right]\boldsymbol{\rho}_{(r-1)}(T_f) \tag{11}$$

where $\boldsymbol{\rho}_{(r)}(T_f) = \mathbf{a} - \mathbf{x}_{(r-1)}(T_f), r = 1, 2, \ldots$.

The operation of partial derivation, being performed at all iterations, is rather complicated. Here the derivative matrix:

$$\tilde{\Pi} = \left(\frac{\partial \boldsymbol{\rho}_{(r-1)}(T_f)}{\partial \psi_{(r-1)}(T_0)}\right) \tag{12}$$

should be evaluated.

This matrix can be received either via finite-difference formulas or via a variational integration of the system. Modification of the method can also be used. The following method is the simplest. The formula (11) is substituted for:

$$\psi_{(r)}(T_0) = \psi_{(r-1)}(T_0) - \gamma_{(r)}\tilde{\Pi}^{-1}\boldsymbol{\rho}_{(r-1)}(T_f) \tag{13}$$

where $\gamma_r\ (0 \le \gamma_r \le 1)$ is selected to meet the constraint:

$$\Delta_u\left(\psi_{(r)}(T_0)\right) < \Delta_H\left(\psi_{(r-1)}(T_0)\right) \tag{14}$$

First, the value $\gamma_{(r)} = 1$ is tried. Then values $1/2, 1/4$, and so on until (14) is true. The selection of $\gamma_{(r)}$ is being performed during all iterations.

The entire solving process is terminated when $\Delta_u\left(\psi_{(r)}(T_0)\right) < \varepsilon_u$, where $\varepsilon_u$ is a given accuracy.

The main advantages of Newton's method and its modifications are a simple realization [there is no need to integrate adjoint system (2)], a fast convergence (if the initial choice of $\psi(T_0)$ is good), and a high accuracy solution.

The main disadvantage is the dependency of a convergence upon the choice of $\psi(T_0)$. In the worst case (absence of a good heuristic plan of MS operation), these methods can be divergent. In addition, if the dimensionality of the vector $\rho(T_f)$ is high, then computational difficulties can arise during calculation as well as an inversion of matrices (12).

The method of penalty functionals. To use this method for the considered two-point boundary problem, we should determine the extended quality functional:

$$\tilde{J}_{ob.p} = J_{ob} + \frac{1}{2}\sum_{i=1}^{\tilde{n}} C_i\left(a_i - x_i(T_f)\right)^2 \tag{15}$$

where $C_i$ are positive coefficients. If the coefficients are sufficiently large, the minimal value of the functional is received in the case of $\rho_i(T_f) = 0$. Therefore, the following algorithm can be used for the solving of the boundary problem. The control program $u_{(r)}(t)$ is searched during all iterations with fixed $C_i$ (Newton's method can be used here, for example). If the accuracy of end conditions is insufficient, then the larger values of $C_i$ are tried. Otherwise, the algorithm is terminated and a solution is received. Although the method seems deceptively simple, it does not provide an exact solution. Therefore, it is advisable to combine it with other methods.

The gradient methods. There are different variants of the gradient methods including generalized gradient (subgradient) methods. All gradient methods use the following recurrence formula:

$$\psi_{(r)}(T_0) = \psi_{(r-1)}(T_0) + \gamma_{(r)}\overline{\Delta}_{(r-1)} \tag{16}$$

Here $r = 1, 2, 3, \ldots$ is an iteration number; $\overline{\Delta}_{(r-1)}$ is a vector defining a direction of a shift from the point $\psi_{(r-1)}(T_0)$. For example, the gradient of the function $\Delta_u$ can be used:

$$\Delta_{(r-1)} = \text{grad}\Delta_u\left(\psi_{(r-1)}(T_0)\right) = \|\frac{\partial\Delta_u}{\partial\psi_{<1,(r-1)>}}, \ldots, \frac{\partial\Delta_u}{\partial\psi_{<\tilde{n},(r-1)>}}\|^{\mathrm{T}}$$

The multiplier $\gamma_{(r)}$ determines the value of the shift in direction of $\Delta_{(r-1)}$. In the subgradient, methods vectors $\Delta_{(r-1)}$ in (16) are some subgradients of the function $\Delta_u$.

In the MS OPC problems, a feasible subgradient can be expressed as:

$$\Delta_{<i,(r-1)>} = a_i - x_{<i,(r-1)>}(T_f)$$

Then, the main recurrence formula can be written as:

$$\psi_{<i,(r)>}(T_0) = \psi_{<i,(r-n)>}(T_0) + \gamma_{<i,r>}\left(a_i - x_{<i,r>}(T_f)\right) \tag{17}$$

where $\gamma_{<i,r>}$ can be selected according to some rule, for example:

$$\gamma_{<i,r>} = \begin{cases} \frac{1}{2}\gamma_{<i,(r-1)>}, & \text{if } \text{sign}dx_i \ne \text{sign}dx1_i \ne \text{sign}dx2_i; \\ \gamma_{<i,(r-1)>}, & \text{if not,} \end{cases}$$

Here $dx_i = a_i - x_{<i,r>}$ is a residual for end conditions at the iteration $r$. Similarly, $dx1_i = a_i - x_{<i,(r-1)>}$ and $dx2_i = a_i - x_{<i,(r-1)>}$ are residual at iterations $(r-1)$ and $(r-2)$. The main advantage of these algorithms over classical gradient algorithms is a simpler calculation of the direction vector during all iterations. However, this results in slower convergence (sometimes in divergence) of the general gradient (subgradient) methods as compared with classical ones. The convergence of all gradient methods depends on the initial approximation $\psi_{(0)}(T_0)$.

The analysis of the existing methods of optimal program control demonstrates that only the combined use of different methods compensates for their disadvantages. Therefore, the vector $\psi_{(0)}(T_0)$ should be determined in two phases ([36,37]). In the first phase, the MS OPC problem is considered without strict end conditions at the time $t = T_f$. The solution of the problem with a free right end is some approximation $\widetilde{\psi}_{(r)}(T_0)$ $(r = 1, 2, \dots)$.

Then, in the second phase, the received vector is used as the initial approximation $\widetilde{\widetilde{\psi}}_{(0)}(T_0) = \widetilde{\psi}_{(r)}(T_0)$ for Newton's method, the penalty functional method, or the gradient method. Thus, in the second phase, the problem of MS OPC construction can be solved over a finite number of iterations.

Let us now consider one of the most effective methods, namely Krylov and Chernousko's method for OPC problem with a free right end [39].

Step 1. An initial solution (an allowable program control) $\mathbf{u}_d(t)$, $\forall t \in (T_0, T_f]$, $\mathbf{u}_d(t) \in M$ is selected.

Step 2. The main system of Equation (1) is integrated under the end conditions $\mathbf{h}_0(\mathbf{x}(T_0)) \leq 0$. This results in the trajectory $\mathbf{x}_d(t)$ $\forall t \in (T_0, T_f]$.

Step 3. The adjoint system (2) is integrated over the time interval from $t = T_f$ to $t = T_0$ under the end conditions:

$$\psi_{<i,d>}(T_f) = \frac{1}{2} \frac{\partial \left( a_i - x_{<i,d>}(T_f) \right)^2}{\partial x_{<i,d>}}, \quad i = 1, \dots, \tilde{n} \tag{18}$$

where the constraints (18) are transversality conditions for the optimal control problem with a free end. The integration results in functions $\psi_{<i,d>}$ of time and particularly in $\psi_{<i,d>}(T_0)$.

Step 4. The control $\mathbf{u}_{(r)}(t)$ is searched for subject to:

$$H\left( \mathbf{x}_{(r)}(t), \mathbf{u}_{(r+1)}(t), \psi_{(r)}(t) \right) = \max_{\vec{u}_{(r)} \in M} H\left( \mathbf{x}_{(r)}(t), \psi_{(r)}(t), \mathbf{u}_{(r)}(t) \right) \tag{19}$$

where $r = 1, 2, \dots$ is an iteration number. An initial solution belongs to the iteration $r = 0$. Apart from the maximization of the Hamiltonian (19), the main and the adjoint systems of Equations (1) and (2) are integrated from $t = T_0$ to $t = T_f$.

Notably, several problems of mathematical programming are solved for each time point (the maximal number of the problems is equal to the number of MS OPC models). These problems define components of Hamilton's function. This is the end of the first iteration $(r = 1)$. If the conditions

$$\left| J_{ob}^{(r)} - J_{ob}^{(r-1)} \right| \leq \varepsilon_1 \tag{20}$$

$$\left\| \mathbf{u}_{(r)}(t) - \mathbf{u}_{(r-1)}(t) \right\| \leq \varepsilon_2 \tag{21}$$

are satisfied, where constants $\varepsilon_1$ and $\varepsilon_2$ define the degree of accuracy, then the optimal control $\mathbf{u}_{(r)}^*(t) = \mathbf{u}_{(r)}(t)$ and the vector $\widetilde{\psi}_{(r)}(T_0)$ are received at the first iteration. If not, we repeat **Step 3** and so on.

In a general case (when the model $M$ is used), the integration step for differential Equations (1) and (2) is selected according to the feasible accuracy of approximation (substitution of initial equations for finite difference ones) and according to the restrictions related with the correctness of the maximum principle. If we linearize the MS motion model ($M_{<g,\Theta>}$), then all the components of the model $M$ ($M_{<0,\Theta>}$, $M_{<k,\Theta>}$, $M_{<p,\Theta>}$, $M_{<n,\Theta>}$, $M_{<e,\Theta>}$, $M_{<c,\Theta>}$, $M_{<v,\Theta>}$) will be finite-dimensional, non-stationary, linear dynamical systems or bi-linear $M_{<k,\Theta>}$ dynamic systems. In this case, the simplest of Euler's formulas can be used for integration.

The advantage of the method is its simple programming. Krylov-Chernousko's method is less dependent on the initial allowable control, as compared with the above-mentioned methods. In addition, control inputs may be discontinuous functions. This is noteworthy as most MS resources are allocated in a relay mode.

However, the simple successive approximations method (SSAM) can diverge.

Several modifications of SSAM were proposed to ensure convergence and the monotony of some SSAM modifications was proved in [40] for two-point linear boundary problems with a convex area of allowable controls and convex goal function. Let us consider one variant of SSAM convergence improvement.

In this case, Equation (22) is applied over some part $\sigma' = (t', t'']$ of the interval $\sigma = (T_0, T_f]$, and not over the whole interval, i.e.,

$$\mathbf{u}_{(r+1)}(t) = \tilde{\tilde{N}}\left(\mathbf{u}_{(r)}(t)\right), \, t \in (t', t'']; \, \mathbf{u}_{(r+1)}(t) = \mathbf{u}_{(r)}(t), \, t \notin (t', t''] \tag{22}$$

where the operator $\tilde{\tilde{N}}$ produces [see formula (19)] a new approximation of a solution for each allowable control $\mathbf{u}(t)$.

The interval $(t', t'']$ is selected as a part of $(T_0, T_f]$, in order to meet the condition:

$$J_{ob}^{(r+1)} < J_{ob}^{(r)}$$

where $J_{ob}^{(r+1)}$, $J_{ob}^{(r)}$ are the values of the quality functional for the controls $\mathbf{u}_{(r+1)}(t)$, $\mathbf{u}_{(r)}(t)$, respectively.

The selection of time points $t'$ and $t''$ is performed in accordance with problem specificity. In our case, the set of possible points $(\tilde{t}'_{<e,(r+1)>}, \tilde{t}''_{<e,(r+1)>})$, $e = 1, \ldots, E_r$ for the iteration $(r + 1)$ is formed during iteration $r$ during the maximization of Hamiltonian function (19). The set includes the time points at which the operations of model $M$ are interrupted. This idea for interval $(t', t'']$ determination was used in a combined method and algorithm of MS OPC construction. The method and the algorithm are based on joint use of the SSAM and the "branch and bounds" methods.

## 4. Combined Method and Algorithm for Short-Term Scheduling in MSs

The basic *technical idea* of the presented approach on the basis of previous works ([13,19,33,34,41]) is the combination of optimal control and mathematical programming. Optimal control is not used for solving the combinatorial problem, but to enhance the existing mathematical programming algorithms regarding non-stationarity, flow control, and continuous material flows. As the control variables are presented as binary variables, it might be possible to incorporate them into the assignment problem. We apply methods of discrete optimization to combinatorial tasks within certain time intervals and use the optimal program control with all its advantages (i.e., accuracy of continuous time, integration of planning and control, and the operation execution parameters as time functions) for the flow control within the operations and for interlinking the decomposed solutions.

The basic computational idea of this approach is that operation execution and machine availability are dynamically distributed in time over the planning horizon. As such, not all operations and machines are involved in decision-making at the same time. Therefore, it becomes natural to transit from large-size allocation matrices with a high number of binary variables to a scheduling problem that is dynamically decomposed. The solution at each time point for a small dimensionality is calculated with mathematical programming. Optimal control is used for modeling the execution of the operations and interlinking the mathematical programming solutions over the planning horizon with the help of the maximum principle. The maximum principle provides the necessary conditions such that the optimal solutions of the instantaneous problems give an optimal solution to the overall problem ([21,27,42]).

We describe the proposed method and algorithm for two classes of models below:

- models $M_{<o>}$ of operation program control;

- models $M_{<v>}$ of auxiliary operations program control.

These models were described in [13,19,33–38]. Along with the initial problem of program control (marked by the symbol $\Gamma$), we consider a relaxed one (marked by P). The latter problem has no constraints of interruption disability operations $D_{\text{æ}}^{(i)}$ (where $D_{\text{æ}}^{(i)}$ is an operation æ with object $i$. See, for example, [36,37]). Let the goal function of problem $\Gamma$ be:

$$
J_p = \left\{ J_{ob} + \sum_{i=1}^{n} \sum_{\text{æ}=1}^{s_i} \sum_{j=1}^{m} \sum_{\lambda=1}^{l} \left[ z_{i\text{æ}j\lambda}^{(0,1)} z_{i\text{æ}j\lambda}^{(0,3)} + \frac{(a_{i\text{æ}})^2}{2} - z_{i\text{æ}j\lambda}^{(0,2)} \right]^2 \left( z_{i\text{æ}j\lambda}^{(0,1)} \right)^2 \right\} \Bigg|_{t=T_f}
\tag{23}
$$

where the auxiliary variables $z_{i\text{æ}j\lambda}^{(0,1)}(t)$, $z_{i\text{æ}j\lambda}^{(0,2)}(t)$, $z_{i\text{æ}j\lambda}^{(0,3)}(t)$ are used for operations with interruption prohibition. More detailed information about these auxiliary variables and the methods of their utilization can be found in [20].

The following theorem expresses the characteristics of the relaxed problem of MS OPC construction.

**Theorem 1.** *Let P be a relaxed problem for the problem $\Gamma$. Then*

(a)   *If the problem P does not have allowable solutions, then this is true for the problem $\Gamma$ as well.*
(b)   *The minimal value of the goal function in the problem P is not greater than the one in the problem $\Gamma$.*
(c)   *If the optimal program control of the problem P is allowable for the problem $\Gamma$, then it is the optimal solution for the problem $\Gamma$ as well.*

The proof of the theorem follows from simple considerations.
**Proof.**

(a)   If the problem P does not have allowable solutions, then a control $\mathbf{u}(t)$ transferring dynamic system (1) from a given initial state $\mathbf{x}(T_0)$ to a given final state $\mathbf{x}(T_f)$ does not exist. The same end conditions are violated in the problem $\Gamma$.
(b)   It can be seen that the difference between the functional $J_p$ in (23) and the functional $J_{ob}$ in the problem P is equal to losses caused by interruption of operation execution.
(c)   Let $\mathbf{u}_p^*(t)$, $\forall\, t \in (T_0,\, T_f]$ be an MS optimal program control in P and an allowable program control in $\Gamma$; let $\mathbf{x}_p^*(t)$ be a solution of differential equations of the models $M_{<o>}$, $M_{<v>}$ subject to $\mathbf{u}(t)=\mathbf{u}_p^*(t)$. If so, then $\mathbf{u}_p^*(t)$ meets the requirements of the local section method (maximizes Hamilton's function) for the problem $\Gamma$. In this case, the vectors $\mathbf{u}_p^*(t)$, $\mathbf{x}_p^*(t)$ return the minimum to the functional (1).

The scheme of computation can be stated as follows.

Step 1. An initial solution $\mathbf{u}_g(t)$, $t \in (T_0,\, T_f]$ (an arbitrary allowable control, in other words, allowable schedule) is selected. The variant $\mathbf{u}_g(t) \equiv 0$ is also possible.

Step 2. The main system of Equation (1) is integrated subject to $\mathbf{h}_0(\mathbf{x}(T_0)) \leq 0$ and $\mathbf{u}(t)=\mathbf{u}_g(t)$. The vector $\mathbf{x}^{(0)}(t)$ is received as a result. In addition, if $t = T_f$, then the record value $J_p = J_p^{(0)}$ can be calculated, and the transversality conditions (5) are evaluated.

Step 3. The adjoint system (2) is integrated subject to $\mathbf{u}(t)=\mathbf{u}_g(t)$ and (5) over the interval from $t = T_f$ to $t = T_0$. For time $t = T_0$, the first approximation $\boldsymbol{\psi}_i^{(0)}(T_0)$ is received as a result. Here the iteration number $r = 0$ is completed.

Step 4. From the point $t = T_0$ onwards, the control $u^{(r+1)}(t)$ is determined ($r = 0, 1, 2, \ldots$ is the number of iteration) through the conditions (19). In parallel with the maximization of the Hamiltonian, the main system of equations and the adjoint system are integrated. The maximization involves the solving of several mathematical programming problems at each time point.

The branching of the problem $\Gamma$ occurs during the process of Hamiltonian maximization at some time $\tilde{t} \in (T_0, T_f]$ if the operation $D_{\text{æ}}^{(i)}$ is being interrupted by the priority operation $D_{\varsigma}^{(\omega)}$. In this case, the problem $\Gamma$ is split into two sub-problems $(P_{\text{æ}}^{(i)}, P_{\varsigma}^{(\omega)})$.

Within the problem $P_{\text{æ}}^{(i)}$, the operation $D_{\text{æ}}^{(i)}$ is executed in an interrupt-disable mode. For other operations, this restriction is removed and the relaxed scheduling problem is solved via the method of successive approximations. Let us denote the received value of the goal function (23) by $J_{p0}^{(1)}$. Within the problem $P_{\varsigma}^{(\omega)}$, the previously started operation $D_{\text{æ}}^{(i)}$ is terminated, and $D_{\varsigma}^{(\omega)}$ begins at time $\tilde{t}$. The resources released by the operation $D_{\text{æ}}^{(i)}$ and not seized by $D_{\varsigma}^{(\omega)}$ can be used for other operations if any are present. Reallocation of resources should be performed according to (19). Subsequently, after completion of operation $D_{\varsigma}^{(\omega)}$, the relaxed scheduling problem is solved. In this case, the value of the goal function (23) is denoted by $J_{p1}^{(1)}$. The record $J_p$ is updated if $J_{p0}^{(1)} < J_p$ or/and $J_{p1}^{(1)} < J_p^{(1)}$ [we assume that the functional (23) is minimized]. If only $J_{p0}^{(1)} < J_p^{(1)}$, then $J_p = J_{p0}^{(1)}$. Similarly, if only $J_{p1}^{(1)} < J_p^{(1)}$, then $J_p = J_{p1}^{(1)}$. If both inequalities are true, then $J_p = \min\{J_{p0}^{(1)}, J_{p1}^{(1)}\}$. In the latter case, the conflict resolution is performed as follows: if $J_{p0}^{(1)} < J_{p1}^{(1)}$, then during the maximization of (19) $D_{\text{æ}}^{(i)}$ is executed in an interrupt-disable mode. Otherwise, the operation $D_{\text{æ}}^{(i)}$ is entered into the Hamiltonian at priority $D_{\varsigma}^{(\omega)}$ at arrival time $\tilde{t}$.

After conflict resolution, the allocation of resources is continued, complying with (19) and without interruption of operations until a new contention is similarly examined. The considered variant of dichotomous division in conflict resolution can be extended to the case of $\bar{k}$-adic branching, where $\bar{k}$ is the number of interrupted operations at some time $\tilde{t}$.

The iterative process of the optimal schedule search is terminated under the following circumstances: either the allowable solution of the problem $\Gamma$ is determined during the solving of a relaxed problem or at the fourth step of the algorithm after the integration we receive:

$$\left| J_p^{(r+1)} - J_p^{(r)} \right| < \varepsilon_1 \tag{24}$$

where $\varepsilon_1$ is a given value, $r = 0, 1, \ldots$ If the condition (24) is not satisfied, then the third step is repeated, etc.

The developed algorithm is similar to those considered in [20,37]. Here, the set of time points in which the Hamiltonian is to be maximized is formed on the previous iteration. These are points of operations interruption. Computational investigations of the proposed algorithm showed that the rate of convergence is predominantly influenced by the choice of the initial adjoint vector $\psi(t_0)$. In its turn, $\psi(t_0)$ depends on the first allowable control that is produced at the first step. During the scheduling process, the machines are primarily assigned to operations of high dynamic priority.

To illustrate the main ideas of the algorithm, let us consider a simple example of a scheduling problem with two jobs, three operations, and a single machine, $(T_0, T_f] = (0, 14]$, $a_{i\text{æ}}^{(o)} = 2$ $(I = 1,2; \text{æ} = 1, 2, 3)$; $\Theta_{i\text{æ}j}(t) = 1 \; \forall \; t$; $\varepsilon_{11}(t) = 1$ at $t \in (0, 14]$, $\varepsilon_{21}(t) = 0$ for $0 \le t < 1$, $\varepsilon_{21}(t) = 1$ for $t \ge 1$. In this case, the model can be described as follows:

$$M = \left\{ \vec{u} \, \middle| \, \dot{x}_{i\text{æ}}^{(o)} = \varepsilon_{i1} u_{i\text{æ}1}^{(o)}; \, \dot{z}_{i\text{æ}1}^{(o,1)} = u_{i\text{æ}1}^{(o)}; \dot{z}_{i\text{æ}1}^{(o,2)} = z_{i\text{æ}1}^{(o,1)}; \right.$$

$$\dot{z}_{i\text{æ}1}^{(o,3)} = w_{i\text{æ}1}^{(o)}; u_{i\text{æ}1}^{(o)}(t), w_{i\text{æ}}^{(o)}(t) \in \{0,1\}, \sum_{\text{æ}=1}^{3} \sum_{i=1}^{2} u_{i\text{æ}1}^{(o)}(t) \le 1; u_{i21}^{(o)}\left(a_{i1}^{(o)} - x_{i1}^{(o)}\right) = 0; u_{i31}^{(o)}\left(a_{i2}^{(o)} - x_{i2}^{(o)}\right) = 0;$$

$$w_{i\text{æ}1}^{(o)}\left(a_{i\text{æ}}^{(o)} - z_{i\text{æ}}^{(o,1)}\right) = 0; T_0 = 0: x_{i\text{æ}}^{(o)}(t_0) = z_{i\text{æ}1}^{(o,1)}(t_0) = z_{i\text{æ}1}^{(o,2)}(t_0) = z_{i\text{æ}1}^{(o,3)}(t_0) = 0; T_f = 14: x_{i\text{æ}}^{(o)}(t_f) = 2; \tag{25}$$

$$z_{i\text{æ}1}^{(o,l)}(t_f) \in R^1, l = 1, 2, 3 \Big\}.$$

The scheduling quality measure is defined by the expression:

$$J_p = \sum_{i=1}^{2} \sum_{\text{æ}=1}^{3} \left[ \left( a_{i\text{æ}}^{(o)} - x_{i\text{æ}}^{(o)} \right)^2 + \left( z_{i\text{æ}1}^{(0,1)} z_{i\text{æ}1}^{(0,3)} + \frac{\left( a_{i\text{æ}}^{(o)} \right)^2}{2} - z_{i\text{æ}1}^{(0,2)} \right)^2 \left( z_{i\text{æ}1}^{(0,1)} \right)^2 \right] \Bigg|_{t=T_f} \tag{26}$$

$$- \sum_{i=1}^{2} \sum_{\text{æ}=1}^{3} \int_{0}^{14} \gamma_{i\text{æ}}(\tau) u_{i\text{æ}1}^{(o)}(\tau) d\tau,$$

where $\gamma_{i\text{æ}}(t)$ are given time functions denoting the most preferable operation intervals: $\gamma_{11} = 15\gamma_+(6 - t)$, $\gamma_{12} = 10\gamma_+(9 - t)$, $\gamma_{13} = 10\gamma_+(11 - t)$, $\gamma_{21} = 20\gamma_+(8 - t)$, $\gamma_{22} = 15\gamma_+(8 - t)$, $\gamma_{23} = 30\gamma_+(11 - t)$. Here, $\gamma_+(\alpha) = 1$ if $\alpha \geq 0$ and $\gamma_+(\alpha) = 0$ if $\alpha < 0$. The integration element in (27) can be interpreted similarly to previous formulas through penalties for operations beyond the bounds of the specified intervals. The scheduling problem can be formulated as follows: facilities-functioning schedule [control program $\vec{u}(t)$] returning a minimal value to the functional (27) under the conditions (26) and the condition of interruption prohibition should be determined.

The resulting schedule is shown in Figure 3.

The upper part of Figure 3 shows the initial feasible schedule implementing First-in-first-out processing. The second and third cases illustrate a conflict resolution for $D_1^{(1)}$ interruption. The optimal schedule is presented in the third and the forth rows.

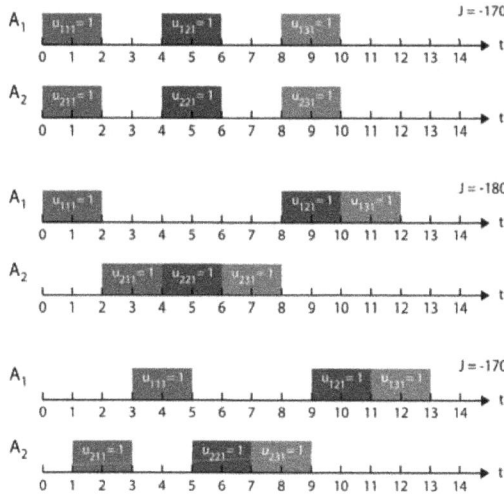

**Figure 3.** Optimal schedule results.

## 5. Qualitative and Quantitative Analysis of the MS Scheduling Problem

The representation of the scheduling problem in terms of a dynamic system (1) control problem allows applications of mathematical analysis tools from modern control theory (38,43,44).

For example, qualitative analysis based on control theory as applied to the dynamic system (1) provides the results listed in Table 1. This table confirms that our interdisciplinary approach to the description of MS OPC processes provides a fundamental base for MS problem decisions (or control and management) that have never been previously formalized and have high practical importance. The table also presents possible directions of practical implementation (interpretation) for these results in the real scheduling of communication with MS. For example, criteria for controllability and attainability in MS OPC problems can be used for MS control process verification for a given time interval.

The attainability sets (AS) of dynamic model (1) have an important place in qualitative analysis of MS control processes. These sets allow detailed analysis of computational procedures to two-point boundary problems with fixed ends of the state trajectory $\mathbf{x}(t)$ and a fixed time interval $(T_0, T_f]$.

An AS is a fundamental characteristic of any dynamic system. The AS approach determines a range of execution policies in the presence of disturbances over which the system can be guaranteed to meet certain goals. The union of these execution policies (i.e., feasible schedules) is called an AS in the state space. The union of possible performance outcomes from the given execution policies is called an AS in the performance space [43]. The AS in the state space depicts the possible states of a schedule subject to variations of the parameters (both planned and perturbation-driven) in the nodes and channels (e.g., different capacities, lot-sizes, etc.). Let us introduce the notation for an AS. $\mathbf{D_x}(t, T_0, \mathbf{x}(T_0), U(\mathbf{x}(T_0)))$ is an AS in the state space, $\mathbf{D}_J(t, T_0, \mathbf{x}(T_0), U(\mathbf{x}(T_0)))$ is an AS in the performance indicator's space, and $\overline{\mathbf{D}}_J^z(t, T_0, \mathbf{x}(T_0), \Xi, U(\mathbf{x}(T_0)))$ is an approximated AS under disturbances at moment $t$. To interconnect schedule execution and performance analysis to the AS in the *state space*, an AS in the *performance space* can be brought into correspondence (see Figure 4).

**Table 1.** Application of optimal control approaches to scheduling.

| No | Results and Their Implementation | The Main Results of Qualitative Analysis of MS Control Processes | The Directions of Practical Implementation of the Results |
|---|---|---|---|
| 1 | | Analysis of solution existence in the problems of MS control | Adequacy analysis of the control processes description in control models |
| 2 | | Conditions of controllability and attainability in the problems of MS control | Analysis MS control technology realizability on the planning interval. Detection of main factors of MS goal and information-technology abilities. |
| 3 | | Uniqueness condition for optimal program controls in scheduling problems | Analysis of possibility of optimal schedules obtaining for MS functioning |
| 4 | | Necessary and sufficient conditions of optimality in MS control problems | Preliminary analysis of optimal control structure, obtaining of main expressions for MS scheduling algorithms |
| 5 | | Conditions of reliability and sensitivity in MS control problems | Evaluation of reliability and sensitivity of MS control processes with respect to perturbation impacts and to the alteration of input data contents and structure |

In projecting these two ASs onto each other, a certain range of the schedule execution policies and the corresponding variation of the performance indicators can be determined. A continuous time representation allows investigation of the changes in execution at each time point. Therefore, at each time point, an AS can be calculated and related to output performance. The *justification of the choice* of the AS method is related to its dynamic nature ([43,44]). An AS may be favourable for obtaining estimations of performance attainability and consideration of the perturbations and attainability abilities of the schedules as *time functions*. In [13,19,33–38], we proposed different methods and algorithms of AS calculations. These results permit improvement of algorithm convergence in MS scheduling.

The proposed model interprets MS scheduling as a response to planning goal changes, demand fluctuations, and resource availability. In this interpretation, the problem is to schedule MS in order to achieve the planned goals (e.g., MS service level).

The model is scalable to other management levels of MS, i.e., orders and operations can be presented as MS configuration elements and orders, respectively. The transformation of parameters and goal criteria is also possible, i.e., the lead time can be considered as the MS cycle time. Hence, the MS strategic configuration and tactical planning can be optimized.

Let us analyse some particular features of the models (1) (see Figure 1). During the conducted experiments, it was revealed that the following model parameters influenced the improvement of the general quality index:

- the total number of operations on a planning horizon;
- a dispersion of volumes of operations;
- a ratio of the total volume of operations to the number of processes;
- a ratio of the amount of data of operation to the volume of the operation (relative operation density).

On the basis of the obtained etalon solutions, we can methodically justify the usage and quality of certain heuristics for certain variants of initial data (see Figure 5).

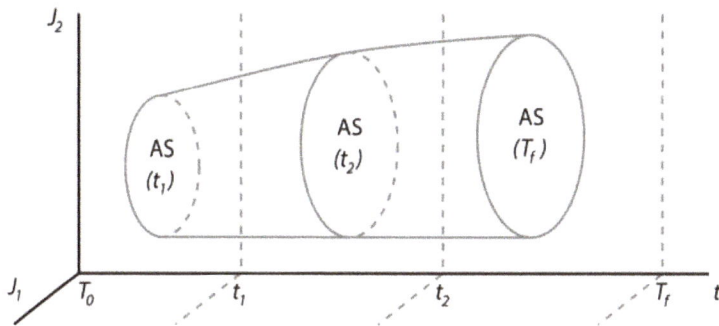

**Figure 4.** Representation of dynamic changes in schedule performance by attainable sets.

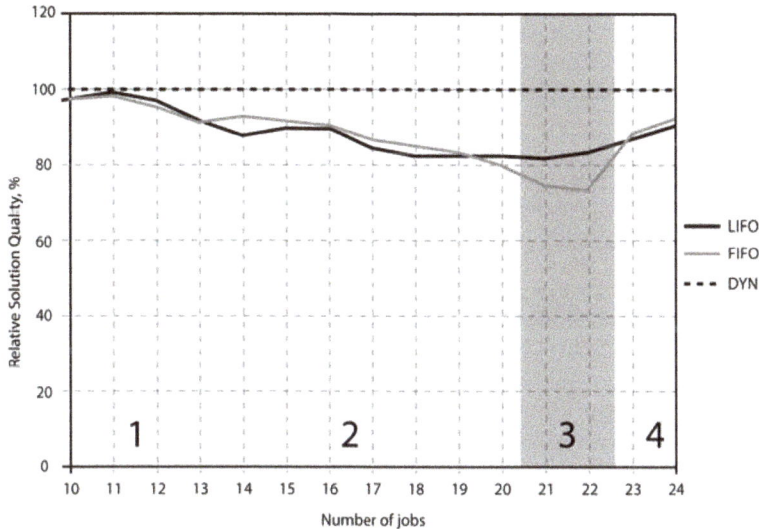

**Figure 5.** Comparison of heuristic algorithms' quality.

Having calculated optimal solutions for several points, it is possible to validate the decision to use either dynamic or heuristic planning algorithms. In Figure 4, the relative solution quality gained by the optimal control algorithm DYN is assumed to be 100%. The relative quality index of the heuristic solutions is calculated as a fraction of the optimal one, i.e., it can be observed that, in the case of a number of processes between 10 and 12, the quality of the heuristic and optimal solutions does not differ by more than 4%. In area 2, the DYN algorithm is preferable to the heuristics. If still using the heuristics, the FIFO algorithm is preferable to the Last-in-first-out one. The largest benefit from using the DYN algorithm is achieved in area 3. In this area, the LIFO algorithm is preferable to the FIFO algorithm.

Finally, the proposed model and algorithm allows for the achievement of better results in many cases in comparison with heuristics algorithms. However, this point is not the most important. The most important point is that this approach allows the interlinking of planning and scheduling models within an adaptation framework. Therefore, the suggested results are important in the Industry 4.0 domain. Hence, the proposed modeling complex does not exist as a "thing in itself" but works in the integrated decision-support system and guides the planning and scheduling decisions in dynamics on the principles of optimization and adaptation.

## 6. Conclusions

Optimal controls as functions of the system and control state allow for the generation of optimal decisions in consideration of a system's evolution in time in the presence of perturbations which result in different system states. Optimal control approaches take another perspective as mathematical programming methods which represent schedules as trajectories.

Computational algorithms with regard to state, control, and conjunctive variable spaces exist in the literature. We have demonstrated that the advantages of optimal control methods are applicable to the treatment of large scale problems with complex constraints, the consideration of non-stationary process execution dynamics, the representation in differential equations of complex interrelations between process execution, capacity evolution, and machine setups. In addition, the advantages of optimal control also include accuracy of continuous time and accurate presentation of continuous flows (e.g., in process industry or energy systems) with the help of continuous state variables.

An important observation is that schedule presentation in terms of optimal control makes it possible to incorporate the rich variety of control theoretic axioms with regard to feedback adaptive control (mostly applied in the framework of production-inventory control models) as well as the use of control tools of qualitative performance analysis, such as attainable (reachable) sets. Limitations of control applications include conceptual and algorithmic restrictions such as continuous process applications and specific (i.e., non-generalized) forms of constructing algorithms with necessary requirements concerning optimality, convergence, and numerical stability.

In this study, we exemplified an application of optimal control to manufacturing scheduling. Fundamentally, this application dynamically decomposes the assignment matrix in time using differential equations, and then solves (by tendency) polynomial problems of small dimensionality at each point of time, with a subsequent integration of these partial solutions using the maximum principle by integrating main and adjoint equation systems. The small dimensionality at each point of time results from dynamic decomposition of job execution described by precedence relation constraints, i.e., at each point of time, we consider only operations that can be assigned to machines at this point of time, excluding those operations that already have been completed as well as those that cannot start because the predecessors have not yet been completed. Algorithmically, we solved these dimensionally small problems at subsequent point of times, integrate main and adjoint systems by the maximum principle, and considered how a particular assignment decision changes the schedule performance metric (e.g., tardiness). If an improvement is observed, the algorithm takes this assignment and moves further to next point of time and continues in this manner until $T_f$.

With regard to the limitations of this study, the schedule algorithm analysis by attainable sets was used to analyse the schedule outputs at the end of the planning horizon. At the same time, attainable sets (more precisely, their geometric approximations) can also be applied prior to schedule optimization to analyse if any feasible schedule exists for the given problem settings (resource capacities, etc.). This issue has not yet been considered.

In light of the revealed methodical shortcomings and application limitations of optimal control methods based on the maximum principle, the following future research avenues can be stated. First, concrete application cases need to be considered for which specific control models and algorithms will be developed. The construction of models and computational procedures within proved axioms of control theory is important. Second, application of qualitative performance analysis methods for control policy dynamic investigation under uncertainty, such as attainable sets, must be named. These tools might be helpful with regard to an analysis of production schedule robustness, supply chain resilience, and Industry 4.0 system flexibility. Third, computational methods themselves require further investigation and modification for concrete application cases. Therefore, a closer collaboration between control and industrial engineers is critical for future applications of control methods in operations and future supply chain management.

**Author Contributions:** B.S. and D.I. conceived and designed the experiments; B.S. performed the experiments; A.D. and B.S. analyzed the data; A.D. and D.I. contributed reagents/materials/analysis tools; all authors wrote the paper.

**Acknowledgments:** The research described in this paper is partially supported by the Russian Foundation for Basic Research (grants 16-07-00779, 16-08-00510, 16-08-01277, 16-29-09482-ofi-i, 17-08-00797, 17-06-00108, 17-01-00139, 17-20-01214, 17-29-07073-ofi-i, 18-07-01272, 18-08-01505), grant 074-U01 (ITMO University), state order of the Ministry of Education and Science of the Russian Federation No. 2.3135.2017/4.6, state research 0073-2018-0003, International project ERASMUS +, Capacity building in higher education, No. 73751-EPP-1-2016-1-DE-EPPKA2-CBHE-JP, Innovative teaching and learning strategies in open modelling and simulation environment for students-centered engineering education.

**Conflicts of Interest:** The authors declare no conflict of interest.

## Abbreviations

| | |
|---|---|
| AS | Attainability Sets |
| OPC | Optimal Program Control |
| MS | Manufacturing System |
| SSAM | Successive Approximations Method |

## Appendix A. List of Notations

$\mathbf{x}$ is a vector of the MS general state
$\mathbf{u}$ is a vector of the generalized control
$\mathbf{h}_0$, $\mathbf{h}_1$ are known vector function that are used for the state $\mathbf{x}$ and conditions
$\mathbf{q}^{(1)}$, $\mathbf{q}^{(2)}$ are known spatio-temporal, technical and technological constraints
$J_\vartheta$ are indicators characterizing MS schedule quality
$\mathcal{E}$ is a known vector-function of perturbation influences
$\mathbf{\psi}$ is a vector of conjugate system state
$\mathbf{H}$ is a Hamilton's function
$J_{ob}$ is a scalar form of the vector quality measure
$M_g$ is a dynamic model of MS motion control,
$M_k$ is a dynamic model of MS channel control,
$M_o$ is a dynamic model of MS operational control,
$M_f$ is a dynamic model of MS flow control,
$M_p$ is a dynamic model of MS resource control,
$M_e$ is a dynamic model of MS operation parameters control,
$M_c$ is a dynamic model of MS structure dynamic control,
$M_v$ is a dynamic model of MS auxiliary operation control,
$t$ is a current value of time,
$T_0$ is start instant of time of the planning (scheduling) horizon,
$T_f$ is end instant of time of the planning (scheduling) horizon,
$\psi_l$ is a component of adjoint vector $\mathbf{\psi}(t)$,

$x_l$ is an element of a general vector $\mathbf{x}(t)$,

$\lambda_\alpha(t)$ is a dynamic Lagrange multiplier which corresponds to the components of vector $\mathbf{q}_\alpha^{(1)}$,

$\rho_\beta(t)$ is a dynamic Lagrange multiplier which is corresponds to the components of vector $\mathbf{q}_\alpha^{(2)}$,

$z_{iju}^{(o)}$ is an auxiliary variable which characterizes the execution of the operation

$\alpha$ is a current number of constraints $\mathbf{q}_\alpha^{(1)}$,

$\beta$ is a current number of constraints $\mathbf{q}_\alpha^{(2)}$,

$\psi_l(T_0)$ is a component of adjoint vector $\boldsymbol{\psi}(t)$ at the moment $t = T_0$,

$\psi_l(T_f)$ is a component of adjoint vector $\boldsymbol{\psi}(t)$ at the moment $t = T_f$,

$H(\mathbf{x}(t), \mathbf{u}(t), \boldsymbol{\psi}(t)) = \boldsymbol{\Psi}^T \mathbf{f}(\mathbf{x}, \mathbf{u}, t)$ is a Hamilton's function,

$\boldsymbol{\Phi}(\boldsymbol{\psi}(T_0))$ is an implicit function of boundary condition,

$\mathbf{a}$ is a vector given quantity value which corresponds to the vector $\mathbf{x}(t)$,

$\Delta_u$ is a function of boundary conditions,

$\boldsymbol{\rho}(T_f)$ is a discrepancy of boudary conditions,

$\boldsymbol{\psi}_{(r)}(T_0)$ is a vector of adjoint system at the moment, $t = T_0$

$r$ is a current number of iteration during schedule optimization,

$\tilde{\Pi}$ is a derivative matrix,

$\varepsilon_u$ is a given accuracy of Newton's method iterative procedure,

$C_i$ is a penalty coefficient,

$\Delta_{<i,(r-1)>}$ is a component of the function $\Delta_{(r-1)}$ gradient on the iteration $r - 1$,

$\gamma_{<i,r>}$ is a step of gradient (subgradient) method (algorithm) on the iteration $r$

$\tilde{\tilde{\boldsymbol{\psi}}}_{(0)}(T_0)$ is a adjoint vector at the moment $t = T_0$ on the iteration "0",

$\tilde{\tilde{\boldsymbol{\psi}}}_{(r)}(T_0)$ is a adjoint vector at the moment $t = T_0$ on the iteration "$r$",

$\sigma'$ is a some part of the interval $\sigma = (T_0, T_f]$,

$\tilde{\tilde{N}}$ is a assumed operator of new approximation procedure,

$\tilde{T}'_{<e,(r+1)>}, \tilde{T}''_{<e,(r+1)>}$ are time moments of operation interruption,

$\mathbf{u}_p^*(t)$ is a vector of generalized control in relaxed problem of MS OPC construction,

$\mathbf{x}_p^*(t)$ is a vector of general state in relaxed problem of MS OPC construction,

$\mathbf{u}_g(t)$ is a vector of an arbitrary allowable control (allowable schedule),

$D_\text{æ}^{(i)}$ is a operation "æ" with object "$i$",

$D_\zeta^{(\omega)}$ is a operation "$\zeta$" with object "$\omega$",

$P_\text{æ}^{(i)}, P_\zeta^{(\omega)}$ is a current number of the branch and bound method subproblems,

$J_{p0}^{(1)}$ is a value of scalar form of MS vector quality measure for the first subproblem,

$J_{p1}^{(1)}$ is a value of scalar form of MS vector quality measure for the second subproblem,

$\mathbf{D}_x$ is an attainable set,

$\vartheta$ is a current index of MS control model,

$g$ is an index of MS motion control model,

$k$ is an index of MS channel control model,

$o$ is an index of MS operation control model,

$f$ is an index of MS flow control model,

$p$ is an index of MS resource control model,

$e$ is an index of MS operation parameters control model,

$c$ is an index of MS structure dynamic control model,

$v$ is an index of MS auxiliary operation control model,

$l$ is a current number of MS elements and subsystems,

$u$ is a scalar allowable control input,

$\Theta$ is a current number of model,

æ is a number of operation "æ", $\zeta$ is a number of operation "$\zeta$", $\omega$ is a number of operation "$\omega$",

$i$ is a current number of external object (customer),

$j$ is a current number of internal object resource,

$\varepsilon_1, \varepsilon_2$ are known constants which characterize the accuracy of iterative solution of boundary-value problem.

$\tilde{\delta}$ is a step of integration

## References

1. Blazewicz, J.; Ecker, K.; Pesch, E.; Schmidt, G.; Weglarz, J. *Scheduling Computer and Manufacturing Processes*, 2nd ed.; Springer: Berlin, Germany, 2001.
2. Pinedo, M. *Scheduling: Theory, Algorithms, and Systems*; Springer: New York, NY, USA, 2008.

3. Dolgui, A.; Proth, J.-M. *Supply Chains Engineering: Useful Methods and Techniques*; Springer: Berlin, Germany, 2010.

4. Werner, F.; Sotskov, Y. (Eds.) *Sequencing and Scheduling with Inaccurate Data*; Nova Publishers: New York, NY, USA, 2014.

5. Lauff, V.; Werner, F. On the Complexity and Some Properties of Multi-Stage Scheduling Problems with Earliness and Tardiness Penalties. *Comput. Oper. Res.* **2004**, *31*, 317–345. [CrossRef]

6. Jungwattanakit, J.; Reodecha, M.; Chaovalitwongse, P.; Werner, F. A comparison of scheduling algorithms for flexible flow shop problems with unrelated parallel machines, setup times, and dual criteria. *Comput. Oper. Res.* **2009**, *36*, 358–378. [CrossRef]

7. Dolgui, A.; Kovalev, S. Min-Max and Min-Max Regret Approaches to Minimum Cost Tools Selection 4OR-Q. *J. Oper. Res.* **2012**, *10*, 181–192. [CrossRef]

8. Dolgui, A.; Kovalev, S. Scenario Based Robust Line Balancing: Computational Complexity. *Discret. Appl. Math.* **2012**, *160*, 1955–1963. [CrossRef]

9. Sotskov, Y.N.; Lai, T.-C.; Werner, F. Measures of Problem Uncertainty for Scheduling with Interval Processing Times. *OR Spectr.* **2013**, *35*, 659–689. [CrossRef]

10. Choi, T.-M.; Yeung, W.-K.; Cheng, T.C.E. Scheduling and co-ordination of multi-suppliers single-warehouse-operator single-manufacturer supply chains with variable production rates and storage costs. *Int. J. Prod. Res.* **2013**, *51*, 2593–2601. [CrossRef]

11. Harjunkoski, I.; Maravelias, C.T.; Bongers, P.; Castro, P.M.; Engell, S.; Grossmann, I.E.; Hooker, J.; Méndez, C.; Sand, G.; Wassick, J. Scope for industrial applications of production scheduling models and solution methods. *Comput. Chem. Eng.* **2014**, *62*, 161–193. [CrossRef]

12. Bożek, A.; Wysocki, M. Flexible Job Shop with Continuous Material Flow. *Int. J. Prod. Res.* **2015**, *53*, 1273–1290. [CrossRef]

13. Ivanov, D.; Dolgui, A.; Sokolov, B. Robust dynamic schedule coordination control in the supply chain. *Comput. Ind. Eng.* **2016**, *94*, 18–31. [CrossRef]

14. Giglio, D. Optimal control strategies for single-machine family scheduling with sequence-dependent batch setup and controllable processing times. *J. Sched.* **2015**, *18*, 525–543. [CrossRef]

15. Lou, S.X.C.; Van Ryzin, G. Optimal control rules for scheduling job shops. *Ann. Oper. Res.* **1967**, *17*, 233–248. [CrossRef]

16. Maimon, O.; Khmelnitsky, E.; Kogan, K. *Optimal Flow Control in Manufacturing Systems*; Springer: Berlin, Germany, 1998.

17. Ivanov, D.; Sokolov, B. Dynamic coordinated scheduling in the manufacturing system under a process modernization. *Int. J. Prod. Res.* **2013**, *51*, 2680–2697. [CrossRef]

18. Pinha, D.; Ahluwalia, R.; Carvalho, A. Parallel Mode Schedule Generation Scheme. In Proceedings of the 15th IFAC Symposium on Information Control Problems in Manufacturing INCOM, Ottawa, ON, Canada, 11–13 May 2015.

19. Ivanov, D.; Sokolov, B. Structure dynamics control approach to supply chain planning and adaptation. *Int. J. Prod. Res.* **2012**, *50*, 6133–6149. [CrossRef]

20. Ivanov, D.; Sokolov, B.; Dolgui, A. Multi-stage supply chains scheduling in petrochemistry with non-preemptive operations and execution control. *Int. J. Prod. Res.* **2014**, *52*, 4059–4077. [CrossRef]

21. Pontryagin, L.S.; Boltyanskiy, V.G.; Gamkrelidze, R.V.; Mishchenko, E.F. *The Mathematical Theory of Optimal Processes*; Pergamon Press: Oxford, UK, 1964.

22. Athaus, M.; Falb, P.L. *Optimal Control: An Introduction to the Theory and Its Applications*; McGraw-Hill: New York, NY, USA; San Francisco, CA, USA, 1966.

23. Lee, E.B.; Markus, L. *Foundations of Optimal Control Theory*; Wiley & Sons: New York, NY, USA, 1967.

24. Moiseev, N.N. *Element of the Optimal Systems Theory*; Nauka: Moscow, Russia, 1974. (In Russian)

25. Bryson, A.E.; Ho, Y.-C. *Applied Optimal Control*; Hemisphere: Washington, DC, USA, 1975.

26. Gershwin, S.B. *Manufacturing Systems Engineering*; PTR Prentice Hall: Englewood Cliffs, NJ, USA, 1994.

27. Sethi, S.P.; Thompson, G.L. *Optimal Control Theory: Applications to Management Science and Economics*, 2nd ed.; Springer: Berlin, Germany, 2000.

28. Dolgui, A.; Ivanov, D.; Sethi, S.; Sokolov, B. Scheduling in production, supply chain Industry 4.0 systems by optimal control: fundamentals, state-of-the-art, and applications. *Int. J. Prod. Res.* **2018**, forthcoming.

29. Bellmann, R. *Adaptive Control Processes: A Guided Tour*; Princeton University Press: Princeton, NJ, USA, 1972.

30. Maccarthy, B.L.; Liu, J. Addressing the gap in scheduling research: A review of optimization and heuristic methods in production scheduling. *Int. J. Prod. Res.* **1993**, *31*, 59–79. [CrossRef]
31. Sarimveis, H.; Patrinos, P.; Tarantilis, C.D.; Kiranoudis, C.T. Dynamic modeling and control of supply chains systems: A review. *Comput. Oper. Res.* **2008**, *35*, 3530–3561. [CrossRef]
32. Ivanov, D.; Sokolov, B. Dynamic supply chains scheduling. *J. Sched.* **2012**, *15*, 201–216. [CrossRef]
33. Ivanov, D.; Sokolov, B. *Adaptive Supply Chain Management*; Springer: London, UK, 2010.
34. Ivanov, D.; Sokolov, B.; Käschel, J. Integrated supply chain planning based on a combined application of operations research and optimal control. *Central. Eur. J. Oper. Res.* **2011**, *19*, 219–317. [CrossRef]
35. Ivanov, D.; Dolgui, A.; Sokolov, B.; Werner, F. Schedule robustness analysis with the help of attainable sets in continuous flow problem under capacity disruptions. *Int. J. Prod. Res.* **2016**, *54*, 3397–3413. [CrossRef]
36. Kalinin, V.N.; Sokolov, B.V. Optimal planning of the process of interaction of moving operating objects. *Int. J. Differ. Equ.* **1985**, *21*, 502–506.
37. Kalinin, V.N.; Sokolov, B.V. A dynamic model and an optimal scheduling algorithm for activities with bans of interrupts. *Autom. Remote Control* **1987**, *48*, 88–94.
38. Ohtilev, M.Y.; Sokolov, B.V.; Yusupov, R.M. *Intellectual Technologies for Monitoring and Control of Structure-Dynamics of Complex Technical Objects*; Nauka: Moscow, Russia, 2006.
39. Krylov, I.A.; Chernousko, F.L. An algorithm for the method of successive approximations in optimal control problems. *USSR Comput. Math. Math. Phys.* **1972**, *12*, 14–34. [CrossRef]
40. Chernousko, F.L.; Lyubushin, A.A. Method of successive approximations for solution of optimal control problems. *Optim. Control Appl. Methods* **1982**, *3*, 101–114. [CrossRef]
41. Ivanov, D.; Sokolov, B.; Dolgui, A.; Werner, F.; Ivanova, M. A dynamic model and an algorithm for short-term supply chain scheduling in the smart factory Industry 4.0. *Int. J. Prod. Res.* **2016**, *54*, 386–402. [CrossRef]
42. Hartl, R.F.; Sethi, S.P.; Vickson, R.G. A survey of the maximum principles for optimal control problems with state constraints. *SIAM Rev.* **1995**, *37*, 181–218. [CrossRef]
43. Chernousko, F.L. *State Estimation of Dynamic Systems*; Nauka: Moscow, Russia, 1994.
44. Gubarev, V.A.; Zakharov, V.V.; Kovalenko, A.N. *Introduction to Systems Analysis*; LGU: Leningrad, Russia, 1988.

*algorithms*

MDPI

Article
# Entropy-Based Algorithm for Supply-Chain Complexity Assessment

Boris Kriheli [1] and Eugene Levner [2,*]

[1]   School of Economics, Ashkelon Academic College, Ashkelon 84101, Israel; borisk@hit.ac.il
[2]   Department of Computer Science, Holon Institute of Technology, Holon 58102, Israel
*   Correspondence: levner@hit.ac.il; Tel.: +972-2-583-2290

Received: 28 February 2018; Accepted: 21 March 2018; Published: 24 March 2018

**Abstract:** This paper considers a graph model of hierarchical supply chains. The goal is to measure the complexity of links between different components of the chain, for instance, between the principal equipment manufacturer (a root node) and its suppliers (preceding supply nodes). The information entropy is used to serve as a measure of knowledge about the complexity of shortages and pitfalls in relationship between the supply chain components under uncertainty. The concept of conditional (relative) entropy is introduced which is a generalization of the conventional (non-relative) entropy. An entropy-based algorithm providing efficient assessment of the supply chain complexity as a function of the SC size is developed.

**Keywords:** industrial supply chain; supply chain complexity; information entropy

---

## 1. Introduction

This paper presents the entropy-based optimization model for estimating structural complexity of supply chain (SC), and, in particular, complexity of relationship between the principal equipment manufacturer (a root node) and its suppliers (preceding supply nodes). The information entropy is used as a measure of decision maker's knowledge about the risks of shortages and pitfalls in relations between the supply chain components under uncertainty, the main attention being paid to relationships between the principal equipment manufacturer (a root node of the corresponding graph model) and its suppliers. A concept of the conditional (relative) entropy is introduced which is a generalization of the conventional (non-relative) entropy that provides more precise estimation of complexity in supply chains as it takes into account information flows between the components of different layers. A main advantage of the suggested entropy-based approach is that it can essentially simplify the hierarchical tree-like model of the supply chain, at the same time retaining the basic knowledge about main sources of risks in inter-layer relationships.

Processes of production, storage, transportation, and utilization of products in SCs may lead to many negative effects on the environment, such as emissions of pollutants into the air and soil; discharges of pollutants into surface and groundwater basins; pollution of soil and water with waste products of production, all these phenomena in their entirety deteriorating the relationships between manufacturing components and their suppliers. These and many other risky situations for manufacturing lead to the uncertainty in SC's phases, i.e., engineering, procurement, production, and distribution [1]. The entropic approach developed in this paper aims at defining the best (minimal but sufficient) level of information going through the several supply chain phases. This paper analyzes the structural complexity of the supply chain affected by technological, organizational and environmental adverse events in the SCs, consequences of which lead, as a result, to violations of correct functioning of the SC. The detailed definition and analysis of the supply-chain structural complexity can be found in [2–10].

Similar to many other researchers (see, e.g., [11–13]), in order to accurately measure the risks of pitfalls and cases of failed supplies, this paper tends to determine the probabilities of undesirable events and their negative impacts on material, financial, and information flows. The economic losses are caused by failures to supply, shipments of inappropriate products, incomplete deliveries, delays in deliveries, etc., and the harmful effect of the adverse events may be expressed in monetary form by relevant penalties.

The main differences of the present work in comparison with close earlier papers (see, e.g., [7,12,14,15]) also exploiting the entropy approach for the SC complexity analysis are the following:

- This paper develops a new graph-theoretic model as a tool for selecting the most vulnerable disruption risks inside the SC which, in turn, can essentially decrease the size of the initial SC model without sacrificing essential knowledge about the risks;
- This paper introduces the conditional entropy as a tool for integrated analysis of the SC complexity under uncertainty, provides more precise estimation of the supply chain complexity taking into account links between the nodes of different layers;
- This paper suggests a new fast entropy-based algorithm for minimizing the SC size.

This paper is structured as follows. The related definitions from graph theory are presented in the next section. The definition of information entropy and the detailed problem description are given in Section 3. Section 4 describes the entropy-based algorithm permitting to reduce the SC model size without a loss of essential information. Section 5 describes the numerical example. Section 6 concludes the paper.

## 2. Basic Definitions

Wishing to avoid any ambiguity in further discussions, we begin with key definitions of risk and ambiguity in relationships between the manufacturer and its supplier. There is a wide specter of different definitions of risk and uncertainty. In this study, we follow Knight's [16] view and his numerous followers. The uncertainty is the absence of certainty in our knowledge, or, in other words, a situation wherein it is impossible to precisely describe future outcomes. In the Knightian sense, the risk is measurable uncertainty, which is possible to calculate.

Similar to many other risk evaluators, we assume that the notion of risk can be described as the expected value of an undesirable outcome, that is, the product of two characteristics, the probability of an undesirable event (that is, a negative deviation of the delayed supply or failure to reach the planned supply target), and the impact or severity, that is, an expected loss in the case of the disruption affecting the supply of products across organizations in a supply network. In a situation with several possible accidents, we admit that the total risk is the sum of the risks for different accidents (see, e.g., [11,17,18]).

In the model considered below, an "event" is the observable discrete change in the state of the SC or its components. A "risk driver" is a factor, a driving force that may be a cause of the undesirable unforeseen event, such as disruptions, breakdowns, defects, mistakes in the design and planning, shortages of material in supply in the SC, etc. In this paper, we study the situations with an "observable uncertainty" where there is an objective opportunity to register, for a pre-specified period of time, the adverse events in the relationships between the components in the SC. Such a registration list, called a "risk protocol", provides us the information whether or not the events are undesirable and, in the case if the event is undesirable what are its risk drivers and possible loss (see [12,19]). Such statistics in the risk protocols permits the decision maker to quantitatively evaluate a contribution of each driver and the total (entropy-based) observable information in the SC.

There exists a wide diversity of risk types, risk drivers, and options for their mitigation in the SC. Their taxonomy lies beyond the scope of this paper. Many authors noticed that if a researcher tries to analyze potential failures/disruptions of all the suppliers in a SC or their absolute majority, he/she encounters a simply impractical and unrealistic problem demanding an astronomic amount of time and budget. Moreover, the supply chain control of the root node of the supply chain is much more important than the control of any of its successors ([6,8]).

Consider a tree-type graph representing the hierarchical structure of an industrial supply chain Define the "parent layer", called also the "main layer", as consisting of a single node, as follows: $L_0 = \{n_0\}$ where $n_0$ is the single node of the parent layer. Called also the original equipment manufacturer (OEM), or the root node. OEM is a firm (or company) that creates an end product, for, instance, assembles and creates an automobile.

Define a *layer* $L_s$ (also denoted as layer $s$) as the set of nodes which are on the same distance $s$ from the root node $n_0$ in the underlying graph of the SC.

Layer 1 (also called Tier 1) are the companies supplying components directly to the OEM that set up the chain. In a typical supply chain, companies in Tier 2 supply the companies in Tier 1; Tier 3 supplies Tier 2, and so on. Tiered supply chains are common in industries such as aerospace or automotive manufacturing where the final product consists of many complex components and sub-assemblies.

Define a "cut" (also called a "cross-section") $C_s$ as a union of all the layers $L_0, L_1, L_2, \ldots L_s$, from 0 to $s$. It is evident that $C_0 = L_0$, $C_{s-1} \subset C_s$, $C_s = \left\{ C_{(s-1)}, L_s \right\} s = 1, 2, \ldots, S$.

Assume that, for each node of the SC, the list of risk drivers $F = \{f_1, f_2, \ldots f_N\}$ is known, each being a source of different adverse events in the nodes of the SC. For simplicity, but without loss of generality, assume that any adverse event is caused by a single risk driver (otherwise, one can split such a multi-drive event into several elementary events each one being caused by a single driver). Here N is the total number of all the drivers.

Another basic assumption to be used in this work is that the drive factors are mutually dependent. It means, for example, that an unfavorable technological decision or an environmental pollution, that is, caused by a technology-based driver in some component at Tier 2 may lead to an adverse event in supply operations to a node at Tier 1. A technological mistake at Tier 3 may be a source of a delayed supply to Tier 2, and so on. In general, any factors $f$ happening at tier $s$ may be depending on a factor $f'$ at an earlier tier $s + 1, f = 1, \ldots, N; f' = 1, \ldots, N; s = 1, \ldots, S$. Below, the dependencies will be described with the help of the $N \times N$ matrix of relative probabilities.

The following Markovian property is assumed to take place. Assume that the dependence between any factor in tier $s$, on the one hand, and the factors in the lower tiers $s + 1, s + 2, \ldots, S$ actually exists only for the factors in a pair of neighboring tiers $(s, s + 1)$, where $s = 0,1, \ldots, S$. Moreover, assume that the pitfalls and any defective decisions do not flow downwards, that is, any risk factor in tier $s$ does not depend upon the risk drivers in the nodes of higher layers, numbered $s - 1, s - 2, \ldots, 1$.

In each layer $s$, when computing the probability of risk drivers $f$ occurring in the nodes of the layer $s$, two types of adverse events have a place. First, there are the events (called "primary events" and denoted by $A_f^{prime}(s)$) that have happened in the nodes of layer $s$ and which are caused by the risk driver $f, f = 1, \ldots, N$. Second, there are the events (called "secondary events" and denoted by $A_f^{second}(s + 1, s)$) that have happened in the nodes of the next layer $(s \pm 1)$ but have an indirect impact upon inverse events in s, since the risk factors are dependent. More precisely, different drivers $f'$ in s+1 have impact upon the driver $f$ in layer $s, f = 1, \ldots, N; f' = 1, \ldots, N; s = 1,2, \ldots, S$.

The impact from $f'$ to $f$ is estimated with the help of the transition probability matrix $M$ which is defined below and computed from the data in the risk protocols.

Denote by $A_j(s)$ the following events:

$A_f(s) = \{$risk driver $f$ is the source of various adverse events in supply to all the nodes of layer $s\}$, $f = 1, \ldots, N, s = 0,1, \ldots, S$.

Denote by $p_f(s) = \text{Pr}(A_f(s))$ the probability that the risk driver $f$ is the source of different adverse events in supply in layer $s$,

$p_i(s) = \text{Pr}(A_i(s)) = \text{Pr}\{$the risk driver $f_i$ is the cause of adverse event on the layer $s$ only$\}$

Denote by $p_f^{prime}(s) = \text{Pr}(A_f^{prime}(s))$ the probability that the risk driver $f$ is the source of different adverse events in layer $s$, and which are caused by the risk driver $f$. These probabilities are termed as "primary". Next, denote by $p_f^{second}(s) = \Pr(A_f^{second}(s))$ the probability that the risk driver $f$ is a source of different adverse events in layer $s$ which is a result of the indirect effect on the $f$ by the risk drivers $f'$ that have caused the adverse events in layer $s+1$; these probabilities are termed as "secondary".

Introduce the following notation:

$$p_i^{(1)}(s) = \Pr\left(A_i^{prime}(s)\right) = \Pr\left(A_i^{(1)}\right) = \Pr \text{ \{the risk driver } f_i \text{ is the cause of adverse event on the}$$

layer s only\}

$$p_i^{(2)}(s) = \Pr\left(A_i^{second}(s)\right) = \Pr\left(A_i^{(2)}\right) = \Pr \text{ \{the risk driver } f_i \text{ is the cause of adverse effect on the}$$

layer $s$ as the result of the risk drivers on the layers $s+1$\}.

For simplicity, and without loss of generality, suppose that the list of risk drivers $F = \{f_1, f_2, \ldots f_N\}$ is complete for each layer. Then the following holds

$$\sum_{i=1}^{N} p_i(s) = 1 \text{ for } s = 0, 1, 2, \ldots \quad (1)$$

Denote $\bar{p}(s) = (p_1(s), p_2(s), \ldots p_N(s))$.

It is obvious that

$$A_i(s) = A_i^{(1)}(s) \cup A_i^{(2)}(s) \text{ and } A_i^{(1)}(s) \cap A_i^{(2)}(s) = \varnothing, \; j = 1, 2, \ldots, N.$$

Therefore,

$$p(A_i(s)) = p\left(A_i^{(1)}(s)\right) + p\left(A_i^{(2)}(s)\right)$$

or

$$p_i(s) = p_i^{(1)}(s) + p_i^{(2)}(s) \qquad i = 1, 2, \ldots, N. \quad (2)$$

Then the vector of risk driver probabilities $\bar{p}(s) = (p_1(s), p_2(s), \ldots p_N(s))$ can be decomposed into two vectors as

$$\bar{p}(s) = \bar{p}^{(1)}(s) + \bar{p}^{(2)}(s), \quad (3)$$

where $\bar{p}^{(1)}(s) = \left(p_1^{(1)}(s), p_2^{(1)}(s), \ldots, p_N^{(1)}(s)\right)$ is the vector of drivers' primary probabilities and $\bar{p}^{(2)}(s) = \left(p_1^{(2)}(s), p_2^{(2)}(s), \ldots, p_N^{(2)}(s)\right)$ the vector of drivers' secondary probabilities.

For any layer $s$, define the transition matrix $M^{(2)}(s)$ of conditional probabilities of the risk drivers on layer $s$ that are obtained as the result of risk drivers existing on layer $s+1$

$$M^{(2)}(s) = \left(p_{ij}^{(2)}(s)\right)_{N \times N}, \; s = 0, 1, 2, \ldots, \text{ with}$$

$$p_{ij}^{(2)}(s) = \Pr\left(A_j^{(2)}(s) \big| A_i(s+1)\right) i, \quad j = 1, 2, \ldots, N, \quad s = 0, 1, 2 \ldots \quad (4)$$

Next, define the matrices $ML(s)$ of the primary drivers' probabilities as

$$ML(s) = \left(q_{ij}(s)\right)_{N \times N}, \; s = 0, 1, 2, \ldots,$$

$$q_{ij}(s) = p_j^{(1)}(s), \qquad i, j = 1, 2, \ldots, N, \; s = 0, 1, \ldots$$

$$ML(s) = \begin{pmatrix} p_1^{(1)}(s) & p_2^{(1)}(s) & \cdot & \cdot & p_N^{(1)}(s) \\ p_1^{(1)}(s) & p_2^{(1)}(s) & \cdot & \cdot & p_N^{(1)}(s) \\ \cdot & \cdot & \cdot & \cdot & \cdot \\ \cdot & \cdot & \cdot & \cdot & \cdot \\ p_1^{(1)}(s) & p_2^{(1)}(s) & \cdot & \cdot & p_N^{(1)}(s) \end{pmatrix} \quad (5)$$

Define the complete transition matrices as

$$\hat{M}^{(1)}(s) = \left( \hat{p}_{ij}^{(1)}(s) \right)_{N \times N}, \ s = 0,1,2, \ \ldots, \ \text{with}$$

$$\hat{p}_{ij}^{(1)}(s) = p_j^{(1)}(s) + p_{ij}^{(2)}(s), i,j = 1,2,\ldots,N, \ s = 0,1,2,\ldots \tag{6}$$

From (5) and (6) it follows that

$$\hat{M}^{(1)}(s) = ML(s) + M^{(2)}(s), \ \ s = 0,1,2\ldots \tag{7}$$

$$p_j^{(2)}\left( A_j^{(2)}(s) \right) = \sum_{i=1}^{N} p_i(s+1) \cdot p_{ij}^{(2)}(s), \ \ \ \ j = 1,2,\ldots,N, \ s = 0,1,\ldots$$

or

$$p_j^{(2)}\left( A_j^{(2)}(s) \right) = \sum_{i=1}^{N} p_i(s+1) \cdot p_{ij}^{(2)}(s), \ \ \ \ j = 1,2,\ldots,N, \ s = 0,1,\ldots \tag{8}$$

In the matrix form, Equation (8) can be rewritten as

$$\bar{p}^{(2)}(s) = \bar{p}(s+1) \cdot M^{(2)}(s), \ \ s = 0,1,\ldots \tag{9}$$

The following claim is true.

**Claim**. The following relation holds:

$$\sum_{i=1}^{N} \left( p_i(s+1) \cdot \left( \sum_{j=1}^{N} p_{ij}^{(2)}(s) \right) \right) = 1 - \sum_{j=1}^{N} p_j^{(1)}(s), \ s = 0,1,\ldots \tag{10}$$

The proof is straightforward and skipped here.

## 3. Information Entropy as a Measure of Supply Chain Complexity

Information entropy is defined by Shannon as follows [20]. Given a set of events $E = \{e_1, \ldots, e_n\}$ with a priori probabilities of event occurrence $P = \{p_1, \ldots, p_n\}$, $p_i \geq 0$, such that $p_i + \ldots + p_n = 1$, the entropy function $H$ is defined by

$$H = -\sum_I p_i \log p_i. \tag{11}$$

In order to yield all necessary information on the SC complexity issues, this paper uses the data recording of all adverse events occurred. The enterprise is to collect and store the data about main adverse events that occurred and led to economic losses in the enterprise, compensation cost, as well as the statistical analysis of the recorded data. Such requirement also applies to the registration of information about control of compliance of target and actual environmental characteristics.

Similar to [12], for each node $u$, consider an information database called a 'risk protocol'. This is a registration list of most important events that have occurred in the node during a pre-specified time period. The protocol provides us the information whether or not the events are undesirable and, in the latter case, what are its risk drivers and possible losses.

This data is recorded in tables $TBL_u$, representing lists of events in each node $u$ during a certain time period $T$ (e.g., month, or year). Each row in the table corresponds to an individual event occurring in a given node at a certain time moment (for example, a day). We use symbol $f$ as an index of risk drivers, $F$ as the total number of risk drivers, and $r$ as an index of the event (row). The value $z_{rf}$ at the intersection of column $f$ and row $r$ is equal to 1 if the risk factor $f$ is a source of the adverse event $r$, and 0—otherwise. The last column, $F + 1$, in each row ($r$) contains the magnitude of economic loss caused by the corresponding event $r$.

As far as the tables $TBL_u$, for all the nodes belonging to a certain arbitrary SC layer, says, are derived, all the tables are gathered into the Cut_Table $CT_s$ for the entire SC cut. Let $R_s(u)$ denote the total number of observed adverse (critical) events in a node $u$ of cut $C_s$ during a certain planning period. If such cut contains $n(s)$ nodes, the total number of critical events in it is $N^s = \sum_{u=1,\dots,n(s)} R_s(u)$. Assume that there are $F$ risk drivers. For each risk driver $f$ ($f = 1, \dots, F$), we can compute the number $N_s(u, f)$ of critical events caused by driver $f$ in the node $u$ and the total number $N_s(f)$ of critical events in all nodes of cut $C_s$, as registered in the risk protocols.

The relative frequency $p_s(f)$ of that driver $f$ is the source of different critical events in nodes of $s$ and can be treated as the estimation of the corresponding probability. Then we compute the latter probability as

$$p_s(f) = N_s(f)/N^s \tag{12}$$

Then $\sum_f p_s(f) = 1$.

For the sake of simplicity of further analysis, our model applies to the case when the critical events are independent within the same tier and the losses are additive (these assumptions will be relaxed in our future research). For any node $u$ from $s$, we can define corresponding probabilities $p_s(u, f)$ of the event that a driver $f$ is the source of adverse events in node $u$

$$p_s(u, f) = N_s(u, f)/R_s(u). \tag{13}$$

This paper treats the $p_s(u, f)$ values defined by Equation(3) as probabilities of events participating in calculation of the entropy function in Equation (1).

The main idea of the suggested entropic approach is that the information entropy in this study estimates the average amount of information contained in a stream of critical events of the risk protocol. Thus the entropy characterizes our uncertainty, or the absence of knowledge, about the risks. The idea here is that the less the entropy is, the more information and knowledge about risks is available for the decision makers.

The entropy value can be computed iteratively for each cut of the SC. Assume that the nodes of a cut $C_{s-1}$, are defined at step (iteration) $s - 1$. Denote by $T_s$ all supplier-nodes in the supply layer $s$ of the given tree. Let $L_s(T_s)$ denote the total losses defined by the risk protocol and summed up for all nodes of cut $C_s$ in tiers $T_s$: $L_s(T_s) = \sum_{u \in T_s} c_s(u)$. Further, let $LT$ denote the total losses for all nodes of the entire chain. Thus, $L_s(T_s)$ are contributions of the suppliers of cut $C_s$ into the total losses $LT$.

Then $L_s(T_s)/LT$ define the relative losses in the $s$-truncated supply chain. The relative contribution of lower tiers, that is, of those with larger $s$ values, are, respectively, $(LT - L_s(T_s))/LT$. One can observe that the larger is the share $(LT - L_s(T_s))/LT$ in comparison with $L_s(T_s)/LT$, the less is the available information about the losses in the cut $C_s$. For example, if the ratio $L_s(T_s)/LT = 0.2$, this case give us less information about the losses in cut $C_s$ in comparison with the opposite case of $L_s(T_s)/LT = 0.8$. This argument motivates us to take the ratios $(LT - L_s(T_s))/LT$ as the coefficients (weights) of the entropy (or, of our unawareness) about the economic losses incurred by adverse events affecting the environmental quality. In other words, the latter coefficients weigh the lack of our knowledge about the losses; as far as the number $s$ grows, these coefficients become less and less significant.

Then the total entropy of all nodes $u$ included into the cut $s$ will be defined as

$$H(s) = \sum_u H(u) \tag{14}$$

where the weighted entropy in each node $u$ is computed as

$$H(u) = -U_s \sum_f p_s(u, f) \log p_s(u, f), \tag{15}$$

$$U_s = (LT - L_s(T_s))/LT, \tag{16}$$

Let $x_u = 1$ if node $u$ from $T_s$ is included into $s$, and $x_u = 0$, otherwise.

Then the total entropy of all nodes included into the cut $s$ will be defined as

$$H(s) = \sum_u H(u)\, x_u, \tag{17}$$

where $H(u)$ is defined in (15).

Computations of $p_s(u, f)$ as well as the summation over risk factors $f$ are taken in the risk event protocols for all the events related to nodes $u$ from $T_s$. As far as entropy values are found for each node, the vulnerability to risks over the supply chain is measured as a total entropy of the $s$-truncated supply chain subject to the restricted losses.

Define the weighted entropy for each cut $s$ as

$$H(C_s) = c(s) H^*(C_s) \tag{18}$$

where

$$H^*(C_s) = -\sum_{j=1}^{N} p_j(C_s) \log p_j(C_s) \text{ is the entropy of cut } C_s$$

We assume that the weight $c(s)$ satisfies the following conditions:

(i)   $c(s)$ is decreasing;
(ii)  $c(0) = L$;
(iii) $\lim_{s\to\infty} c(s) = 0$.

Define the "variation of relative entropy" depending upon the cut number is

$$REV(s) = \frac{H(s-1) - \frac{c(s-1)}{c(s)} H(s)}{H(1) - \frac{c(s-1)}{c(s)} H(s)}. \tag{19}$$

The following claim is valid:

**Theorem.** For the process of sequentially computing of the relative entropy variation (REV), for any fixed value $\varepsilon$, there exists the layer number s* for which it holds: $|REV(s*)| < \varepsilon$.

**Proof.** For simplicity, we assume that the entropy of any layer depends only upon the information of the neighbor layers, that is,

$$H^*(L_s | L_{s+1}, L_{s+2}, \ldots L_k) - H^*(L_s | L_{s+1}), \quad s = 0, 1, 2, \ldots, k-1$$

Let us exploit the following Formula for the entropy of combined system (see [21]):

$$H(X_1, X_2, \ldots X_s) = H(X_1) + H(X_2 | X_1) + H(X_3 | X_1, X_2) + \ldots + H(X_s | X_1, X_2, \ldots, X_{s-1}),$$

Applying it for the entropy $H^*(C_s)$ of cut $C_s$. We have

$$\begin{aligned}
H^*(C_s) &= H^*(L_s, L_{s-1}, L_{s-2}, \ldots, L_0) = \\
&= H^*(L_s) + H^*(L_{s-1} | L_s) + H^*(L_{s-2} | L_{s-1}, L_s) + \ldots + H^*(L_0 | L_1, L_2, \ldots L_s) = \\
&= H^*(L_s) + H^*(L_{s-1} | L_s) + H^*(L_{s-2} | L_{s-1}) + \ldots + H^*(L_0 | L_1)
\end{aligned} \tag{20}$$

Using the latter Formula for cut $C_{s-1}$, we obtain

$$H^*(C_{s-1}) = H^*(L_{s-1}) + H^*(L_{s-2} | L_{s-1}) + H^*(L_{s-3} | L_{s-2}) + \ldots + H^*(L_0 | L_1) \tag{21}$$

From Formulas (20) and (21), we obtain that

$$H^*(C_s) = H^*(C_{s-1}) + H^*(L_s) - H^*(L_{s-1}) + H^*(L_{s-1}|L_s) \tag{22}$$

Here $H^*(L_{s-1}|L_s)$ denotes the conditional entropy of the layer $s-1$ under the condition that the probabilities and entropy of layer $s$ are found.

Denote, for convenience, $H(C_s) = H(s)$ and $H^*(C_s) = H^*(s)$.

Using the definition of the weighted entropy and Formula (21), we obtain

$$H(s) = c(s)H^*(s) = c(s)(H^*(C_{s-1}) + H^*(L_s) - H^*(L_{s-1}) + H^*(L_{s-1}|L_s)).$$

Using the Formula of the conditional entropy, the definitions of events $A_i(s)$, probabilities $p_i(s)$ and matrices $M^{(2)}(s)$, we can write that

$$H^*(L_{s-1}|L_s) = -\sum_{i=1}^{N} \left( P(A_i(s)) \cdot \sum_{j=1}^{N} P(A_j(s-1)|A_i(s)) \cdot \log P(A_j(s-1)|A_i(s)) \right)$$

$$= -\sum_{i=1}^{N} \left( p_i(s) \cdot \sum_{j=1}^{N} \hat{p}_{ij}^{(1)}(s-1) \cdot \log\left(\hat{p}_{ij}^{(1)}(s-1)\right) \right)$$

$$H^*(L_{s-1}|L_s) = -\sum_{i=1}^{N} \left( p_i(s) \cdot \sum_{j=1}^{N} \hat{p}_{ij}^{(1)}(s-1) \cdot \log\left(\hat{p}^{(1)}(s-1)\right) \right) \tag{23}$$

Using Formula (22) we can write

$$H(s) = c(s) \left( H^*(C_{s-1}) + H^*(L_s) - H^*(L_{s-1}) - \sum_{i=1}^{N} \left( p_i(s) \cdot \sum_{j=1}^{N} \hat{p}_{ij}^{(1)}(s-1) \cdot \log\left(\hat{p}_{ij}^{(1)}(s-1)\right) \right) \right) \tag{24}$$

$$s = 1, 2, \ldots$$

$$H(s-1) - H(s) =$$
$$= H(s-1) - c(s) \left( H^*(C_{s-1}) + H^*(L_s) - H^*(L_{s-1}) - \sum_{i=1}^{N} \left( p_i(s) \cdot \sum_{j=1}^{N} \hat{p}_{ij}^{(1)}(s-1) \cdot \log\left(\hat{p}_{ij}^{(1)}(s-1)\right) \right) \right) =$$
$$= \left(1 - \frac{c(s)}{c(s-1)}\right) H(s-1) - c(s)(H^*(L_s) - H^*(L_{s-1})) + c(s) \sum_{i=1}^{N} \left( p_i(s) \cdot \sum_{j=1}^{N} \hat{p}_{ij}^{(1)}(s-1) \cdot \log\left(\hat{p}_{ij}^{(1)}(s-1)\right) \right)$$

We obtain that

$$H(s-1) - H(s) = \left(1 - \frac{c(s)}{c(s-1)}\right) H(s-1) - c(s)(H^*(L_s) - H^*(L_{s-1})) + c(s) \cdot \sum_{i=1}^{N} \left( p_i(s) \cdot \sum_{j=1}^{N} \hat{p}_{ij}^{(1)}(s-1) \cdot \log\left(\hat{p}_{ij}^{(1)}(s-1)\right) \right) \tag{25}$$

$$s = 1, 2, \ldots$$

Since the following relations are valid,

$$0 < 1 - \frac{c(s)}{c(s-1)} < 1,$$

$$0 < -\sum_{i=1}^{N} \left( p_i(s) \cdot \sum_{j=1}^{N} \hat{p}_{ij}^{(1)}(s-1) \cdot \log\left(\hat{p}_{ij}^{(1)}(s-1)\right) \right) < \log N$$

$$0 < H^*(L_s) < \log N, \quad 0 < H^*(L_{s-1}) < \log N$$

$$\lim_{s \to \infty} H(s-1) = 0, \quad \lim_{s \to \infty} c(s) = 0$$

we obtain that

$$\lim_{s \to \infty} (H(s-1) - H(s)) = 0$$

Therefore, for any accuracy level $\varepsilon$, we can select the number of a cut $s_1$ for which

$$\frac{H(s_1 - 1) - H(s_1)}{H(0) - H(s_1)} < \varepsilon$$

The truncated part of the SC containing only layers of the cut $(s_1 - 1)$ possesses the required level of the entropy variation. The theorem is proved.

The theorem permits the decision maker to define the decreased size of the SC model, such that the decreased number of the layers in the SC model is sufficient for planning and coordinating the knowledge about the risks in the relations between the SC components, without the loss of essential information about the risks.

## 4. Entropy-Based Algorithm for Complexity Assessment

This section summarizes theoretical findings of the previous sections for obtaining a decreased SC model on which planning and coordination of supplies can be done without loss of essential information.

Input data of the algorithm:

- the given number $N$ of risk drivers,
- weight functions $c(s)$ selected by the decision maker,
- probabilities $p_f^{prime}(s) = \Pr\left(A_f^{\,prime}(s)\right)$ that the risk driver $f$ is the direct source of the supply failure/delay in layer $s$, which are caused by risk driver $f$,
- probabilities $p_f^{second}(s) = \Pr\left(A_f^{\,second}(s)\right)$ that the risk driver $f$ is the source of the supply failure/delay in layer $s$, which is a result of the indirect effect on the $f$ by the risk drivers $f'$ of supply delay in layer $s + 1$; these probabilities are termed as *secondary*.
- transition probability matrices $M^{(2)}(s) = \left(p_{ij}^{(2)}(s)\right)_{N \times N}$, $s = 0, 1, 2, \ldots, k$.

**Step 1.** Using the entropy Formulas (11)–(17), calculate entropy of the layer 0:

$$H^*(L_0) = -\sum_{j=1}^{N} p_j(0) \log(p_j(0))$$
$$H(0) = H^*(L_0)$$

**Step 2.** Using Formulas (2)–(9), compute the matrix
$\hat{M}^{(1)}(0)$, and vector $\bar{p}(1) = (p_1(1), p_2(1), \ldots p_N(1))$
**Step 3.** Compute the corrected vector of probabilities for the layer $L_0$, using Formula
$\bar{p}^c(0) - \bar{p}(1) \cdot \hat{M}^{(1)}(0)$ and corrected entropy for layer $L_0$

$$HC^*(L_0) = -\sum_{j=1}^{N} p_j^c(0) \log\left(p_j^c(0)\right)$$

**Step 4.** For $s = 2, 3, \ldots$, using matrix $\hat{M}^{(1)}(s)$, vector $\bar{p}(s) - (p_1(s), p_2(s), \ldots p_N(s))$ compute sequentially the corrected vectors of probabilities for the layers $L_{s-1}$: $\bar{p}^c(s - 1) = \bar{p}(s) \cdot \hat{M}^{(1)}(s - 1)$
**Step 5.** Compute $HC^*(L_{s-1}) = -\sum_{j=1}^{N} p_j^c(s - 1) \log\left(p_j^c(s - 1)\right)$
**Step 6.** Compute

$$H(0) - H(1) =$$
$$= \left(1 - \frac{c(1)}{c(0)}\right) H(0) - c(1)(H^*(L_1) - HC^*(L_0)) + c(1) \sum_{i=1}^{N} \left(p_i(1) \cdot \sum_{j=1}^{N} \hat{p}_{ij}^{(1)}(0) \cdot \log\left(\hat{p}_{ij}^{(1)}(0)\right)\right)$$

**Step 7.** For $s - 1, 2, \ldots,$ compute

$$H(s-1) - H(s) =$$

$$= \left(1 - \frac{c(s)}{c(s-1)}\right) H(s-1) - c(s)(H^*(L_s) - HC^*(L_{s-1})) + c(s) \sum_{i=1}^{N} \left( p_i(s) \cdot \sum_{j=1}^{N} \hat{p}_{ij}^{(1)}(s-1) \cdot \log\left(\hat{p}_{ij}^{(1)}(s-1)\right) \right)$$

As the stopping rule use the following rule:

Stop at the cut $s_1$ for which $\frac{H(s_1-1) - H(s_1)}{H(0) - H(s_1)} < \varepsilon$ holds.

Then the reduced SC model contains only the cut $(s_1 - 1)$.

## 5. Numerical Example

Input data:

- the number or risk factor drivers in each layer, $N = 3$;
- level of accuracy $\varepsilon = 0.01$;
- the weight function $c(s) = \frac{1}{(s+1)^2}$ (selected by the decision maker).
- probabilities $p_f{}^{prime}(s) = \Pr\left(A_f{}^{prime}(s)\right)$

$$\bar{p}_f^{prime}(0) = (0.3457 \quad 0.0835 \quad 0.0918)$$
$$\bar{p}_f^{prime}(1) = (0.1644 \quad 0.3017 \quad 0.0542)$$
$$\bar{p}_f^{prime}(2) = (0.1256 \quad 0.1602 \quad 0.2156)$$
$$\bar{p}_f^{prime}(3) = (0.0845 \quad 0.2001 \quad 0.3025)$$
$$\bar{p}_f^{prime}(4) = (0.2623 \quad 0.1056 \quad 0.2369)$$
$$\bar{p}_f^{prime}(5) = (0.2014 \quad 0.2032 \quad 0.1356)$$
$$\bar{p}_f^{prime}(6) = (0.1422 \quad 0.2258 \quad 0.1047)$$
$$\bar{p}_f^{prime}(7) = (0.1056 \quad 0.3241 \quad 0.2658)$$
$$\bar{p}_f^{prime}(8) = (0.1599 \quad 0.3056 \quad 0.1422)$$
$$\bar{p}_f^{prime}(9) = (0.2014 \quad 0.3068 \quad 0.0856)$$
$$\bar{p}_f^{prime}(10) = (0.2145 \quad 0.0241 \quad 0.2536)$$

- probabilities $p_f{}^{second}(s) = \Pr\left(A_f{}^{second}(s)\right)$

$$\bar{p}_f^{second}(0) = (0.3014 \quad 0.0725 \quad 0.1051)$$
$$\bar{p}_f^{second}(1) = (0.1851 \quad 0.2532 \quad 0.0414)$$
$$\bar{p}_f^{second}(2) = (0.2098 \quad 0.1308 \quad 0.1280)$$
$$\bar{p}_f^{second}(3) = (0.0837 \quad 0.2011 \quad 0.1281)$$
$$\bar{p}_f^{second}(4) = (0.1272 \quad 0.1013 \quad 0.1667)$$
$$\bar{p}_f^{second}(5) = (0.2334 \quad 0.0687 \quad 0.1577)$$
$$\bar{p}_f^{second}(6) = (0.1393 \quad 0.2709 \quad 0.1171)$$
$$\bar{p}_f^{second}(7) = (0.0824 \quad 0.1803 \quad 0.0417)$$
$$\bar{p}_f^{second}(8) = (0.1379 \quad 0.1143 \quad 0.1401)$$
$$\bar{p}_f^{second}(9) = (0.1456 \quad 0.1703 \quad 0.0903)$$
$$\bar{p}_f^{second}(10) = (0.2350 \quad 0.0383 \quad 0.2345)$$

- transition probability matrices $M^{(2)}(s) = \left(p_{ij}^{(2)}(s)\right)_{N \times N}, s = 1, 2, \ldots, 10,$

$$M^{(2)}(1) =$$

$$\begin{matrix} 0.3124 & 0.3320 & 0.3556 \\ 0.3456 & 0.4158 & 0.2386 \\ 0.4258 & 0.0256 & 0.5486 \end{matrix}$$

$$M^{(2)}(2) =$$

| | | |
|---|---|---|
| 0.2587 | 0.3568 | 0.3845 |
| 0.6587 | 0.0254 | 0.3159 |
| 0.4872 | 0.0246 | 0.4882 |

$$M^{(2)}(3) =$$

| | | |
|---|---|---|
| 0.1054 | 0.4503 | 0.4443 |
| 0.0865 | 0.6514 | 0.2621 |
| 0.4536 | 0.1458 | 0.4006 |

$$M^{(2)}(4) =$$

| | | |
|---|---|---|
| 0.3548 | 0.2149 | 0.4303 |
| 0.2826 | 0.5239 | 0.1935 |
| 0.2596 | 0.6823 | 0.0581 |

$$M^{(2)}(5) =$$

| | | |
|---|---|---|
| 0.1036 | 0.5483 | 0.3481 |
| 0.0598 | 0.2136 | 0.7266 |
| 0.1269 | 0.2456 | 0.6275 |

$$M^{(2)}(6) =$$

| | | |
|---|---|---|
| 0.3265 | 0.4251 | 0.2484 |
| 0.6421 | 0.2386 | 0.1193 |
| 0.0425 | 0.1243 | 0.8332 |

$$M^{(2)}(7) =$$

| | | |
|---|---|---|
| 0.0934 | 0.4652 | 0.4414 |
| 0.5219 | 0.1361 | 0.3420 |
| 0.4582 | 0.0348 | 0.5070 |

$$M^{(2)}(8) =$$

| | | |
|---|---|---|
| 0.2458 | 0.1965 | 0.5577 |
| 0.0563 | 0.7581 | 0.1856 |
| 0.4103 | 0.1784 | 0.4113 |

$$M^{(2)}(9) =$$

| | | |
|---|---|---|
| 0.4397 | 0.1535 | 0.4068 |
| 0.3627 | 0.4863 | 0.1510 |
| 0.2642 | 0.5529 | 0.1829 |

$$M^{(2)}(10) =$$

| | | |
|---|---|---|
| 0.7412 | 0.0985 | 0.1603 |
| 0.0525 | 0.4368 | 0.5107 |
| 0.1056 | 0.7296 | 0.1648 |

The algorithm solves this example as follows.

Using Formulas (3)–(6) compute:

The probabilities of risk factor drivers on the layers $L_0, L_1, L_2, \ldots$

$$\bar{p}(0) = (0.6471 \quad 0.1560 \quad 0.1969)$$
$$\bar{p}(1) = (0.3495 \quad 0.5549 \quad 0.0956)$$
$$\bar{p}(2) = (0.3354 \quad 0.2910 \quad 0.3736)$$
$$\bar{p}(3) = (0.1682 \quad 0.4012 \quad 0.4306)$$
$$\bar{p}(4) = (0.3895 \quad 0.2069 \quad 0.4036)$$
$$\bar{p}(5) = (0.4348 \quad 0.2719 \quad 0.2933)$$
$$\bar{p}(6) = (0.2815 \quad 0.4967 \quad 0.2218)$$
$$\bar{p}(7) = (0.1880 \quad 0.5044 \quad 0.3075)$$
$$\bar{p}(8) = (0.2978 \quad 0.4199 \quad 0.2823)$$
$$\bar{p}(9) = (0.3470 \quad 0.4771 \quad 0.1759)$$
$$\bar{p}(10) = (0.4495 \quad 0.0624 \quad 0.4881)$$

The complete transition matrices

$$\hat{M}^{(1)}(s) = \left( \hat{p}_{ij}^{(1)}(s) \right)_{3 \times 3}$$

$$\hat{M}^{(1)}(s) = (p_{ij}(s)), \quad s = 1, 2, \ldots, 10$$

$\hat{M}(:, :, 1) =$

$$
\begin{matrix}
0.4124 & 0.4227 & 0.1649 \\
0.8076 & 0.0130 & 0.1794 \\
0.5732 & 0.0114 & 0.4153
\end{matrix}
$$

$\hat{M}(:, :, 2) =$

$$
\begin{matrix}
0.1978 & 0.6567 & 0.1455 \\
0.1029 & 0.8599 & 0.0372 \\
0.6779 & 0.2258 & 0.0963
\end{matrix}
$$

$\hat{M}(:, :, 3) =$

$$
\begin{matrix}
0.6331 & 0.3176 & 0.0494 \\
0.1494 & 0.0275 & 0.8230 \\
0.3925 & 0.5261 & 0.0814
\end{matrix}
$$

$\hat{M}(:, :, 4) =$

$$
\begin{matrix}
0.1104 & 0.8743 & 0.0153 \\
0.1262 & 0.0455 & 0.8284 \\
0.2454 & 0.1270 & 0.6276
\end{matrix}
$$

$\hat{M}(:, :, 5) =$

$$
\begin{matrix}
0.4019 & 0.3097 & 0.2884 \\
0.7173 & 0.2444 & 0.0382 \\
0.0672 & 0.0197 & 0.9131
\end{matrix}
$$

$\hat{M}(:, :, 6) =$

$$
\begin{matrix}
0.1399 & 0.6222 & 0.2379 \\
0.4887 & 0.1686 & 0.3427 \\
0.6885 & 0.0584 & 0.2531
\end{matrix}
$$

$$\hat{M}\,(:,:,7) =$$

$$
\begin{array}{ccc}
0.5105 & 0.1190 & 0.3705 \\
0.3377 & 0.4472 & 0.2152 \\
0.0493 & 0.8089 & 0.1418
\end{array}
$$

$$\hat{M}\,(:,:,8) =$$

$$
\begin{array}{ccc}
0.1930 & 0.2289 & 0.5781 \\
0.0701 & 0.8885 & 0.0414 \\
0.3581 & 0.2238 & 0.4180
\end{array}
$$

$$\hat{M}\,(:,:,9) =$$

$$
\begin{array}{ccc}
0.4487 & 0.2271 & 0.3242 \\
0.2554 & 0.4309 & 0.3137 \\
0.1151 & 0.7706 & 0.1143
\end{array}
$$

$$\hat{M}\,(:,:,10) =$$

$$
\begin{array}{ccc}
0.6751 & 0.1230 & 0.2019 \\
0.0287 & 0.3383 & 0.6330 \\
0.0856 & 0.8210 & 0.0934
\end{array}
$$

Omitting intermediate calculations, at Steps 6 and 7, the algorithm computes values $H(s)$ and $REV(s)$ for each cut $s$, as presented in Table 1.

**Table 1.** Computational results.

| $S$ | 0 | 1 | 2 | 3 | 4 | 5 | 6 |
|-----|------|------|------|------|------|------|------|
| $H(s)$ | 2.5693 | 1.2861 | 0.5972 | 0.1735 | 0.0953 | 0.0378 | 0.0235 |
| $REV(s)$ | - | 1 | 0.5368 | 0.3301 | 0.0609 | 0.0448 | 0.0111 |

We observe that $REV(6)$ value is less then $\varepsilon = 0.01$, therefore we can reduce our SC size by taking the truncated model with $s = 6$. The results of computations are graphically presented in Figure 1.

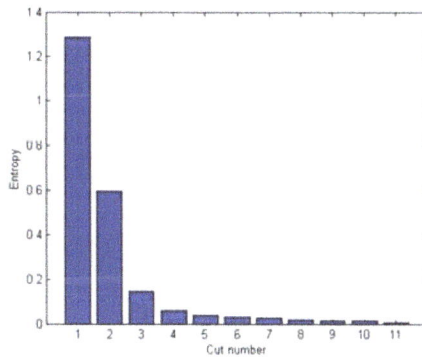

**Figure 1.** Results of entropy computations.

## 6. Conclusions

A main contribution of this paper is the entropy-based method for a quantitative assessment of the information and knowledge relevant to the analysis of the SC size and complexity. Using the entropy approach, the suggested model extracts a sufficient amount of useful information from the risk protocols generated at different SC units. Assessment of the level of entropy for the risks allows the decision maker, step by step, to assess the entropy in the SC model and, consequently, to increase the amount of useful knowledge. As a result, we arrive at a reduced graph model of the supply chain such that it contains essentially the same amount of information about controllable parameters as the complete SC graph but much smaller in size.

An attractive direction for further research is to incorporate the human factor and, in particular, the effect of human fatigue (see [22]) on performance of industrial supply chains under uncertainty in dynamic environments.

**Acknowledgments:** There are no grants or other sources of funding of this study.

**Author Contributions:** Boris Kriheli contributed to the overall idea, model and algorithm, as well as the detailed writing of the manuscript; Eugene Levner contributed to the overall ideas and discussions on the algorithm, as well as the preparation of the paper.

**Conflicts of Interest:** The authors declare no conflict of interest.

## References

1. Fera, M.; Fruggiero, F.; Lambiase, A.; Macchiaroli, R.; Miranda, S. The role of uncertainty in supply chains under dynamic modeling. *Int. J. Ind. Eng. Comput.* **2017**, *8*, 119–140. [CrossRef]
2. Calinescu, A.; Efstathiou, J.; Schirn, J.; Bermejo, J. Applying and assessing two methods for measuring complexity in manufacturing. *J. Oper. Res. Soc.* **1998**, *49*, 723–733. [CrossRef]
3. Sivadasan, S.; Efstathiou, J.; Calinescu, A.; Huatuco, L.H. Advances on measuring the operational complexity of supplier–customer systems. *Eur. J. Oper. Res.* **2006**, *171*, 208–226. [CrossRef]
4. Sivadasan, S.; Efstathiou, J.; Frizelle, G.; Shirazi, R.; Calinescu, A. An information-theoretic methodology for measuring the operational complexity of supplier-customer systems. *Int. J. Oper. Prod. Manag.* **2002**, *22*, 80–102. [CrossRef]
5. Sivadasan, S.; Smart, J.; Huatuco, L.H.; Calinescu, A. Reducing schedule instability by identifying and omitting complexity-adding information flows at the supplier–customer interface. *Int. J. Prod. Econ.* **2013**, *145*, 253–262. [CrossRef]
6. Battini, D.; Persona, A. Towards a Use of Network Analysis: Quantifying the Complexity of Supply Chain Networks. *Int. J. Electron. Cust. Relatsh. Manag.* **2007**, *1*, 75–90. [CrossRef]
7. Isik, F. An Entropy-based Approach for Measuring Complexity in Supply Chains. *Int. J. Prod. Res.* **2010**, *48*, 3681–3696. [CrossRef]
8. Allesina, S.; Azzi, A.; Battini, D.; Regattieri, A. Performance Measurement in Supply Chains: New Network Analysis and Entropic Indexes. *Int. J. Prod. Res.* **2010**, *48*, 2297–2321. [CrossRef]
9. Modraka, V.; Martona, D. Structural Complexity of Assembly Supply Chains: A Theoretical Framework. *Proced. CIRP* **2013**, *7*, 43–48. [CrossRef]
10. Ivanov, D. Entropy-Based Supply Chain Structural Complexity Analysis. In *Structural Dynamics and Resilience in Supply Chain Risk Management*; International Series in Operations Research & Management Science; Springer: Berlin, Germany, 2018; Volume 265, pp. 275–292.
11. Kogan, K.; Tapiero, C.S. *Supply Chain Games: Operations Management and Risk Valuation*; Springer: New York, NY, USA, 2007.
12. Levner, E.; Ptuskin, A. An entropy-based approach to identifying vulnerable components in a supply chain. *Int. J. Prod. Res.* **2015**, *53*, 6888–6902. [CrossRef]
13. Aven, T. *Risk Analysis*; Wiley: New York, NY, USA, 2015.
14. Harremoës, P.; Topsøe, F. Maximum entropy fundamentals. *Entropy* **2001**, *3*, 191–226. [CrossRef]
15. Herbon, A.; Levner, E.; Hovav, S.; Shaopei, L. Selection of Most Informative Components in Risk Mitigation Analysis of Supply Networks: An Information-gain Approach. *Int. J. Innov. Manag. Technol.* **2012**, *3*, 267–271.
16. Knight, F.H. *Risk, Uncertainty, and Profit*; Hart, Schaffner & Marx: Boston, MA, USA, 1921.
17. Zsidisin, G.A.; Ellram, L.M.; Carter, J.R.; Cavinato, J.L. An Analysis of Supply Risk Assessment Techniques. *Int. J. Phys. Distrib. Logist. Manag.* **2004**, *34*, 397–413. [CrossRef]
18. Tapiero, C.S.; Kogan, K. Risk and Quality Control in a Supply Chain: Competitive and Collaborative Approaches. *J. Oper. Res. Soc.* **2007**, *58*, 1440–1448. [CrossRef]

19. Tillman, P. An Analysis of the Effect of the Enterprise Risk Management Maturity on Sharehokder Value during the Economic Downturn of 2008–2010. Master's Thesis, University of Pretoria, Pretoria, South Africa, 2011.
20. Shannon, C.E. A Mathematical Theory of Communication. *Bell Syst. Tech. J.* **1948**, *27*, 379–423. [CrossRef]
21. Wentzel, E.S. *Probability Theory*; Mir Publishers: Moscow, Russia, 1982.
22. Fruggiero, F.; Riemma, S.; Ouazene, Y.; Macchiaroli, R.; Guglielmim, V. Incorporating the human factor within the manufacturing dynamics. *IFAC-PapersOnline* **2016**, *49*, 1691–1696. [CrossRef]

*algorithms*

MDPI

*Article*

# PHEFT: Pessimistic Image Processing Workflow Scheduling for DSP Clusters

Alexander Yu. Drozdov [1], Andrei Tchernykh [2,3,*], Sergey V. Novikov [1], Victor E. Vladislavlev [1] and Raul Rivera-Rodriguez [2]

[1] Moscow Institute of Physics and Technology, Moscow 141701, Russia;
alexander.y.drozdov@gmail.com (A.Y.D.); serg.v.novikov@gmail.com (S.V.N.);
victor.vladislavlev@gmail.com (V.E.V.)
[2] Computer Science Department, CICESE Research Center, 22860 Ensenada, Baja California, Mexico;
rrivera@cicese.mx
[3] School of Electrical Engineering and Computer Science, South Ural State University,
Chelyabinsk 454080, Russia
* Correspondence: chernykh@cicese.mx or chernykhan@susu.ru; Tel.: +52-646-178-6994

Received: 27 February 2018; Accepted: 9 April 2018; Published: 22 May 2018

**Abstract:** We address image processing workflow scheduling problems on a multicore digital signal processor cluster. We present an experimental study of scheduling strategies that include task labeling, prioritization, resource selection, and digital signal processor scheduling. We apply these strategies in the context of executing the Ligo and Montage applications. To provide effective guidance in choosing a good strategy, we present a joint analysis of three conflicting goals based on performance degradation. A case study is given, and experimental results demonstrate that a pessimistic scheduling approach provides the best optimization criteria trade-offs. The Pessimistic Heterogeneous Earliest Finish Time scheduling algorithm performs well in different scenarios with a variety of workloads and cluster configurations.

**Keywords:** DSP microprocessor; multicore; multiprocessors; scheduling; workflow; resource management; job allocation

---

## 1. Introduction

In this paper, we address the multi-criteria analysis of image processing with communication workflow scheduling algorithms and study the applicability of Digital Signal Processor (DSP) cluster architectures.

The problem of scheduling jobs with precedence constraints is a fundamental problem in scheduling theory [1,2]. It arises in many industrial and scientific applications, particularly, in image and signal processing, and has been extensively studied. It has been shown to be NP-hard and includes solving a complex task allocation problem that depends not only on workflow properties and constraints, but also on the nature of the infrastructure.

In this paper, we consider a DSP compatible with TigerSHARC TS201S [3,4]. This processor was designed in response to the growing demands of industrial signal processing systems for real-time processing of real-world data, performing the high-speed numeric calculations necessary to enable a broad range of applications. It is optimized for both floating point and fixed point operations. It provides ultra-high performance; static superscalar processing optimized for memory-intensive digital signal processing algorithms from fully implemented 5G stations; three-dimensional ultrasound scanners and other medical imaging systems; radio and sonar; industrial measurement; and control systems.

It supports low overhead DMA transfers between internal memory, external memory, memory-mapped peripherals, link ports, host processors, and other DSPs, providing high performance for I/O algorithms.

Flexible instruction sets and high-level language-friendly DSP support the ease of implementation of digital signal processing with low communications overhead in scalable multiprocessing systems. With software that is programmable for maximum flexibility and supported by easy-to-use, low-cost development tools, DSPs enable designers to build innovative features with high efficiency.

The DSP combines very wide memory widths with execution six floating-point and 24 64-bit fixed-point operations for digital signal processing. It maintains a system-on-chip scalable computing design, including 24 M bit of on-chip DRAM, six 4 K word caches, integrated I/O peripherals, a host processor interface, DMA controllers, LVDS link ports, and shared bus connectivity for Glueless Multiprocessing without special bridges and chipsets.

It typically uses two methods to communicate between processor nodes. The first one is dedicated point-to-point communication through link ports. Other method uses a single shared global memory to communicate through a parallel bus.

For full performance of such a combined architecture, sophisticated resource management is necessary. Specifically, multiple instructions must be dispatched to processing units simultaneously, and functional parallelism must be calculated before runtime.

In this paper, we describe an approach for scheduling image processing workflows using the networks of a DSP-cluster (Figure 1).

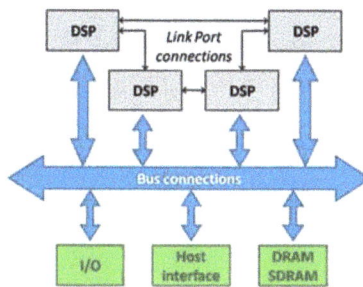

**Figure 1.** Digital signal processor (DSP) cluster.

## 2. Model

### 2.1. Basic Definitions

We address an offline (deterministic) non-preemptive, clairvoyant workflow scheduling problem on a parallel cluster of DSPs.

DSP-clusters consist of $m$ integrated modules ($IM$) $IM_1, IM_2, \ldots, IM_m$. Let $k_i$ be the size of $IM_i$ (number of DSP-processors). Let $n$ workflow jobs $J_1, J_2, \ldots, J_n$ be scheduled on the cluster.

A workflow is a composition of tasks subject to precedence constraints. Workflows are modeled as a Directed Acyclic Graph (DAG) $G_j = (V_j, E_j)$, where $V_j$ is the set of tasks, and $E_j = \{(i,k) \mid i,k \in V_j,\ i \neq k\}$, with no cycles.

Each arc $(i,k)$ is associated with a communication time $d_{i,k}$ representing the communication delay, if $i$ and $k$ are executed on different processors. Task $i$ must be completed, and data must be transmitted during $d_{i,k}$ prior to when execution of task $k$ is initiated. If $i$ and $k$ are executed on the same processor, no data transmission between them is needed; hence, communication delay is not considered.

Each workflow task $i$ is a sequential application (thread) and described by the tuple $(r'_i, p'_i)$, with release date $r'_i$, and execution time $p'_i$.

Due to the offline scheduling model, the release date of a workflow $r_j = 0$. However, the release date of a task $r'_i$ is not available before the task is released. Tasks are released only after all dependencies have been satisfied and data are available. At its release date, a task can be allocated to a DSP-processor for an uninterrupted period of time $p'_i$. $c_j$ is completion time of the job $j$.

Total workflow processing time $p_j^G$ and critical path execution cost $p_j$ are unknown until the job has been scheduled. We allow multiprocessor workflow execution; hence, tasks of $J_j$ can be run on different DSPs.

### 2.2. Performance Metrics

Three criteria are used to evaluate scheduling algorithms: makespan, critical path waiting time, and critical path slowdown. Makespan is used to qualify the efficiency of scheduling algorithms. To estimate the quality of workflow executions, we apply two workflow metrics: critical path waiting time and critical path slowdown.

Let $C_{max} = \max_{i=1..n}\{C_i\}$ be the maximum completion time (makespan) of all tasks in the schedule $C^*_{max}(I)$. The waiting time of a task $tw_i = c'_i - p'_i - r'_i$ is the difference between the completion time of the task, its execution time, and its release date. Note that a task is not preemptable and it is immediately released when the input data it needs from predecessors are available. However, note that we do not require that a job is allocated to processors immediately at its submission time as in some online problems.

Waiting time of a critical path is the difference between the completion time of the workflow, length of its critical path and data transmission time between all tasks in the critical path. It takes into account waiting times of all tasks in the critical path and communication delay.

The critical path execution time $p_j$ depends on the schedule that allocates tasks on the processor. The minimal value of $p_j$ includes only execution time of the tasks that belong to the critical path. The maximal value includes maximal data transmission times between all tasks in the critical path.

The waiting time of a critical path is defined as $cpw_j = c_j - p_j$. Critical path slowdown $cps_j = 1 + cpw_j/p_j$ is the relative critical path waiting time and evaluates the quality of the critical path execution. A slowdown of one indicates zero waiting times for critical path tasks, while a value greater than one indicates that the critical path completion is increased by increasing the waiting time of critical path tasks. Mean critical path waiting time is $cpw = 1/n \sum_{j=1}^{n} cpw_j$, and mean critical path slowdown is $cps = 1/n \sum_{j=1}^{n} cps_j$.

### 2.3. DSP Cluster

DSP-clusters consist of $m$ integrated modules ($IM$). Each $IM_i$ contains $k_i$ DPS-processors with their own local memory. Data exchange between DPS-processors of the same $IM$ is performed through local ports. The exchange of data between DPS-processors from different $IM$ is performed via external memory, which needs a longer transmission time than through the local ports. The speed of data transfer between processors depends on their mutual arrangement in the cluster.

Let $f_{ij}$ be the data rate coefficient from the processor of the $IM_i$ to the processor of $IM_j$. We neglect the communication delay $\varepsilon$ inside DSP; however, we take into account the communication delay between DSP-processors of the same $IM$. Data rate coefficients of this communication are represented as a matrix $D$ of the size $k_i \times k_i$. We assume that the transmission rates between different $IM$ are equal to $\alpha \gg \varepsilon$. Table 1 shows a complete matrix of data rate coefficients for a DSP-cluster with four $IMs$.

**Table 1.** Data rate coefficients of the cluster of DSP with four integrated modules (IMs).

| IM | 1 | 2 | 3 | 4 |
|----|---|---|---|---|
| 1 | $\varepsilon$ | $\alpha$ | $\alpha$ | $\alpha$ |
| 2 | $\alpha$ | $\varepsilon$ | $\alpha$ | $\alpha$ |
| 3 | $\alpha$ | $\alpha$ | $\varepsilon$ | $\alpha$ |
| 4 | $\alpha$ | $\alpha$ | $\alpha$ | $\varepsilon$ |

The values of the matrix $D$ depend on the specific communication topology of the $IMs$. In Figure 2, we consider three examples of the $IM_i$ communication topology for $k_i = 4$.

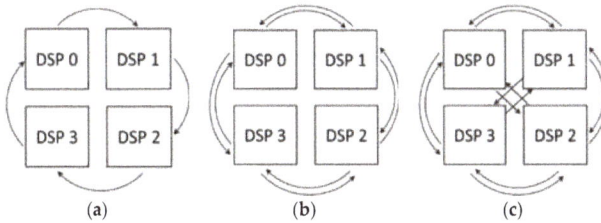

**Figure 2.** Examples of communication topology of DSP-processors: (**a**) uni-directional; (**b**) bi-directional; (**c**) all to all.

Figure 2a shows uni-directional DSP communication. Let us assume that the transfer rate between processors connected by an internal link port is equal to $\alpha = 1$. The corresponding matrix of data rate coefficients is presented in Table 2a. Figure 2b shows bi-directional DSP communication. The corresponding matrix of data rate coefficients is presented in Table 2b. Figure 2c shows all-to-all communication of DSP. Table 2c shows the corresponding data rate coefficients.

**Table 2.** Data rate coefficient matrix $D$ for three communication topologies between DSP-processors.

| (a) Uni-Directional | | | | (b) Bi-Directional | | | | (c) All to All | | | |
|---|---|---|---|---|---|---|---|---|---|---|---|
| 0 | 1 | 2 | 3 | 0 | 1 | 2 | 1 | 0 | 1 | 1 | 1 |
| 3 | 0 | 1 | 2 | 1 | 0 | 1 | 2 | 1 | 0 | 1 | 1 |
| 2 | 3 | 0 | 1 | 2 | 1 | 0 | 1 | 1 | 1 | 0 | 1 |
| 1 | 2 | 3 | 0 | 1 | 2 | 1 | 0 | 1 | 1 | 1 | 0 |

For the experiments, we take into account two models of the cluster (Figure 3). In the cluster $A$, ports connect only neighboring DSPs, as shown in Figure 3a. In the cluster $B$, DSPs are connected to each other, as shown in Figure 3b.

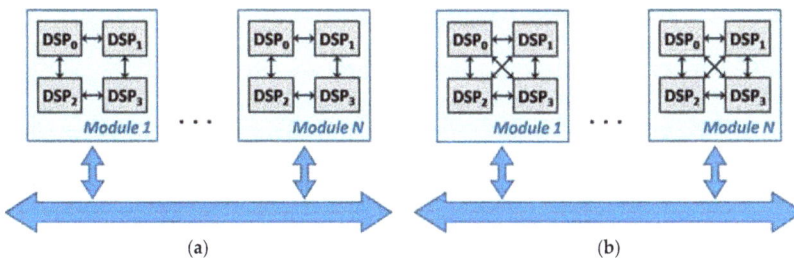

**Figure 3.** DSP cluster configuration. (**a**) Cluster $A$; (**b**) Cluster $B$.

*IMs* are interconnected by a bus. In the current model, for each connection, different data transmission coefficients are used. Data transfer within the same DSP has a coefficient of 0, between adjacent DSPs in a single *IM* has a coefficient of 1, and between *IMs*, has a data transmission coefficient of 10.

## 3. Related Work

State of the art studies tackle different workflow scheduling problems by focusing on general optimization issues; specific workflow applications; minimization of critical path execution time; selection of admissible resources; allocation of suitable resources for data-intensive workflows; Quality of Service (QoS) constraints; and performance analysis, among other factors. [5–17].

Many heuristics have been developed for scheduling DAG-based task graphs in multiprocessor systems [18–20]. In [21], the authors discussed clustering DAG tasks into chains and allocating them to single machines. In [22], two strategies were considered: Fairness Policy based on Finishing Time (FPFT) and Fairness Policy based on Concurrent Time (FPCT). Both strategies arranged DAGs in ascending order of their slowdown value, selected independent tasks from the DAG with the minimum slowdown, and scheduled them using Heterogeneous Earliest Finishing Time first (HEFT) [23] or Hybrid.BMCT [24]. FPFT recalculates the slowdown of a DAG each time the task of a DAG completes execution, while FPCT recalculates the slowdown of all DAGs each time any task in a DAG completes execution.

HEFT is considered as an extension of the classical list scheduling algorithm to cope with heterogeneity and has been shown to produce good results more often than other comparable algorithms. Many improvements and variations to HEFT have been proposed considering different ranking methods, looking ahead algorithms, clustering, and processor selection, for example [25].

The multi-objective workflow allocation problem has rarely been considered so far. It is important, especially in scenarios that contain aspects that are multi-objective by nature: Quality of Service (QoS) parameters, costs, system performance, response time, and energy, for example [14].

## 4. Proposed DSP Workflow Scheduling Strategies

The scheduling algorithm assigns to each graph's task start execution time. The time assigned to the stop task is the main result metric of the algorithm. The lower the time, the better the scheduling of the graph.

The algorithm uses a list of ready for scheduling tasks and a waiting list of scheduling tasks. If all predecessors of the task are scheduled, then it is inserted into the waiting list. If all incoming data are ready, then the task is inserted into the ready list, otherwise, into the waiting list. Available DPS-processors are placed into the appropriate list.

The list of tasks that are ready to be started is maintained. Independent tasks with no predecessors and with predecessors that completed their execution and available input data are entered into the list. Allocation policies are responsible for selecting a suitable DSP for task allocation.

We introduce five task allocation strategies: PESS (Pessimistic), OPTI (Optimistic), OHEFT (Optimistic Heterogeneous Earliest Finishing Time), PHEFT (Pessimistic Heterogeneous Earliest Finishing Time), and BC (Best Core). Table 3 briefly describes the strategies.

OHEFT and PHEFT are based on HEFT, a workflow scheduling strategy used in many performance evaluation studies.

HEFT schedules DAGs in two phases: job labeling and processor selection. In the job labeling phase, a rank value (upward rank) based on mean computation and communication costs is assigned to each task of a DAG. The upward rank of a task $i$ is recursively computed by traversing the graph upward, starting from the exit task, as follows: $rank_u(i) = \overline{w_i} + max_{j \in succ(i)}\left(\overline{d_{i,j}} + rank_u(i)\right)$, where $succ(i)$ is the set of immediate successors of task $i$; $\overline{d_{i,j}}$ is the average communication cost of $arc(i,j)$ over all processor pairs; and $\overline{w_i}$ is the average of the set of computation costs of task $i$.

Although HEFT is well-known, the study of different possibilities for computing rank values in a heterogeneous environment is limited. In some cases, the use of the mean computation and communication as the rank value in the graph may not produce a good schedule [26].

In this paper, we consider two methods of calculating the rank: best and worst. The best version assumes that tasks are allocated to the same DSP. Hence, no data transmission is needed. Alternatively, the worst version assumes that tasks are allocated to the DSP from different nodes, so data transmission is maximal. To determine the critical path, we need to know the execution time of each task of the graph and the data transfer time, considering every combination of DSPs, where the two given tasks may be executed taking into account the data transfer rate between the two connected nodes.

Tasks labeling prioritizes workflow tasks. Labels are not changed nor recomputed on completion of predecessor tasks. This also distinguishes our model from previous research (see, for instance [22]). Task labels are used to identify properties of a given workflow. We distinguish four labeling approaches: Best Downward Rank (BDR), Worst Downward Rank (WDR), Best Upward Rank (BUR), and Worst Upward Rank (WUR).

BDR estimates the length of the path from considered task to a root passing a set of immediate predecessors in a workflow without communication costs. WDR estimates the length of the path from considered task to a root passing a set of immediate predecessors in a workflow with worst communications. The descending order of BDR and WDR supports scheduling tasks by the depth-first approach.

BUR estimates the length of the path from the considered task to a terminal start task passing a set of the immediate successors in a workflow without communication costs.

**Table 3.** Task allocation strategies.

| | Description |
|---|---|
| Rand | Allocates task $T_k$ to a DSP with the number randomly generated from a uniform distribution in the range [1,$m$] |
| BC (best core) | Allocates task $T_k$ to the DSP that can start the task as early as possible considering communication delay of all input data |
| PESS (pessimistic) | Allocates task $T_k$ from the ordered list to the DSP according to Worst Downward Rank (WDR). |
| OPTI (optimistic) | Allocates task $T_k$ from the ordered list to the DSP according to Best Downward Rank (BDR). |
| PHEFT (pessimistic HEFT) | Allocates task $T_k$ from the ordered list to the DSP according to Worst Upward Rank (WUR) |
| OHEFT (optimistic HEFT) | Allocates task $T_k$ from the ordered list to the DSP according to Best Upward Rank (BUR) |

WUR estimates the length of the path from the considered task to a terminal task passing a set of immediate successors in a workflow with the worst communication costs. The descending order of BUR and WUR supports scheduling tasks on the critical path first. The upward rank represents the expected distance of any task to the end of the computation. The downward rank represents the expected distance of any task from the start of the computation.

## 5. Experimental Setup

This section presents the experimental setup, including workload and scenarios, and describes the methodology used for the analysis.

*5.1. Parameters*

To provide a performance comparison, we used workloads from a parametric workload generator that produces workflows such as Ligo and Montage [27,28]. They are a complex workflow of parallelized computations to process larger-scale images.

We considered three clusters with different numbers of DSPs and two architectures of individual DSPs (Table 4). Their clock frequency was considered to be equal.

**Table 4.** Experimental settings.

| Description | Settings |
|---|---|
| Workload type | 220 Montage workflows, 98 Ligo workflows |
| DSP clusters | 3 |
| Cluster 1 | 5 *IMs* in a cluster *B*, 4 DSP per module |
| Cluster 2 | 2 *IMs* in a cluster *A*, 4 DSP per module |
| Cluster 3 | 5 *IMs* in a cluster *A*, 4 DSP per module |
| Data transmission coefficient $K$ | 0—within the same DSP<br>1—between connected DSPs in a *IM*;<br>20—between DSP of different *IMs* |
| Metrics | $C_{max}, cpw, cps$ |
| Number of experiments | 318 |

*5.2. Methodology of Analysis*

Workflow scheduling involves multiple objectives and may use multi-criteria decision support. The classical approach is to use a concept of Pareto optimality. However, it is very difficult to achieve the fast solutions needed for DSP resource management by using the Pareto dominance.

In this paper, we converted the problem to a single objective optimization problem by multiple-criteria aggregation. First, we made criteria comparable by normalizing them to the best values found during each experiment. To this end, we evaluated the performance degradation of each strategy under each metric. This was done relative to the best performing strategy for the metric, as follows:

$$(\gamma - 1) \cdot 100, \text{ with } \gamma = \frac{strategy\ metric\ value}{best\ found\ metric\ value}.$$

To provide effective guidance in choosing the best strategy, we performed a joint analysis of several metrics according to the methodology used in [14,29]. We aggregated the various objectives to a single one by averaging their values and ranking. The best strategy with the lowest average performance degradation had a rank of 1.

Note that we tried to identify strategies that performed reliably well in different scenarios; that is, we tried to find a compromise that considered all of our test cases with the expectation that it also performed well under other conditions, for example, with different DSP-cluster configurations and workloads. For example, the rank of the strategy could not be the same for any of the metrics individually or any of the scenarios individually.

**6. Experimental Results**

*6.1. Performance Degradation Analysis*

Figure 4 and Table 5 show the performance degradation of all strategies for $C_{max}$, $cpw$, and $cps$. Table 5 also shows the mean degradation of the strategies and ranking when considering all averages and all test cases.

A small percentage of degradation indicates that the performance of a strategy for a given metric is close to the performance of the best performing strategy for the same metric. Therefore, small degradations represent better results.

We observed that Rand was the strategy with the worst makespan, with up to 318 percent performance degradation compared with the best-obtained result. PHEFT strategy had a small percent of degradation, almost in all metrics and test cases. We saw that *cps* had less variation compared with $C_{max}$ and *cpw*. It yielded to lesser impact on the overall score. The makespan of PHEFT and OHEFT were near the lower values.

Because our model is a simplified representation of a system, we can conclude that these strategies might have similar efficiency in real DSP-cluster environments when considering the above metrics. However, there exist differences between PESS and OPTI, comparing *cpw*. In PESS strategy, the critical path completion time did not grow significantly. Therefore, tasks in the critical path experienced small waiting times. Results also showed that for all strategies, small mean critical path waiting time degradation corresponded to small mean critical path slowdown.

BC and Rand strategies had rankings of 5 and 6. Their average degradations were within 67% and 18% of the best results. While PESS and OPTI had rankings of 3 and 4, with average degradations within 8% and 11%.

PHEFT and OHEFT showed the best results. Their degradations were within 6% and 7%, with rankings of 1 and 2.

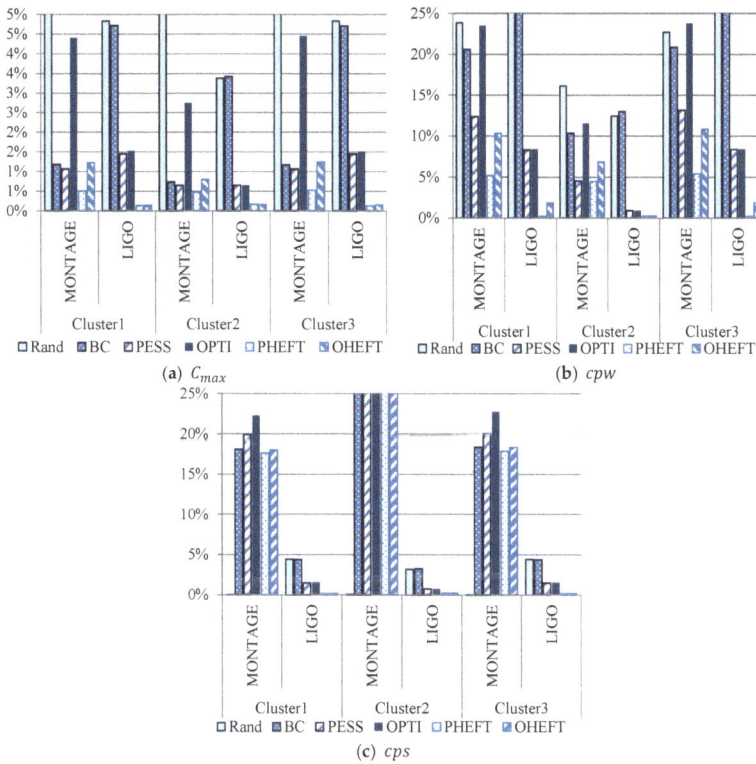

**Figure 4.** Performance degradation.

**Table 5.** Rounded performance degradation and ranking.

| | Criteria | Strategy | | | | | |
|---|---|---|---|---|---|---|---|
| | | **Rand** | **BC** | **PESS** | **OPTI** | **PHEFT** | **OHEFT** |
| Montage | $C_{max}$ | 3.189 | 0.010 | 0.009 | 0.039 | 0.005 | 0.011 |
| | $cpw$ | 0.209 | 0.173 | 0.100 | 0.196 | 0.050 | 0.093 |
| | $cps$ | 0.001 | 0.305 | 0.320 | 0.348 | 0.302 | 0.306 |
| | Mean | 1.133 | 0.163 | 0.143 | 0.194 | 0.119 | 0.137 |
| | Rank | 6 | 4 | 3 | 5 | 1 | 2 |
| Ligo | $C_{max}$ | 0.044 | 0.043 | 0.012 | 0.012 | 0.001 | 0.002 |
| | $cpw$ | 0.580 | 0.542 | 0.059 | 0.059 | 0.002 | 0.013 |
| | $cps$ | 0.040 | 0.040 | 0.012 | 0.013 | 0.002 | 0.002 |
| | Mean | 0.221 | 0.208 | 0.028 | 0.028 | 0.002 | 0.005 |
| | Rank | 6 | 5 | 3 | 4 | 1 | 2 |
| All test cases | $C_{max}$ | 1.616 | 0.027 | 0.011 | 0.025 | 0.003 | 0.006 |
| | $cpw$ | 0.394 | 0.357 | 0.079 | 0.128 | 0.026 | 0.053 |
| | $cps$ | 0.020 | 0.173 | 0.166 | 0.180 | 0.152 | 0.154 |
| | Mean | 0.677 | 0.186 | 0.085 | 0.111 | 0.060 | 0.071 |
| | Rank | 6 | 5 | 3 | 4 | 1 | 2 |

*6.2. Performance Profile*

In the previous section, we presented the average performance degradations of the strategies over three metrics and test cases. Now, we analyze results in more detail. Our sampling data were averaged over a large scale. However, the contribution of each experiment varied depending on its variability or uncertainty [30–32]. To analyze the probability of obtaining results with a certain quality and their contributors on average, we present the performance profiles of the strategies. Measures of result deviations provide useful information for strategies analysis and interpretation of the data generated by the benchmarking process.

The performance profile $\delta(\tau)p\tau$ is a non-decreasing, piecewise constant function that presents the probability that a ratio $\gamma$ is within a factor $\tau$ of the best ratio [33]. The function $\delta(\tau)$ is the cumulative distribution function. Strategies with larger probabilities $\delta(\tau)$ for smaller $\tau$ will be preferred.

Figure 5 shows the performance profiles of the strategies according to total completion time, in the interval $\tau = [1 \ldots 1.2]$, to provide objective information for analysis of a test set.

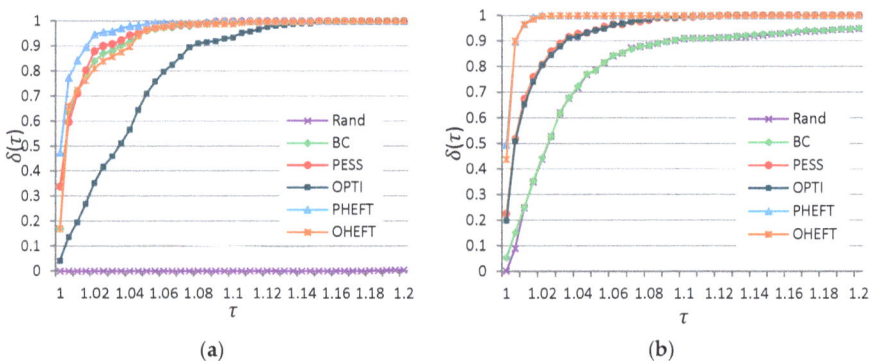

**Figure 5.** $C_{max}$ performance profile, $\tau = [1 \ldots 1.2]$. (**a**) Montage; (**b**) Ligo.

Figure 5a displays results for Montage workflows. PHEFT had the highest probability of being the better strategy. The probability that it was the winner on a given problem within factors of 1.02 of the best solution was close to 0.9. If we chose to be within a factor of 1.1 as the scope of our interest, then strategies except Rand and OPTI would have sufficed with a probability of 1. Figure 5b displays results for Ligo workflows. Here, PHEFT and OHEFT were the best strategies, followed by OPTI and PESS.

Figure 6 shows *cpw* performance profiles of six strategies for Montage and Ligo workflows considering $\tau = [1\ldots1.2]$. In both cases, PHEFT had the highest probability of being the better strategy for *cpw* optimization. The probability that it was the winner on a given problem within factors of 1.1 of the best solution was close to 0.85 and 1 for Montage and Ligo, respectively.

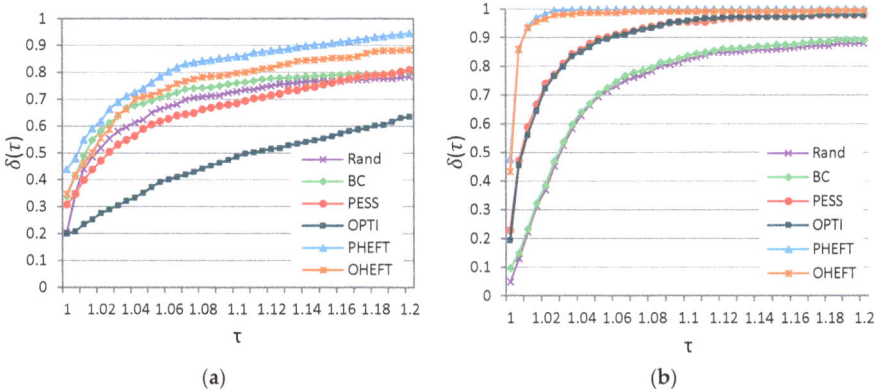

**Figure 6.** *cpw* performance profile, $\tau = [1\ldots1.2]$. (**a**) Montage; (**b**) Ligo.

Figure 7 shows the mean performance profiles of all metrics, scenarios and test cases, considering $\tau = [1\ldots1.2]$. There were discrepancies in performance quality. If we want to obtain results within a factor of 1.02 of the best solution, then PHEFT generated them with probability 0.8, while Rand with a probability of 0.47. If we chose $\tau = 1.2$, then PHEFT produced results with a probability of 0.9, and Rand with a probability of 0.76.

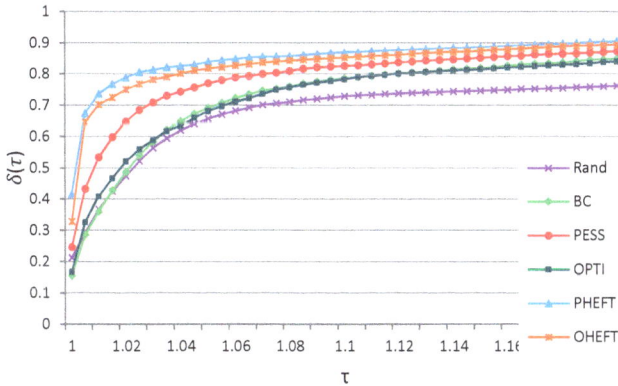

**Figure 7.** Mean performance profile over all metrics and test cases, $\tau = [1\ldots1.2]$.

## 7. Conclusions

Effective image and signal processing workflow management requires the efficient allocation of tasks to limited resources. In this paper, we presented allocation strategies that took into account both infrastructure information and workflow properties. We conducted a comprehensive performance evaluation study of six workflow scheduling strategies using simulation. We analyzed strategies that included task labeling, prioritization, resource selection, and DSP-cluster scheduling.

To provide effective guidance in choosing the best strategy, we performed a joint analysis of three metrics (makespan, mean critical path waiting time, and critical path slowdown) according to a degradation methodology and multi-criteria analysis, assuming the equal importance of each metric.

Our goal was to find a robust and well-performing strategy under all test cases, with the expectation that it would also perform well under other conditions, for example, with different cluster configurations and workloads.

Our study resulted in several contributions:

(1)   We examined overall DSP-cluster performance based on real image and signal processing data, considering Ligo and Montage applications;
(2)   We took into account communication latency, which is a major factor in DSP scheduling performance;
(3)   We showed that efficient job allocation depends not only on application properties and constraints but also on the nature of the infrastructure. To this end, we examined three configurations of DSP-clusters.

We found that an appropriate distribution of jobs over the clusters using a pessimistic approach had a higher performance than an allocation of jobs based on an optimistic one.

There were two differences to PHEFT strategy, compared to its original HEFT version. First, the data transfer cost within a workflow was set to maximal values for a given infrastructure to support pessimistic scenarios. All data transmissions were assumed to be made between different integrated modules and different DSPs to obtain the worst data transmission scenario with the maximal data rate coefficient.

Second, PHEFT had reduced time complexity compared to HEFT. It did not need to consider every combination of DSPs, where the two given tasks were executed, and did not need to take into account the data transfer rate between the two nodes to calculate a rank value (upward rank) based on mean computation and communication costs. Low complexity is important for industrial signal processing systems and real-time processing.

We conclude that for practical purposes, the scheduler PHEFT can improve the performance of workflow scheduling on DSP clusters. Although, more comprehensive algorithms can be adopted.

**Author Contributions:** All authors contributed to the analysis of the problem, designing algorithms, performing the experiments, analysis of data, and writing the paper.

**Acknowledgments:** This work was partially supported by RFBR, project No. 18-07-01224-a.

**Conflicts of Interest:** The authors declare no conflict of interest.

## References

1.   Conway, R.W.; Maxwell, W.L.; Miller, L.W. *Theory of Scheduling*; Addison-Wesley: Reading, MA, USA, 1967.
2.   Błażewicz, J.; Ecker, K.H.; Pesch, E.; Schmidt, G.; Weglarz, J. *Handbook on Scheduling: From Theory to Applications*; Springer: Berlin, Germany, 2007.
3.   Myakochkin, Y. 32-bit superscalar DSP-processor with floating point arithmetic. *Compon. Technol.* **2013**, *7*, 98–100.
4.   TigerSHARC Embedded Processor ADSP-TS201S. Available online: http://www.analog.com/en/products/processors-dsp/dsp/tigersharc-processors/adsp-ts201s.html#product-overview (accessed on 15 May 2018).
5.   Muchnick, S.S. *Advanced Compiler Design and Implementation*; Morgan Kauffman: San Francisco, CA, USA, 1997.
6.   Novikov, S.V. Global Scheduling Methods for Architectures with Explicit Instruction Level Parallelism. Ph.D. Thesis, Institute of Microprocessor Computer Systems RAS (NIISI), Moscow, Russia, 2005.

7.  Wieczorek, M.; Prodan, R.; Fahringer, T. Scheduling of scientific workflows in the askalon grid environment. *ACM Sigmod Rec.* **2005**, *34*, 56–62. [CrossRef]

8.  Bittencourt, L.F.; Madeira, E.R.M. A dynamic approach for scheduling dependent tasks on the xavantes grid middleware. In Proceedings of the 4th International Workshop on Middleware for Grid Computing, Melbourne, Australia, 27 November–1 December 2006.

9.  Jia, Y.; Rajkumar, B. Scheduling scientific workflow applications with deadline and budget constraints using genetic algorithms. *Sci. Program.* **2006**, *14*, 217–230.

10. Ramakrishnan, A.; Singh, G.; Zhao, H.; Deelman, E.; Sakellariou, R.; Vahi, K.; Blackburn, K.; Meyers, D.; Samidi, M. Scheduling data-intensive workflows onto storage-constrained distributed resources. In Proceedings of the 7th IEEE Symposium on Cluster Computing and the Grid, Rio De Janeiro, Brazil, 14–17 May 2007.

11. Szepieniec, T.; Bubak, M. Investigation of the dag eligible jobs maximization algorithm in a grid. In Proceedings of the 2008 9th IEEE/ACM International Conference on Grid Computing, Tsukuba, Japan, 29 September–1 October 2008.

12. Singh, G.; Su, M.-H.; Vahi, K.; Deelman, E.; Berriman, B.; Good, J.; Katz, D.S.; Mehta, G. Workflow task clustering for best effort systems with Pegasus. In Proceedings of the 15th ACM Mardi Gras conference, Baton Rouge, LA, USA, 29 January–3 February 2008.

13. Singh, G.; Kesselman, C.; Deelman, E. Optimizing grid-based workflow execution. *J. Grid Comput.* **2005**, *3*, 201–219. [CrossRef]

14. Tchernykh, A.; Lozano, L.; Schwiegelshohn, U.; Bouvry, P.; Pecero, J.-E.; Nesmachnow, S.; Drozdov, A. Online Bi-Objective Scheduling for IaaS Clouds with Ensuring Quality of Service. *J. Grid Comput.* **2016**, *14*, 5–22. [CrossRef]

15. Tchernykh, A.; Ecker, K. Worst Case Behavior of List Algorithms for Dynamic Scheduling of Non-Unit Execution Time Tasks with Arbitrary Precedence Constrains. IEICE-Tran Fund Elec. *Commun. Comput. Sci.* **2008**, *8*, 2277–2280.

16. Rodriguez, A.; Tchernykh, A.; Ecker, K. Algorithms for Dynamic Scheduling of Unit Execution Time Tasks. *Eur. J. Oper. Res.* **2003**, *146*, 403–416. [CrossRef]

17. Tchernykh, A.; Trystram, D.; Brizuela, C.; Scherson, I. Idle Regulation in Non-Clairvoyant Scheduling of Parallel Jobs. *Disc. Appl. Math.* **2009**, *157*, 364–376. [CrossRef]

18. Deelman, E.; Singh, G.; Su, M.H.; Blythe, J.; Gil, Y.; Kesselman, C.; Katz, D.S. Pegasus: A framework for mapping complex scientific workflows onto distributed systems. *Sci. Program.* **2005**, *13*, 219–237. [CrossRef]

19. Blythe, J.; Jain, S.; Deelman, E.; Vahi, K.; Gil, Y.; Mandal, A.; Kennedy, K. Task Scheduling Strategies for Workflow-based Applications in Grids. In Proceedings of the IEEE International Symposium on Cluster Computing and the Grid, Cardiff, Wales, UK, 9–12 May 2005.

20. Kliazovich, D.; Pecero, J.; Tchernykh, A.; Bouvry, P.; Khan, S.; Zomaya, A. CA-DAG: Modeling Communication-Aware Applications for Scheduling in Cloud Computing. *J. Grid Comput.* **2016**, *14*, 23–39. [CrossRef]

21. Bittencourt, L.F.; Madeira, E.R.M. Towards the scheduling of multiple workflows on computational grids. *J. Grid Comput.* **2010**, *8*, 419–441. [CrossRef]

22. Zhao, H.; Sakellariou, R. Scheduling multiple dags onto heterogeneous systems. In Proceedings of the 20th International Parallel and Distributed Processing Symposium, Rhodes Island, Greece, 25–29 April 2006.

23. Topcuouglu, H.; Hariri, S.; Wu, M.-Y. Performance-effective and low-complexity task scheduling for heterogeneous computing. *IEEE Trans. Parallel Distrib. Syst.* **2002**, *13*, 260–274. [CrossRef]

24. Sakellariou, R.; Zhao, H. A hybrid heuristic for dag scheduling on heterogeneous systems. In Proceedings of the 13th IEEE Heterogeneous Computing Workshop, Santa Fe, NM, USA, 26 April 2004.

25. Bittencourt, L.F.; Sakellariou, R.; Madeira, E.R. DAG Scheduling Using a Lookahead Variant of the Heterogeneous Earliest Finish Time Algorithm. In Proceedings of the 18th Euromicro Conference on Parallel, Distributed and Network-Based Processing, Pisa, Italy, 17–19 February 2010.

26. Zhao, H.; Sakellariou, R. *An Experimental Investigation into the Rank Function of the Heterogeneous Earliest Finish Time Scheduling Algorithm*; Springer: Berlin/Heidelberg, Germany, 2003.

27. Pegasus. Available online: http://pegasus.isi.edu/workflow_gallery/index.php (accessed on 15 May 2018).

28. Hirales-Carbajal, A.; Tchernykh, A.; Roblitz, T.; Yahyapour, R. A grid simulation framework to study advance scheduling strategies for complex workflow applications. In Proceedings of the 2010 IEEE International Symposium on Parallel & Distributed Processing, Workshops and Phd Forum (IPDPSW), Atlanta, GA, USA, 19–23 April 2010.
29. Hirales-Carbajal, A.; Tchernykh, A.; Yahyapour, R.; Röblitz, T.; Ramírez-Alcaraz, J.-M.; González-García, J.-L. Multiple Workflow Scheduling Strategies with User Run Time Estimates on a Grid. *J. Grid Comput.* **2012**, *10*, 325–346. [CrossRef]
30. Ramírez-Velarde, R.; Tchernykh, A.; Barba-Jimenez, C.; Hirales-Carbajal, A.; Nolazco, J. Adaptive Resource Allocation in Computational Grids with Runtime Uncertainty. *J. Grid Comput.* **2017**, *15*, 415–434. [CrossRef]
31. Tchernykh, A.; Schwiegelsohn, U.; Talbi, E.-G.; Babenko, M. Towards understanding uncertainty in cloud computing with risks of confidentiality, integrity, and availability. *J. Comput. Sci.* **2016**. [CrossRef]
32. Tchernykh, A.; Schwiegelsohn, U.; Alexandrov, V.; Talbi, E.-G. Towards Understanding Uncertainty in Cloud Computing Resource Provisioning. *Proced. Comput. Sci.* **2015**, *51*, 1772–1781. [CrossRef]
33. Dolan, E.D.; Moré, J.J.; Munson, T.S. Optimality measures for performance profiles. *Siam. J. Optim.* **2006**, *16*, 891–909. [CrossRef]

MDPI

St. Alban-Anlage 66

4052 Basel

Switzerland

Tel. +41 61 683 77 34

Fax +41 61 302 89 18

www.mdpi.com

*Algorithms* Editorial Office

E-mail: algorithms@mdpi.com

www.mdpi.com/journal/algorithms

www.ingramcontent.com/pod-product-compliance
Lightning Source LLC
Chambersburg PA
CBHW051848210326
41597CB00033B/5819